教育部高等学校特色专业建设教材
山东省高等学校特色专业建设教材

MATLAB 语言基础与应用

（第 2 版）

王永龙　张兆忠　张桂红　主　编
袁　红　许春磊　曹伟然　副主编

电子工业出版社
Publishing House of Electronics Industry
北京·BEIJING

内 容 简 介

本书在第 1 版的基础上修订而成，以 MATLAB R2013a 软件为基础，系统讲解了 MATLAB 基本环境和操作方法；分章阐述矩阵与数组的创立和运算、符号运算、M 脚本与函数编写、基本绘图方法与属性设置、数值计算方法、图形用户界面设计、Simulink 仿真系统、在信号系统的应用。在阐述上述内容的过程中给出大量的教学实例，并给出便于重复的过程表述。

本书可作为高等院校理工科各专业的高年级专科生、本科生、研究生学习 MATLAB 的教材或参考资料，也可以作为希望在这一领域进行研究和应用的广大科学技术工作者的参考书。

未经许可，不得以任何方式复制或抄袭本书之部分或全部内容。
版权所有，侵权必究。

图书在版编目（CIP）数据

MATLAB 语言基础与应用/王永龙，张兆忠，张桂红主编. —2 版. —北京：电子工业出版社，2016.1
ISBN 978-7-121-28124-2

Ⅰ．①M… Ⅱ．①王… ②张… ③张… Ⅲ．①Matlab 软件－高等学校－教材 Ⅳ．①TP317

中国版本图书馆 CIP 数据核字（2016）第 024700 号

策划编辑：张贵芹
责任编辑：桑　昀
印　　刷：北京七彩京通数码快印有限公司
装　　订：北京七彩京通数码快印有限公司
出版发行：电子工业出版社
　　　　　北京市海淀区万寿路 173 信箱　邮编 100036
开　　本：787×1 092　1/16　印张：25　字数：640 千字
版　　次：2010 年 9 月第 1 版
　　　　　2016 年 1 月第 2 版
印　　次：2025 年 1 月第 11 次印刷
定　　价：49.80 元

凡所购买电子工业出版社图书有缺损问题，请向购买书店调换。若书店售缺，请与本社发行部联系，联系及邮购电话：（010）88254888。

质量投诉请发邮件至 zlts@phei.com.cn，盗版侵权举报请发邮件至 dbqq@phei.com.cn。
服务热线：（010）88258888。

前　　言

自《MATLAB 语言基础与应用》于 2010 年 10 月出版以来，一方面由于 MATLAB 版本不断更新软件的需要，如在数值计算、绘图功能、编程手段和工具箱等方面都有很大改进；另一方面由于教学的需要，如开设 MATLAB 课程的专业越来越多、学生层次也趋向多样化等，促使作者对《MATLAB 语言基础与应用》一书进行修订。

本书在第 1 版的基础上有较大的修改，主要修改如下：

（1）4.2 节三维绘图内容细化更明确，增加了函数 patch 示例和 easy 三维绘图函数列表，使得内容更加完善。

（2）原第 8 章数值计算方法调整为第 5 章，增加了 5.6 节非线性方程求根，便于学生对于 MATLAB 中一些内置函数的基础算法有更加全面的了解，便于后面章节数值计算函数的调用。

（3）原第 5 章图形用户界面调整为第 6 章图形用户界面设计（GUI），增加了常用 GUI 组件创建和设置，便于读者对于 GUI 组件有更加全面的掌握。

（4）原第 7 章 Simulink 仿真调整为第 7 章 Simulink 仿真系统，对 Simulink 基础进行了细化，基础知识介绍更加系统。同时，增加了仿真系统中的子系统，使借助 Simulink 仿真系统解决较复杂问题得到简化。

（5）增加了第 8 章 MATLAB 在数字信号中的应用，满足电子信息科学与技术、电子信息工程专业学生的要求和教学过程课程间衔接的需要。

（6）原第 6 章偏微分方程工具箱本书不再进行介绍。

（7）本书对示例中的执行语句进行了英汉双语注释，尤其是前 4 章注释非常详细，希望为读者在后面章节的学习打下良好基础，对读者能够自己读懂 MATLAB 自带函数解释有所帮助。

本书着重体现以下 4 个方面的特点。

（1）内容结构简单，适合教学。内容的增减、结构的调整都是基于学生学习和教学中课程衔接的需要。

（2）课堂实用性强，课程衔接容易。增加了一些应用函数的编写，能够加深读者对 MATLAB 的理解，也便于 MATLAB 语言在其他课程中的应用及扩展。

（3）英汉双语注释，便于读者借助 MATLAB 帮助系统自学。

（4）为方便教学和学习，本书提供所有例题的 M 文件源代码及插图的 png 格式文件。

本书由 8 章构成。第 1 章绪论，主要介绍 MATLAB 的特点和 MATLAB R2013a 版的系统环境。第 2 章矩阵、数组、符号运算，讲解矩阵、数组、符号表达式的创建及各种运算操作。第 3 章编程，对示例注释进行细致修改，便于读者对后面章节示例给出的程序有更深入的理解。第 4 章绘图，对二维绘图、三维绘图、高维图形可视化、动画制作及图形属性设置进行系统介绍。第 5 章数值计算方法，讲解了线性方程、多项式插值与最小二乘曲线拟合、微积分、矩阵特征值与特征向量、常微分方程、非线性方程的数值解方法。第 6 章图形用户界面设计（GUI），阐述了如何通过函数 guide 设计图形用户交互界面，如何将函数脚本文件编译为可独立运行的 exe 文件。第 7 章 Simulink 仿真系统，介绍了 Simulink 基础，讲解了 Simulink 模型操作和仿真系统设置，给出具体示例，叙述了子系统的创建、子系统的条件执行及封装。第 8 章 MATLAB 在数字信号中的应用，阐述了 MATLAB 在数字信号中的应用，为学生将 MATLAB 在信号处理与系统课程中的拓展与应用提供了很好的桥梁。

本书第 1～4、6、7 章由王永龙编写，第 5 章由张兆忠编写，第 8 章由袁红编写，全书英文注释由张桂红编写修改。本书第 1～3 章由许春磊修订，第 4、6、7 章由曹伟然修订。全书由王永龙修改定稿。

本书在编写修订过程中，得到张贵芹编辑的全力支持，在此向她致以感谢！本书还使用了南京大学的网络资源，在此向南京大学物理学院的相关人员致以衷心的感谢！对为本书改版给予大力支持的江兆林院长、刘建华书记致以深深的敬意！

由于作者水平有限，书中难免出现错误或表述不妥之处，恳请广大读者给予批评指正，作者联系邮箱 wylong322@163.com。

作　者
2015 年 12 月于临沂

目　　录

第1章　绪论 ··· 1
　1.1　MATLAB 是什么 ··· 1
　1.2　MATLAB 启动与退出 ·· 3
　　　1.2.1　MATLAB 启动 ·· 3
　　　1.2.2　MATLAB 退出 ·· 6
　1.3　MATLAB 桌面 ·· 7
　　　1.3.1　命令窗口 ··· 8
　　　1.3.2　命令历史窗口 ·· 14
　　　1.3.3　工作空间窗口 ·· 14
　　　1.3.4　当前路径浏览器 ··· 15
　1.4　MATLAB 高级功能 ··· 15
　1.5　帮助系统 ·· 16
　　　1.5.1　帮助命令 ·· 16
　　　1.5.2　帮助浏览器窗口 ··· 17
　　　1.5.3　网络帮助 ·· 18
　小结 ··· 18
　习题 ··· 19

第2章　矩阵、数组、符号运算 ·· 20
　2.1　创建矩阵方法 ··· 20
　　　2.1.1　直接输入法 ·· 21
　　　2.1.2　矩阵生成命令 ··· 22
　2.2　构建数组方法 ··· 31
　　　2.2.1　数组生成命令 ··· 31
　　　2.2.2　矢量生成命令 ··· 31
　2.3　矩阵数组的运算操作 ··· 35
　　　2.3.1　四则运算 ·· 37
　　　2.3.2　初等数学运算 ··· 42
　　　2.3.3　矩阵运算操作函数 ··· 43
　2.4　多项式 ··· 47
　　　2.4.1　多项式表述 ·· 47
　　　2.4.2　多项式操作 ·· 48
　2.5　符号表达式的生成 ·· 51
　　　2.5.1　创建符号对象 ··· 52

	2.5.2 操作符号对象	58
2.6	微积分	62
2.7	求解符号方程	66
	2.7.1 解代数方程	66
	2.7.2 解微分方程	69
2.8	积分变换	70
2.9	实例应用	71
	2.9.1 解多项式	71
	2.9.2 解线性方程组	73
	2.9.3 求平行六面体体积	74
	2.9.4 特征值与特征向量	74
	2.9.5 多元数据	76
	2.9.6 电路问题	77
	2.9.7 稀疏矩阵绘图	78
小结		79
习题		79

第3章 编程 83

3.1	Top-Down 设计模式	83
3.2	伪代码	85
3.3	顺序结构	86
3.4	分支结构	93
	3.4.1 关系算符和逻辑算符	93
	3.4.2 if 结构	94
	3.4.3 switch 结构	100
	3.4.4 try/catch 结构	107
3.5	循环结构	109
	3.5.1 for 结构	109
	3.5.2 while 结构	111
3.6	函数编写	114
	3.6.1 MATLAB 函数	115
	3.6.2 MATLAB 中变量传递	117
	3.6.3 选择变量相关函数	120
	3.6.4 全局变量和永久变量	127
	3.6.5 子函数和私人函数	130
小结		134
习题		134

第4章 绘图 137

4.1	二维绘图	137
	4.1.1 函数 plot	137
	4.1.2 图形参数设置	144

	4.1.3 特殊二维图形绘制函数	151
	4.1.4 easy 二维绘图函数	155
4.2	三维绘图	157
	4.2.1 函数 plot3	157
	4.2.2 函数 patch	159
	4.2.3 三维网格图和曲面图函数	160
	4.2.4 函数 contour 和 contour3	168
	4.2.5 函数 quiver	170
	4.2.6 easy 三维绘图函数	172
	4.2.7 三维图形的参数设置	174
4.3	高维图形可视化	180
4.4	动画制作示例	184
	4.4.1 电影程序编写	185
	4.4.2 函数 movie	186
4.5	应用实例	189
	4.5.1 布朗运动	189
	4.5.2 相干波	190
	4.5.3 带洞的峰面	192
	4.5.4 透视图	193
	4.5.5 能流图	196
4.6	鼠标对图形的操作	207
4.7	图形句柄	209
	4.7.1 图形窗口	210
	4.7.2 核心对象	211
	4.7.3 注释对象	223
	4.7.4 总结	225
小结		226
习题		226
第 5 章 数值计算方法		**228**
5.1	线性方程组数值解法	228
	5.1.1 直接方法	228
	5.1.2 迭代方法	231
5.2	多项式插值与最小二乘曲线拟合	238
	5.2.1 多项式插值	238
	5.2.2 最小二乘曲线拟合	241
5.3	积分与微分	245
	5.3.1 数值积分	245
	5.3.2 数值微分	247
5.4	矩阵的特征值与特征向量	248
	5.4.1 特征值函数	249

5.4.2　幂法和反幂法 249
　　5.4.3　Jacobi 方法 252
　　5.4.4　QR 方法 253
5.5　常微分方程数值解法 255
　　5.5.1　欧拉（Euler）方法 255
　　5.5.2　龙格库塔（Runge-Kutta）方法 258
　　5.5.3　MATLAB 的相关函数 259
5.6　非线性方程求根 260
　　5.6.1　二分法 260
　　5.6.2　牛顿迭代法 261
　　5.6.3　弦截法 263
小结 264
习题 264

第 6 章　图形用户界面设计（GUI） 266
6.1　借助函数 guide 创建 GUI 266
6.2　创建 GUI 示例 268
6.3　GUI 实例 283
6.4　常用 GUI 组件创建与设置 296
6.5　编译独立的应用程序 302
　　6.5.1　编译器的安装与配置 302
　　6.5.2　编译 exe 文件 305
小结 305
习题 306

第 7 章　Simulink 仿真系统 307
7.1　Simulink 基础 307
　　7.1.1　启动 Simulink 307
　　7.1.2　Simulink 模块库浏览器 309
　　7.1.3　Commonly Used Blocks 模块库 310
　　7.1.4　Simulink 模型窗口 311
　　7.1.5　Simulink 建模仿真示例 313
7.2　Simulink 模型操作和仿真系统设置 317
　　7.2.1　Simulink 模型操作 317
　　7.2.2　Simulink 仿真系统设置 318
7.3　系统建模实例 321
7.4　仿真系统中的子系统 329
　　7.4.1　创建子系统 329
　　7.4.2　子系统的条件执行 331
　　7.4.3　封装子系统 333
小结 335
习题 335

第8章 MATLAB 在数字信号中的应用 ································ 336
8.1 时域离散信号和系统 ·· 336
8.1.1 信号、实现信号的基本运算及求解差分方程 ····················· 336
8.1.2 序列运算 ··· 340
8.2 离散时间傅里叶变换（DTFT）与 Z 变换函数 ·················· 348
8.2.1 函数 freqz ··· 348
8.2.2 函数 zplane ··· 349
8.3 离散傅里叶变换及快速傅里叶变换 ·· 352
8.3.1 几个扩展函数 ··· 353
8.3.2 快速傅里叶变换 ··· 355
8.4 IIR 滤波器的设计 ·· 361
8.4.1 滤波器设计函数 ··· 362
8.4.2 双线性变换法及冲激响应不变法设计 IIRDF ····················· 367
8.4.3 MATLAB 自带函数设计各类数字滤波器 ····················· 378
8.4.4 基于数字频带变换法设计数字滤波器 ····················· 382
小结 ·· 387
习题 ·· 387
参考文献 ·· 389

第1章 绪 论

本章含有 5 小节，分别是"MATLAB 是什么"、"MATLAB 启动与退出"、"MATLAB 桌面"、"MATLAB 高级功能"和"帮助系统"。

1.1 MATLAB 是什么

MATLAB 软件是在 1980 年前后，由新墨西哥大学计算机系主任 Clever Moler 博士在讲授线性代数课程时，发现用其他高级语言编程极为不便后构思开发的。MATLAB 是利用 EISPACK 和 LINPACK 两大软件包，借助 Fortran 语言编写，集计算、绘图、功能模块等于一身的交互式软件系统。后来 Moler 博士与斯坦福大学的 John Little 工程师等人合作成立 MathWorks 公司，于 1984 年首次推出 MATLAB 商业版本，编译语言也由 Fortran 改为 C 语言。MATLAB 初期就具有丰富多彩的图形图像处理功能、多媒体功能、符号运算功能和开放的体系结构等。MATLAB 商业版以此优秀的品质占据了大部分数学计算软件的市场，原来应用于控制领域的一些封闭式数学软件包（如英国的 UMIST、瑞典的 LUND 和 SIMNON、德国的 KEDDC）纷纷被淘汰或者在 MATLAB 基础上进行重建。随后的时间里获得以下主要发展历史。

1992 年 MathWorks 公司推出支持 Windows3.x 的 MATLAB4.0 版本，增加了 Simulink、Control、Neural Network、Signal Processing 等专用工具箱。

1993 年，推出 MATLAB4.1 版本，主要增加了符号运算功能。

1997 年，MATLAB5.0 版本问世，此时能够处理更多的数据类型，比如单元数据、数据结构体、多维矩阵、对象与类等，同时真正地实现了 32 位运算，提高了计算速度，使得图形绘制更加有效。

2002 年，推出的 MATLAB6.5 版本，Simulink 模块升级为 5.0，性能取得很大提高，同时退出 JIT 程序加速器，使得 MATLAB 的计算速度有了明显提高。

2005 年推出的 MATLAB7.1 版本，Simulink 模块升级为 6.3 版本，用户界面更加友好，并且采用了更加先进的数学程序库，即"LAPACK"和"BLAS"。

2007 年 3 月正式发布 MATLAB R2007a 版本，Simulink6.6 版本，增加了对 64 位 Windows 的支持，新推出.net 工具箱；同年 9 月，正式发布 MATLAB R2007b 版本，Simulink 升级为 7.0 版本，新增 2 个产品物理模型仿真模块（Simscape）和液压模拟模块组（SimHydraulics）。自此以后，MathWorks 公司每年 3 月和 9 月分别发布新版本，对 MATLAB 的新功能、新产品以及主要产品的升级情况进行公布。时至本书修订完成前夕，MATLAB R2015a 版本已经发布，增加了 4 个新产品（Antenna Toolbox、Simulink Test、Robotics System Toolbox、Vision HDL Toolbox），并对 79 个其他产品更新。基于通用性考虑，本书基于 MATLAB R2013a 版本进行阐述。

MATLAB来自于英文"matrix"和"laboratory"两单词,所以说MATLAB是基于矩阵为运算单元的高级计算机语言。MATLAB提供了基本的数学算法,例如矩阵运算、数值分析算法,集成了二维和三维图形功能,以完成相应的数值可视化的工作,并且提供了一种交互式的高级编程语言——M语言,利用M语言可以通过编写脚本或者函数文件实现用户自己的算法。MATLAB基础功能主要包含:数学运算、符号运算、数据获取、建模仿真、数据分析与可视化、图形用户界面开发等。

利用MATLAB语言还开发了相应的专业工具箱及函数供用户直接使用。这些工具箱应用的算法是开放的可扩展的,用户不仅可以查看其中的算法,还可以针对一些算法进行修改,甚至允许开发自己的算法扩充工具箱的功能。目前MATLAB R2013a版本含有的工具箱十分丰富,分别涵盖数据获取、科学计算、数值分析、数值和符号计算、工程与科学绘图、物理建模、控制系统的设计与分析、数字图像处理、数字信号处理、通信系统设计与仿真、金融财务分析、计算生物、代码生成和验证、应用发布、数据库访问与报告、仿真绘图与报告等专业领域。工具箱含有大量的MATLAB语言编写的M文件,拓展了MATLAB语言适应特殊专业的能力。

MATLAB是高级矩阵/数组语言,含有控制流程语句、函数、数据结构、输入与输出等编程基本要素。它即便于创建运行小程序,也能够编写比较大的程序。MATLAB语言具有如下长处。

1. 易用

MATLAB是翻译语言,如同很多版本的BASIC语言,非常容易使用。MATLAB程序常常被当作便笺本来处理简单问题。在其命令窗口直接输入需要计算的数学表达式,按回车键就能直接得到结果。当然,在命令窗口也能方便运行已经编写程序给出结果。对于程序,MATLAB可通过编写/调试器方便创建和修改程序。由于该语言具有如此易用特点,被认为是教学使用的理想语言,学习者也比较容易掌握编写新程序的基本知识。

同时,MATLAB越来越丰富的工具箱,也使得它易用特点得以彰显。易于学习编程的内容有MATLAB整合的编写/调试器、在线的文件与手册、工作空间浏览器,以及实例丰富的联机演示等。

2. 支持操作系统广泛

MATLAB支持多种计算机操作系统软件,比如Windows XP/Windows 7/Windows 8/UNIX/ Linux/Mac等。

3. 丰富的自带函数

MATLAB带有非常丰富的自带函数,这些函数涵盖许多基本技术问题的测试求解功能,比如假设要编写数据统计方面的程序。借助最常用的编程语言C语言或者Fortran语言,必须编写大量的子程序来完成计算,如平均值、标准差、中值等。而这些子函数在MATLAB语言中都已经自带,可以节省大量的时间。

另外,MATLAB软件的工具箱含有大量函数可以帮助用户解决一些专业而复杂的问题。比如可视化工具箱、图像处理工具箱、控制系统工具箱、神经网络工具箱、模糊逻辑工具箱、数字信号工具箱、统计工具箱、小波变换工具箱、偏微分方程工具箱、样条工具箱、优化工具箱和财政金融工具箱等。

4．卓著的绘图功能

不像其他计算语言，MATLAB 还有很多绘图和操作图形的命令。这些函数任何时候在装有 MATLAB 软件的计算机上都一样有效。并且仅在绘制的图形窗口也能够对图形所有基本属性进行设置。

5．图形用户界面

MATLAB 含有的工具箱允许程序员为自己的程序创建交互式的图形用户界面。具有了这样能力，使得经验相对欠缺的用户也能够编译复杂的程序进行数据分析成为可能。

6．MATLAB 编译器

MATLAB 具有独立的编译器。编译器可以将 MATLAB 编写的 M 文件转化为脱离 MATLAB 环境仍可独立运行的 EXE 文件。

当然，MATLAB 也有一些缺点，但是这些缺点随着计算机运行速度的快速提高变得越来越微不足道。

1.2　MATLAB 启动与退出

启动 MATLAB 的方法依靠运行计算机的操作系统，可参考 MathWorks 网站 http://www.mathworks.com，搜寻浏览关于启动 MATLAB 的说明，如 Windows 操作系统下的启动、UNIX 操作系统下启动、Mac 操作系统下启动，以及 Linux 操作系统下启动等。直接启动 MATLAB，需注意启动选项，以及与 MATLAB 连接的工具箱等。

1.2.1　MATLAB 启动

1．Windows 操作系统下启动 MATLAB

找到 Windows 操作系统左下角"开始"按钮，然后选择"开始→所有程序→MATLAB Starter Application"菜单命令单击，或者双击 Windows 桌面上 MATLAB R2013a 的快捷图标""。桌面上的快捷图标一般在安装 MATLAB 时会自动创建，如果没有可以自己创建。

在 Windows 窗口，也可以通过双击 MATLAB 文件，如*.mat、*.mdl 和*.fig 来启动 MATLAB。在确认 MATLAB 没有启动情况下，双击 Windows 窗口中的 MATLAB 文件也可以启动 MATLAB。如果 MATLAB 已经启动，将会重新打开 MATLAB 一新窗口。

2．在 DOS 窗口启动 MATLAB

首先将路径更改到含有 MATLAB.exe 文件处，然后输入 MATLAB 并敲击回车键即可。

3．通过 M 文件或其他类型文件启动 MATLAB

在默认情况下，双击 M 文件将打开 MATLAB 单机编辑器。如果通过双击 M 文件或其他类型文件启动 MATLAB，需要借助 Windows 窗口改变文件类型链接，如下描述。

为了改变 M 文件链接，请首先单击"开始"菜单选择"默认程序"菜单，如图 1.1 所示；然后单击"将文件类型或协议与程序关联"菜单，如图 1.2 所示；然后在打开的新窗口找到扩展名为".m"的文件选中，背景为灰色，如图 1.3 所示；单击右上角"更改程序…"按钮，打开新的窗口，从其中选择 MATLAB Starter Application，如果不含 MATLAB Starter Application 选项，单击"浏览"按钮打开新窗口，在新窗口中选中"MATLAB Starter Application"选项，如图 1.4 所示；最后在图 1.4 上单击"确定"按钮，功能设置完成，如图 1.5 所示。

图 1.1 "开始"菜单窗口

上面的示例说明并不一定适应读者的 Windows 操作系统版本，需要读者自己注意。如果读者遇到不一致的问题，请参阅计算机上的 Windows 文件配置说明。

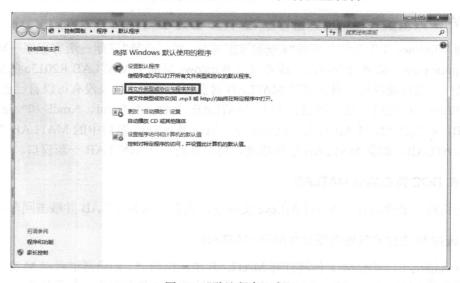

图 1.2 "默认程序"窗口

图 1.3　设置关联窗口

图 1.4　打开方式窗口

图 1.5　M 文件关联设置结果

1.2.2 MATLAB 退出

退出 MATLAB 的方法比较多，可以在命令窗口输入 quit 命令并按回车键，也可以直接单击快捷按钮"✕"，还可以使用组合键"Alt+F4"。

如果想在退出 MATLAB 时给出确认提示，可以进行如下设置。单击 MATLAB 窗口上"Preferences"按钮，如图 1.6 所示；打开"Preferences"设置窗口，然后在左侧分支结构选择"General→Confirmation Dialogs"，如图 1.7 所示，将"Confirm before exiting MATLAB"前面复选框选中，然后单击"OK"按钮。如果再关闭 MATLAB 时，会出现确认对话框如图 1.8 所示。

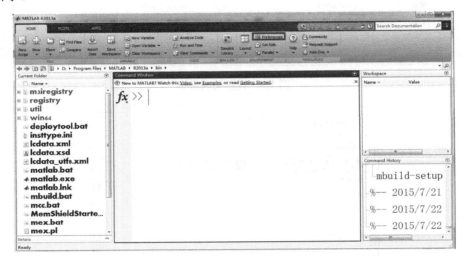

图 1.6 Desktop 操作桌面

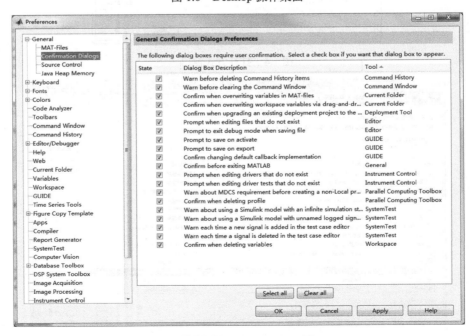

图 1.7 Preferences 属性设置窗口

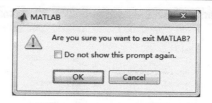

图 1.8 MATLAB 退出确认对话框

1.3 MATLAB 桌 面

MATLAB 桌面主要含有当前文件夹窗口、命令窗口、命令历史窗口、工作空间窗口、当前路径窗口、编辑/调试窗口和图形窗口。前面 5 个窗口整合在 MATLAB 默认桌面；后面 2 个窗口需要另外打开，没有显示在 MATLAB 的默认桌面上。当编写 M 文件和画图时，自然会打开编辑/调试窗口和图形窗口，也可以将它们嵌入 MATLAB 桌面中。

MATLAB 桌面提供了非常友好的环境，可以在命令窗口运行命令、可以直接看到数据、可以编写 M 程序、可以浏览文件、可以获得帮助等。它还提供了非常有用且有效的工具箱，便于用户管理和操作程序。

这里有几种常用办法设置个性桌面：通过右击 MATLAB 桌面菜单栏空白处，弹出"Minimize Toolstrip"快捷命令，单击它可实现隐藏工具栏，也可以通过右击菜单栏空白处，弹出"Restore Toolstrip"快捷命令，单击它可实现显示工具栏；可用通过拖拉窗口边界改变窗口大小；通过单击标题栏上的"   "，然后再单击"   "按钮将选择窗口单独显示还是嵌入在 MATLAB 桌面；可以长按鼠标左键移动窗口进行拖动，放开的位置就是该窗口的新位置；单击工具栏快捷按钮"   "，打开"Preferences"窗口，选中其中"Fonts"和"Colors"对字体及颜色进行设置，分别如图 1.9 和图 1.10 所示。

图 1.9 字体设置窗口

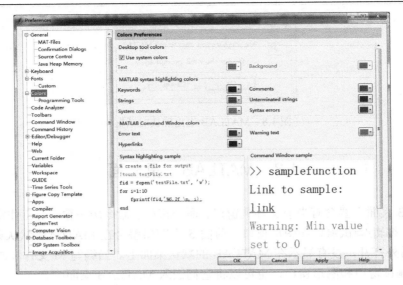

图1.10 字体颜色设置

1.3.1 命令窗口

在命令窗口可以输入变量，也可以输入函数，还可以直接输入数学表达式，按回车键就能给出准确结果；当然也可以输入用户自己编写的 M 文件名，按回车键开始执行。

命令窗口通常用来执行 MATLAB 命令函数，实现多任务，比如：创建变量、回调 M 文件、数值计算、符号计算、绘制二维图形、绘制三维图形等。如果希望再次执行已经运行过的语句，可以借助向上键进行选取，可以借助"Delete"和"Backspace"键进行删除，然后按回车键执行。

EXAMP01001

创建变量，并借助初等数学知识进行计算。

```
% Create a variable "a" as 3, and display the result in Command Window.
%  (创建变量a，并赋值为3)
>> a=3
a = 3
% Create two variables b and c with "6" and "17", respectively,
% but disappear answers.
% (创建两个变量b和c，分别赋值为6和17，但不显示)
>> b=6; c=17;
% Solve (a+b)*c-c, and show answer.
% (计算表达式(a+b)*c-c，并显示结果)
>> (a+b)*c-c
ans = 136
% Calculate sin(a)+cos(b)-tan(c), and display answer.
% (计算sin(a)+cos(b)-tan(c)，并显示结果)
>> sin(a)+cos(b)-tan(c)
ans = -2.3926
% Solve exp(a)*sin(b)/cos(c), and display answer.
% (计算exp(a)*sin(b)/cos(c)，并显示结果)
```

```
>> exp(a)*sin(b)/cos(c)
ans = 20.3959
```

借助 MATLAB，我们不仅能解决初等数学问题，还能求解高等数学问题，如微积分等。

EXAMP01002

借助 MATLAB 计算微积分。

```
% Define symbolic variable x.
% (定义符号变量 x)
>> syms x
% Integrate the function input by user with function "int",
% and display answer.
% (借助"int"函数对用户输入的表达式进行积分，并显示结果)
>> int(x^3+4*x^2-7*x+19,x)
ans = x^4/4 + (4*x^3)/3 - (7*x^2)/2 + 19*x
% Express the above ans in usual mathematical form by "pretty".
% (通过函数"pretty"将上面答案转化为通常数学书写形式)
>> pretty(ans)
    4       3       2
   x    4 x     7 x
   -- + ---- - ---- + 19 x
   4     3      2
% Calculate the definite integral by "int"
% with the integral limits "0" and "5".
% (借助函数"int"计算积分上下限分别为"0"和"5"的定积分)
>> int(x^3+4*x^2-7*x+19,x,0,5)
ans =3965/12
% Define symbolic variables f, x, y, z.
% (定义符号变量 f, x, y, z)
>> syms f x y z
% Define a function f depending on x, y, and z.
% (定义函数 f，其自变量为 x, y, z)
>> f=3*x^3*y*z^2-8*y*z;
% Solve the one order derivative of f with respect to x.
% (求函数 f 对 x 的一阶微分)
>> fx1=diff(f,x)
fx1 =9*x^2*y*z^2
% Solve the two order derivative of function f with respect to x.
% (求 f 对 x 二阶微分)
>> fx2=diff(f,x,2)
fx2 =18*x*y*z^2
% Solve the two order derivative of function f with respect
% to variables x and y.
% (求 f 对 x 和 y 二阶微分)
>> fxy=diff(diff(f,x),y)
fxy =9*x^2*z^2
% Solve the three order derivative of function f with respect
% to x, y and z.
```

```
%  (求f对x, y, z三阶微分)
>> fxyz=diff(diff(diff(f,x),y),z)
fxyz =18*x^2*z
```

借助MATLAB，还可以绘制二维图形和三维图形，下面提供两个示例，所绘制结果如图1.11至图1.14所示。

EXAMP01003

```
%  Define an independent variable x.
%  (定义一个自变量为x)
>> x=1:50;
%  Define a dependent variable y.
%  (定义一个因变量为y, 对应x的每个值为任意数)
>> y=rand(1,50);
%  Plot the plot of x and y as 图.1.11 with "plot".
%  (借助函数"plot"绘制x和y的图形, 如图1.11所示)
>> plot(x,y)
%  Clear all variables in workspace.
%  (清除工作空间中所有变量)
>> clear
%  Define an independent variable x by "linspace".
%  (借助"linspace"函数, 定义一个自变量x)
>> x=linspace(0,2*pi,200);
%  Define a variable y by "sin".
%  (借助"sin"函数定义因变量y)
>> y=sin(x);
%  Open a new figure.
%  (打开一个新图形窗口)
>> figure
%  Plot the curve of x and y as 图1.12 by "plot".
%  (借助"plot"函数绘制x和y的图形, 如图1.12所示)
>> plot(x,y)
%  Clear all variables in workspace.
%  (清除工作空间中所有变量)
>> clear
%  Define a parameter variable t.
%  (定义一个参数变量t)
>> t=0:pi/20:6*pi;
%  Open a new figure.
%  (打开一个新图形窗口)
>> figure
%  Plot a three-dimensional line as 图11.13 by "plot3".
%  (借助"plot3"函数绘制一条三维曲线, 如图1.13所示)
>> plot3(sin(t),cos(2*t),t)
%  Clear all variables in workspace.
%  (清除工作空间中所有变量)
>> clear
%  Define three variables x, y, z by "sphere".
%  (通过"sphere"函数定义三个变量x, y, z)
>> [x,y,z]=sphere(30);
```

```
% Define a new variable x1 depending on x.
% (通过 x 定义一个新变量为 x1)
>> x1=1.5*x;
% Define a new variable y1 depending on y.
% (通过 y 定义一个新变量为 y1)
>> y1=1.5*y;
% Define a new variable z1 depending on z.
% (通过 z 定义一个新变量为 z1)
>> z1=1.5*z;
% Define a new variable x2 depending on x.
% (通过 x 定义一个新变量为 x2)
>> x2=2*x;
% Define a new variable y2 depending on y.
% (通过 y 定义一个新变量为 y2)
>> y2=2*y;
% Define a new variable z2 depending on z.
% (通过 z 定义一个新变量为 z2)
>> z2=2*z;
% Plot a three-dimensional solid sphere by "surf".
% (借助"surf"函数绘制三维实体球)
>> surf(x,y,z);
% Hold the present figure window on.
% (保持当前绘图窗口为当前窗口)
>> hold on
% Plot a mesh three-dimensional sphere by "mesh"
% in the same figure window.
% (借助"mesh"函数,在同一个窗口同时绘制三维球的网格图)
>> mesh(x1,y1,z1);
% Define the characteristic color is "cool" by "colormap".
% (借助"colormap"函数定义色调为"color")
>> colormap(cool)
% Plot a mesh three-dimensional sphere by "mesh"
% in the same figure window.
% (借助"mesh"函数,在同一个窗口同时绘制三维球的网格图)
>> mesh(x2,y2,z2);
% Hold off the present figure window.
% (释放当前图形窗口)
>> hold off
% Hidden off the present figure.
% (隐藏关闭)
>> hidden off
% Set the present axis as square.
% (定义坐标轴都相等,为方坐标轴)
>> axis square
% Set the present axis off.
% (释放坐标轴设置)
>> axis off
```

绘制图形结果如图 1.11 至图 1.14 所示,需要注意的是,每次运行得到的图 1.11 都会不同。

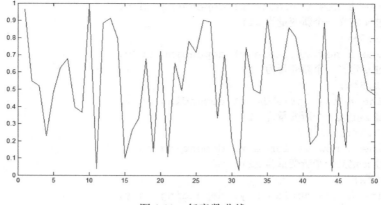

图 1.11 任意数曲线

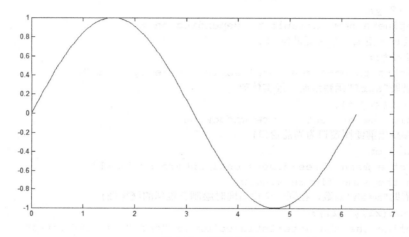

图 1.12 正弦曲线

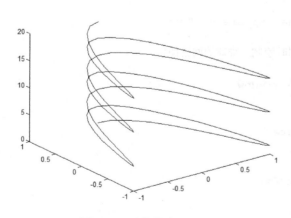
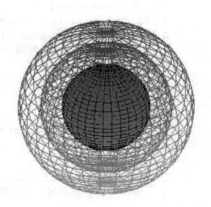

图 1.13 三维曲线　　　　　　　　图 1.14 三重透视球

通过 MATLAB，还可以制作电影，示例如下，其电影剪辑如图 1.15 所示。

EXAMP01004

为逐渐变大的球编写程序。

```
%   Define values of x, y and z by "sphere".
%   (通过函数"sphere",定义 x,y 和 z 变量值)
[x,y,z]=sphere(50);
%   Evaluate the elements of matrix m through "for" structure.
%   (借助 for 循环完成矩阵 m 的矩阵元的赋值)
for i=1:30
    surf((1+(i/30))*x,(1+(i/30))*y,(1+(i/30))*z);
    colormap(hot)
    shading interp
    m=getframe;
end
%   Display the movie in 图1.15.
%   (电影剪辑如图 1.15 所示)
movie(m);
```

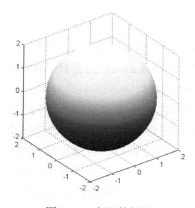

图 1.15　电影剪辑图

为了后面学习 MATLAB 操作方便,下面列出常用操作函数与常用快捷键,分别参见表 1.1 和表 1.2。

表 1.1　常用操作函数

函　数	描　述	函　数	描　述
cd	显示或更改当前路径	clear	清除当前工作空间
dir	显示当前路径下所有文件与文件夹	disp	显示变量内容
clc	清除命令窗口	type	显示特定路径下的文件内容
clf	清除当前图形窗口	exit	退出 MATLAB
cla	清除当前坐标系	quit	退出 MATLAB

表 1.2　常用快捷键

键	组　合　键	描　述
↑	Ctrl+P	回调命令窗口上执行
↓	Ctrl+N	回调命令窗口下执行
←	Ctrl+B	将光标向左移动一个字符
→	Ctrl+F	将光标向右移动一个字符

(续表)

键	组合键	描述
Ctrl+→		将光标向右移动一个单词
Ctrl+←		将光标向左移动一个单词
Home	Ctrl+A	将光标移动到行头
End	Ctrl+E	将光标移动到行尾点
Ctrl+Home		将光标移动到命令窗口初始点
Ctrl+End		将光标移动到命令窗口结尾点
Esc	Ctrl+U	清除命令行
Delete	Ctrl+D	删除光标后1个字符
Backspace	Ctrl+H	删除光标前1个字符

1.3.2 命令历史窗口

在 MATLAB 默认桌面，命令历史窗口显示在右下角。一旦命令历史窗口被误操作关闭，可以单击工具栏快捷按钮"　"，打开 Layout 的悬挂菜单如图 1.16 所示，选中"Command History"选项，重新显示命令历史窗口。

自从 MATLAB 启动后，所有在命令窗口执行的语句都会存储在命令历史窗口中。如果需要重新运行某一命令行，可以双击命令历史窗口中的相应语句，自动就会将运行该语句行，并将该运行语句和运行结果显示在命令窗口。也可以选中欲运行命令行，按住鼠标左键将其拖入命令窗口，单击回车键即可，如需修改也可以修改后再单击回车键运行。并且还可以将能实现特殊功能多行语句选中，单击鼠标右键打开悬挂菜单如图 1.17 所示，然后选择"Create Script（创建脚本程序）"或者"Create Shortcut（创建快捷按钮）"创建 M 文件，或者创建快捷按钮。在命令窗口输入存储文件名就可运行选中的所有命令行并给出结果，当然也可以单击快捷按钮。

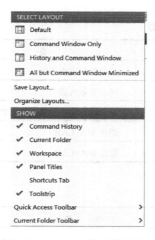

图 1.16　Layout 的悬挂菜单

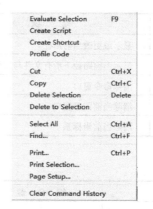

图 1.17　命令历史窗口中的悬挂菜单

1.3.3 工作空间窗口

工作空间窗口罗列了 MATLAB 工作空间中存储的所有变量，或者运行函数中的所有变量。双击工作空间中显示的任意变量,将会打开数组编辑器窗口(Array Editor)显示该变量的数值;并且可以同时打开多个变量;还可以借助工作空间中变量存储的数值进行绘图。

如果同时操作多个文件,比如数组、M 文件或者图形窗口,可以同时给它们命名、也可同时操作它们;还可以通过拖拉鼠标将各文件放到新的位置等。

1.3.4 当前路径浏览器

当前路径浏览器允许用户导航文件,并且用户可以重新设定他们所期望的当前路径。

另外,桌面工具还有帮助浏览器窗口如图 1.18 所示、编辑/调试窗口、图形窗口等。帮助浏览器窗口将在 1.5 节"帮助系统"中进行简单介绍;编辑/调试窗口将在第 3 章"编程"中进行阐述;图形窗口将在第 4 章"绘图"中进行讲解。

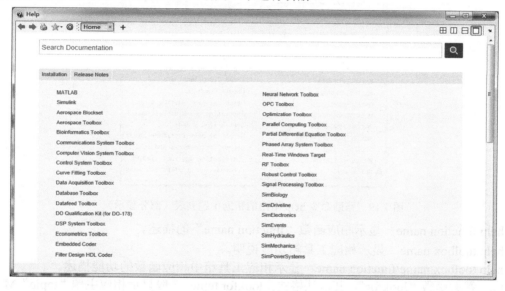

图 1.18　帮助浏览器窗口

1.4　MATLAB 高级功能

MATLAB 软件附带许多工具箱,这些工具箱拓展了 MATLAB 在许多特殊领域的应用功能,为更多的专业用户提供了友好专业的工作界面。工具箱有生物信息工具箱、射频工具箱、通信工具箱、控制系统工具箱、偏微分方程工具箱,等等。这些工具箱为相应领域的科学家和工程师们提供了相关领域的大量计算工具,甚至包含许多有意义的函数程序,这对相应领域的专家学者及工作人员都有很好的借鉴意义。对于想对某一领域进行研究的初学者,MATLAB 可以提供第一手的帮助资料及程序。

由于作者能力所限,本书只对图形用户界面开发工具箱(GUI)、Simulink 仿真工具箱、偏微分方程工具箱(PDE)进行简单介绍,如果想了解其他工具箱功能,请参阅相关书籍资料。

1.5 帮助系统

这里提供了许多途径可以得到读者需要的答案,主要有3个途径:帮助命令、帮助浏览器窗口、网络帮助。

1.5.1 帮助命令

MATLAB 帮助命令常用的有 "help" 和 "lookfor"。

(1)"help" 命令的语法格式。其语法格式:help 在命令窗口显示所有初等帮助主题,每一个主要帮助主题都有直接的搜索路径,如图 1.19 所示,这里只是显示了部分主题。

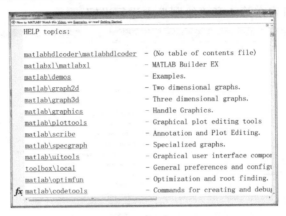

图 1.19 帮助命令 help 给出的帮助主题列表(部分显示)

help function name   显示相应函数 "function name" 的描述。
help toolbox name   显示相应工具箱功能说明。
help toolbox name/function name   显示相应工具箱中相应函数的功能描述。

(2)帮助函数 "lookfor"。其语法格式:lookfor topic 一般显示相应主题 "topic" M 文件帮助说明的第一行,简洁明了。一般对于对 MATLAB 软件函数相对熟悉用户比较有效,能起到很好的提示作用。

lookfor magic –all   搜索显示所有与主题 "magic" 对应的相关 M 文件,如图 1.20 所示。

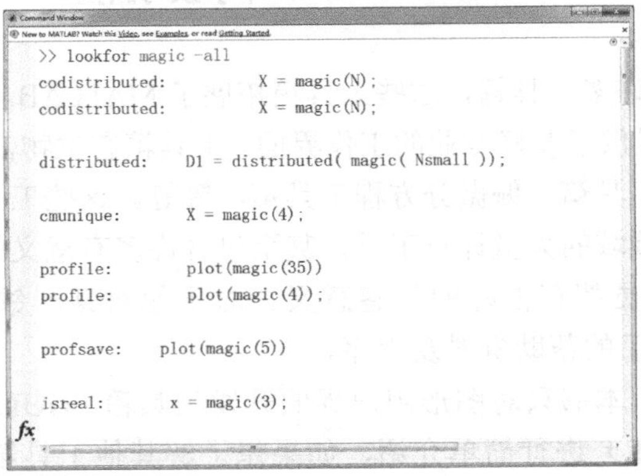

图 1.20 帮助命令 lookfor 给出的帮助主题列表(部分显示)

1.5.2 帮助浏览器窗口

通过帮助浏览器可以找到所有与 MathWorks 产品相关的解说文件以及联机演示，帮助浏览器是整合到 MATLAB 桌面的 HTML 阅读器。

打开帮助浏览器有很多途径，可以在 MATLAB 桌面单击工具栏的"Help"按钮，还可以通过键盘"F1"键打开。如图 1.21 所示为帮助浏览窗口。

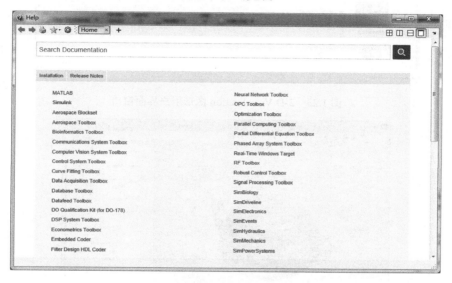

图 1.21　帮助浏览器窗口

单击工具栏的"Help"下拉选项的 Examples，调出 MathWorks 提供的解说演示文件，如图 1.22 所示。选择"3-D Visualization"演示后单击"teapot"按钮，如图 1.23 所示，就可得到窗口如图 1.24 所示。

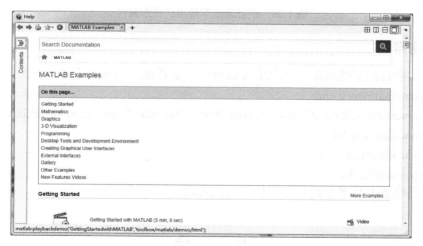

图 1.22　帮助浏览器的联机演示面板

借助联机演示所提供的许多文件，可以学习图形用户界面开发基本方法，并且可得到很多设计经验，值得用户仔细琢磨，尤其图形用户界面软件开发爱好者。

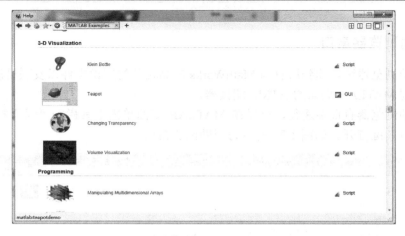

图 1.23　3-D Visualization 图形用户界面窗口

图 1.24　茶壶操作图形用户界面窗口

1.5.3　网络帮助

这里为读者提供了些网址，从上面可以得到许多帮助。网址如下：

http://www/mathwork.com.

http://www.cs.cornell.edu/Courses/cs100j/2001sp/Matlab/Chapman/chapman.htm

http://www.ilovematlab.cn/

http://www.labfans.com/bbs/

http://www.matlabsky.com/

当然，网络还有很多帮助，在于读者自己发现。

小　　结

本章首先简单介绍 MATLAB 的发展过程及其主要特点、启动与退出的基本途径、MATLAB 桌面的主要窗口构成并简单说明。然后对 MATLAB 提供的帮助途径进行较详细阐述，为读者自己进一步学习 MATLAB 提供了途径。

习　　题

1.1　MATLAB 桌面主要由哪些窗口构成？这些窗口主要功能是什么？
1.2　计算下列表达式。
　　（1）3+6
　　（2）sin(1+i)
　　（3）(1+2i)×(2-i)
　　（4）2×4+3
1.3　借助"help"函数了解"meshgrid"函数的意义。
1.4　通过帮助浏览器窗口了解函数 sin 和 cos。
1.5　通过 MATLAB 帮助浏览器窗口和帮助命令学习绘图命令"plot"的用法。
1.6　借助 MATLAB 的"plot"函数完成函数 sin(x), cos(x)和 sin(x)+cos(x)在区间 0 到 2π 区间内绘图。
1.7　借助帮助浏览器，了解一个你感兴趣的工具箱。
1.8　通过帮助 Examples 部分，了解 MATLAB 基本功能。
1.9　计算下面数学表达式。
　　（1）$y_1 = \dfrac{2\sin(0.5\pi)}{1+\sqrt{3}}$

　　（2）$y_2 = 2\pi e^{i\frac{\pi}{3}}$

　　（3）$y_3 = 2 + 5 \times 3.6 - 7.2$

　　（4）$y_4 = 23^4 - 12^2$

第2章 矩阵、数组、符号运算

在 MATLAB 中，矩阵是一方形的数组。一些特殊情形，比如只含有一个元素的矩阵称为标量，对于只含有一行或一列的矩阵称为矢量。MATLAB 为存储数或非数也提供了其他方式，但是最好在一开始就将所有数据定义为矩阵，因为这种操作在 MATLAB 中是最自然的。对于操作矩阵，MATLAB 赋予了我们能够同时操作一个矩阵所有元素的能力，不像其他程序语言一次只能对一个数（也就是一个矩阵元）进行操作。标量和矢量是特殊的矩阵，我们不仅可以通过操作矩阵来解决矩阵的问题，还可以解决所有和标量与矢量相关的问题。尤其是在操作含有多于一个元素的矩阵时，MATLAB 表现出远胜过其他编程软件的长处。

符号运算是本章介绍的另一重要内容。符号运算在数学中扮演着非常重要的角色，尤其在高等数学（微积分、高等代数等）中有较广泛的应用。在当前发达的社会，在处理许多事情时都引入了高等数学的方法，日常生活中的很多现象也满足高等数学基本定律。作为一门被广泛应用于科学与工程技术领域的计算机高级语言，引入符号计算是必需的。MATLAB 中符号计算是借助引入数学工具箱来完成的，数学工具箱出自于另一数学软件 MAPLE。MAPLE 拥有非常强的符号计算能力，自身创建并整合了大量有用而又有效的函数，可以对符号表达式进行各种数学操作。这章将对符号表达式的定义，以及一些常见基本操作函数进行阐述。

在 MATLAB 中，矩阵就是特殊的操作元素，在这章我们将首先介绍如何创建矩阵，然后介绍与矩阵相关的操作命令。由于数值计算的广泛应用，本章还将讲述数值计算的一些基本函数，数值计算方法将在第 5 章详细讲述。随着计算机的发展进步，数据处理软件也得到迅速发展。尽管 MATLAB 软件也提供了创建数据库的一些基本函数，但是限于篇幅本书不讲述这方面内容。另外，本章将对符号表达式的定义，求解等基本函数进行阐述，尤其是微积分和积分变换方面函数进行较详细讲解并给出示例。对符号方程、常微分方程求解函数也进行较详细阐述。

本章中所有例题求解中，开头"…"符号前为题目内容陈述，"%"后为注释，">>"后为执行行语句，一般"="后为输出结果。执行行命令应该在 Command Window 内输入运行，相对应本书附带的程序是以 M 文件模式存在的，没有">>"符号，也没有结果显示行，这些需要注意。

2.1 创建矩阵方法

在学习数学过程中，整数是我们的第一位朋友，然后一步一步逐渐认识了各种各样的数。对各种各样数的理解和认识，与我们对整数的理解和认识是分不开的。矩阵在 MATLAB 语言相当于整数在数学中的地位，所以学习创建矩阵是学习 MATLAB 编程的基石。创建矩阵有多种方法，下面将分别介绍。

2.1.1 直接输入法

对简单矩阵来讲，直接输入法创建矩阵是一种最直接、简单、有效的办法。并且基于已有矩阵，我们还可以通过直接输入法定义一些特定元素生成较大矩阵。

【例 2.1】 通过直接输入法，创建矩阵 $\begin{bmatrix} 1 & 3 \\ 5 & 7 \end{bmatrix}$，并通过定义特定元素，生成 3×5 矩阵如下：

$$A = \begin{bmatrix} 1 & 3 & 0 & 0 & 0 \\ 5 & 7 & 0 & 0 & 0 \\ 0 & 0 & 0 & 0 & 0 \end{bmatrix}$$

并将矩阵 A 的值赋予给另一个矩阵 B。

EXAMP02001

创建矩阵 $A = \begin{bmatrix} 1 & 3 \\ 5 & 7 \end{bmatrix}$，我们可以通过在命令窗口一个接一个地输入所有矩阵元素来实现。

```
% By typing, one enters a matrix in Command Window. Need to note that
% "," is used to separate elements in same row, and ";" is used to
% distinguish different rows.
% (通过键盘在命令窗口输入矩阵。需要注意：借助","隔离同行内元素，借助";"
% 分离矩阵的行)
>> A=[1,3;5,7]
A =
     1     3
     5     7
% Use "space" key to separate different elements in same row, and
% use "enter" key to distinguish different rows.
% (通过"space"键区分同行的不同元素，通过"enter"键去区分不同的行)
>> A=[1 3
5 7]
A =
     1     3
     5     7
% In terms of a matrix A, by defining a specific element to generate
% a 3-by-5 matrix as B.
% (基于矩阵A，通过定义特定元素产生3×5矩阵即为B)
>> A(3,5)=0
A =
     1     3     0     0     0
     5     7     0     0     0
     0     0     0     0     0
>> B=A
B =
     1     3     0     0     0
     5     7     0     0     0
     0     0     0     0     0
```

【例 2.2】 创建 Dürer 矩阵，又称魔术阵，如下：

$$A = \begin{bmatrix} 16 & 2 & 3 & 13 \\ 5 & 11 & 10 & 8 \\ 9 & 7 & 6 & 12 \\ 4 & 14 & 15 & 1 \end{bmatrix}$$

EXAMP02002

```
% Type all elements of matrix A directly in Command window.
% (在命令窗口直接输入矩阵 A 的所有元素)
>>A=[16,2,3,13;5,11,10,8;9,7,6,12;4,14,15,1]
A =
    16     2     3    13
     5    11    10     8
     9     7     6    12
     4    14    15     1
% Create a 4*4 magic matrix by "magic".
% (借助函数"magic"创建一个 4×4 魔术阵)
>> B=magic(4)
A =
    16     2     3    13
     5    11    10     8
     9     7     6    12
     4    14    15     1
```

2.1.2 矩阵生成命令

1. 全零矩阵

全零矩阵函数描述参见表 2.1。

表 2.1 zeros 函数

语法	描述	语法	描述
zeros	生成 1 个 0 标量	zeros(m, n, p, ⋯)	生成 1 个 $m×n×p×⋯$ 的全零数组
zeros(n)	生成 1 个 $n×n$ 的全零矩阵	zeros(size(A))	生成 1 个和矩阵 A 同尺寸的全零矩阵
zeros(m, n)	生成 1 个 $m×n$ 的全零矩阵		

【例 2.3】 创建几个全零矩阵。

EXAMP02003

```
创建几个全零矩阵。
    % Clear Command Window.
    % (清除命令窗口)
    >>clc
    % Clear variables in workspace.
    % (清除工作空间中的变量)
    >> clear
    % Create a 3-by-3 matrix containing random values drawn
    % from a uniform distribution on the unit interval by
```

```
% "rand".
% (借助函数"rand"创建一个 3×3 的任意矩阵)
>> A=rand(3)
A =
    0.8147    0.9134    0.2785
    0.9058    0.6324    0.5469
    0.1270    0.0975    0.9575
% Create a scalar 0 by "zeros".
% (通过函数"zeros"创建一个零标量)
>> zero1=zeros
zero1 =
    0
% Create a 2-by-2 matrix of zeros.
% (创建一个 2×2 的全零矩阵)
>> zero2=zeros(2)
zero2 =
    0    0
    0    0
% Create a 2-by-3 matrix of zeros.
% (创建一个 2×3 的全零矩阵)
>> zero23=zeros(2,3)
zero23 =
    0    0    0
    0    0    0
% Create a 3-by-5-by-2 array of zeros.
% (创建一个 3×5×2 层的全零数组)
>> zero352=zeros(3,5,2)
zero352(:,:,1) =
    0    0    0    0    0
    0    0    0    0    0
    0    0    0    0    0
zero352(:,:,2) =
    0    0    0    0    0
    0    0    0    0    0
    0    0    0    0    0
% Create a same size as A of zeros.
% (创建一个和矩阵 A 同行数同列数的全零矩阵)
>> zeroA=zeros(size(A))
zeroA =
    0    0    0
    0    0    0
    0    0    0
```

> **好习惯**
>
> 在编写程序时，通常开始我们就将所有矩阵定义为全零矩阵，因为这样将会使得所编写程序运行节省一些时间。

2. 全 1 矩阵

全 1 矩阵函数描述参见表 2.2。

表 2.2 ones 函数

语法	描述	语法	描述
ones	生成 1 个标量 1	ones(m, n, p, ⋯)	生成 1 个 $m \times n \times p \times \cdots$ 的全 1 数组
ones(n)	生成 1 个 $n \times n$ 的全 1 矩阵	ones(size(A))	生成 1 个和矩阵 A 同大小的全 1 矩阵
ones(m, n)	生成 1 个 $m \times n$ 的全 1 矩阵		

【例 2.4】 创建几个全 1 矩阵。

EXAMP02004

创建几个全 1 矩阵。

```
% Clear Command Window.
% (清除命令窗口)
>>clc
% Clear variables in workspace.
% (清除工作空间中的变量)
>> clear
% Create a 3-by-3 matrix containing pseudo-random values
% drawn from a normal distribution with mean zero and
% standard deviation one by the function "randn".
% (借助函数"randn"创建 3×3 的任意矩阵)
>> A=randn(3)
A =
    2.7694    0.7254   -0.2050
   -1.3499   -0.0631   -0.1241
    3.0349    0.7147    1.4897
% Create a scalar 1 by "ones".
% (借助函数"ones"创建一个标量)
>> one1=ones
one1 =  1
% Create a 5-by-5 matrix of ones by "ones".
% (借助函数"ones"创建一个 5×5 列全 1 矩阵)
>> one5=ones(5)
one5 =
     1     1     1     1     1
     1     1     1     1     1
     1     1     1     1     1
     1     1     1     1     1
     1     1     1     1     1
% Create a 2-by-7 matrix of ones by "ones".
% (借助函数"ones"创建 2 行 7 列的全 1 矩阵)
>> one27=ones(2,7)
one27 =
     1     1     1     1     1     1     1
     1     1     1     1     1     1     1
% Create a 3-by-6-by-3 array of ones by "ones".
% (借助函数"ones"创建 3 行 6 列 3 层的全 1 数组)
>> one363=ones(3,6,3)
one363(:,:,1) =
```

```
            1     1     1     1     1     1
            1     1     1     1     1     1
            1     1     1     1     1     1
one363(:,:,2) =
            1     1     1     1     1     1
            1     1     1     1     1     1
            1     1     1     1     1     1
one363(:,:,3) =
            1     1     1     1     1     1
            1     1     1     1     1     1
            1     1     1     1     1     1
% Create a matrix of ones as the same size of A.
% (创建与矩阵 A 同行数同列数的全 1 矩阵)
>> oneA=ones(size(A))
oneA =
     1     1     1
     1     1     1
     1     1     1
```

3. 单位矩阵

单位矩阵函数描述参见表 2.3。

表 2.3 eye 函数

语 法	描 述	语 法	描 述
eye	生成 1 个标量 1	eye(m, n)	生成 1 个 $m×n$ 的单位矩阵
eye(n)	生成 1 个 $n×n$ 的单位矩阵	eye(size(A))	生成 1 个和矩阵 A 一样大小的单位矩阵

【例 2.5】 创建几个单位矩阵。

EXAMP02005

创建几个单位矩阵。

```
    % Clear Command Window.
    % (清除命令窗口)
>>clc
    % Clear variables in workspace.
    % (清除工作空间中的变量)
>> clear
    % Create a 3-by-3 matrix containing pseudo-random values drawn from
    % a normal distribution with mean zero and standard deviation one.
    % (创建 3×3 的任意矩阵,矩阵元素符合正态分布)
>> A=randn(3)
A =
   -0.4326    0.2877    1.1892
   -1.6656   -1.1465   -0.0376
    0.1253    1.1909    0.3273
    % Create a unit scalar by "eye".
    % (借助"eye"函数创建单位标量)
>> eye1=eye
```

```
eye1 =  1
% Create a 2-by-2 identity matrix by "eye".
% (借助"eye"函数创建一个 2×2 的单位矩阵)
>> eye2=eye(2)
eye2 =
     1     0
     0     1
% Create a 3-by-5 identity matrix by "eye".
% (借助"eye"函数创建 3×5 的对角单位矩阵)
>> eye35=eye(3,5)
eye35 =
     1     0     0     0     0
     0     1     0     0     0
     0     0     1     0     0
% Create an identity matrix as the same size of A.
% (创建与矩阵 A 同规格的单位矩阵)
>> eyeA=eye(size(A))
eyeA =
     1     0     0
     0     1     0
     0     0     1
```

4. 随机矩阵生成函数

MATLAB 中 rand 函数是产生 0 到 1 的随机数，函数具体用法参见表 2.4。

表 2.4 rand 函数

语 法	描 述
rand	生成 1 个随机数标量
rand(n)	生成 1 个 $n \times n$ 随机数矩阵
rand(m, n)	生成 1 个 $m \times n$ 随机数矩阵
rand(m, n, p)	生成 1 个 $m \times n \times p$ 随机数数组
rand(size(A))	生成 1 个和矩阵 A 同尺寸的随机数矩阵
rand(method, s)	通过"method"来完成生成随机数的取值区间，"method"有以下三种选择，分别为"state"、"seed"和"twister"。而 s 是标量整数，取值范围为从 0 到 $2^{32}-1$，"state"对应取值范围为$[2^{(-53)}, 1-2^{(-53)}]$，执行 2^{1492} 次才有可能重复一次；"seed"对应取值范围为 $[1/(2^{31}-1), 1-1/(2^{31}-1)]$，周期为 $2^{31}-2$；"twister"对应取值范围$[2^{(-53)}, 1-2^{(-53)}]$，周期为$(2^{19937}-1)/2$

【例 2.6】 创建几个随机矩阵。

EXAMP02006

通过 rand 函数创建几个随机矩阵。

```
% Clear Command Window.
% (清除命令窗口)
>>clc
% Clear variables in workspace.
% (清除工作空间中的变量)
```

```
>> clear
% Create a double precision pseudo-random number.
% (产生一个随机数标量)
>> rand
ans =
    0.2311
% Create a 3-by-3 double precision pseudo-random matrix.
% (创建一个 3×3 的随机数矩阵)
>> rand(3)
ans =
    0.6068    0.7621    0.8214
    0.4860    0.4565    0.4447
    0.8913    0.0185    0.6154
% Create a 3-by-5 double precision pseudo-random matrix.
% (创建一个 3×5 的随机数矩阵)
>> rand(3,5)
ans =
    0.7919    0.1763    0.9169    0.0579    0.0099
    0.9218    0.4057    0.4103    0.3529    0.1389
    0.7382    0.9355    0.8936    0.8132    0.2028
% Create a 3-by-5-by-2 double precision pseudo-random array.
% (创建一个 3×5×2 的随机数数组)
>> rand(3,5,2)
ans(:,:,1) =
    0.1987    0.1988    0.4451    0.4186    0.2026
    0.6038    0.0153    0.9318    0.8462    0.6721
    0.2722    0.7468    0.4660    0.5252    0.8381
ans(:,:,2) =
    0.0196    0.8318    0.4289    0.1934    0.5417
    0.6813    0.5028    0.3046    0.6822    0.1509
    0.3795    0.7095    0.1897    0.3028    0.6979
% Create a 3-by-3 magical matrix.
% (创建一个 3×3 的魔术矩阵)
>> A=magic(3)
A =
    8    1    6
    3    5    7
    4    9    2
% Create a matrix with double precision pseudo-random numbers as
% the same size of A.
% (创建一个与矩阵 A 同尺寸的随机矩阵)
>> rand(size(A))
ans =
    0.3784    0.5936    0.8216
    0.8600    0.4966    0.6449
    0.8537    0.8998    0.8180
% Generate uniform values from the interval [5, 99] to construct
% a 5-by-5 matrix.
% (创建一个 5×5 的随机矩阵,矩阵元素取值区间为[5,99])
>> B=5+(99-5)*rand(5)
B =
```

```
       33.1993    95.9007    91.1195    92.2779    65.3216
       96.5908    84.4530    97.7124    14.3555    40.0986
       27.7735     8.3009    77.9867    74.8394    57.2399
       84.6939    23.3938    85.9498    17.0268    41.4749
       14.6084    77.0194    65.6323    40.1215    65.1922
% Generate an integer array with 50 elements ranging from 1 to 100.
% (产生一个含 50 个元素的整数数组,元素随机分布在 1 到 100 之间)
>> C=ceil(100*rand(1,50))
C =
 Columns 1 through 14
  8  31  68  93   5  43  54  59  41  57  97  50  61  39
 Columns 15 through 28
 73   4  52  99   9  96  69  44  50  89  35  12  94  81
 Columns 29 through 42
 45  18  58  81  71   1  97  80  52   7  60  41  92  45
 Columns 43 through 50
 83  47  65  82  95  69  32  78
```

注意:
  创建随机矩阵时,输入的 m、n 和 p 应该是非负整数,如果输入的是负整数,将按 0 处理。

5. 正态分布的随机矩阵生成函数

MATLAB 提供 randn 函数用于产生正态分布的随机矩阵,其具体用法参见表 2.5。

表 2.5 randn 函数

语法	描述
randn	生成正态分布的随机数标量
randn(n)	生成 1 个 $n×n$ 正态分布的随机矩阵
randn(m, n)	生成 1 个 $m×n$ 正态分布的随机矩阵
randn(m, n, p)	生成 1 个 $m×n×p$ 正态分布的随机数组
randn(size(A))	生成 1 个与矩阵 A 同尺寸的正态分布随机矩阵

【例 2.7】 创建几个正态分布随机矩阵。

EXAMP02007

通过 randn 函数创建几个正态分布随机矩阵。

```
    % Clear Command Window.
    % (清除命令窗口)
>>clc
    % Clear variables in workspace.
    % (清除工作空间中的变量)
>> clear
    % Generate a double precision random number.
    % (产生一双精度随机数)
>> randn
```

```
ans =
    0.7258
% Generate a 3-by-3 double precision random matrix.
%  (创建一个 3×3 正态分布的随机矩阵)
>> randn(3)
ans =
   -0.5883    0.1139   -0.0956
    2.1832    1.0668   -0.8323
   -0.1364    0.0593    0.2944
% Generate a 2-by-3 double precision random matrix.
%  (创建一个 2×3 正态分布的随机矩阵)
>> randn(2,3)
ans =
   -1.3362    1.6236    0.8580
    0.7143   -0.6918    1.2540
% Generate a 2-by-5-by-2 double precision random array.
%  (创建一个 2×5×2 的正态分布随机数组)
>> randn(2,5,2)
ans(:,:,1) =
   -1.5937    0.5711    0.6900    0.7119    0.6686
   -1.4410   -0.3999    0.8156    1.2902    1.1908
ans(:,:,2) =
   -1.2025   -0.1567    0.2573    1.4151    0.5287
   -0.0198   -1.6041   -1.0565   -0.8051    0.2193
% Produce a 3-by-3 Hilbert matrix.
%  (生成一个 3×3 希尔伯特矩阵)
>> A=hilb(3)
A =
    1.0000    0.5000    0.3333
    0.5000    0.3333    0.2500
    0.3333    0.2500    0.2000
% Generate a double precision random matrix as
%  the same size of A matrix.
%  (生成一个与矩阵 A 同尺寸的正态分布随机矩阵)
>> randn(size(A))
ans =
   -0.9219   -1.0106    1.6924
   -2.1707    0.6145    0.5913
   -0.0592    0.5077   -0.6436
% Generate a double precision random matrix with mean 1 and
%  standard deviation 2.
%  (生成一个正态分布含有 25 个元素的行数组,该数组平均值为 1,标准差为 2)
>> 1+2*randn(1,25)
ans =
  Columns 1 through 8
    1.7607  -1.0182   0.9610   0.9036   1.0001   0.3643   3.1900  -2.7480
  Columns 9 through 16
    1.8564   2.7913   2.4619   2.1557   1.0806   2.3542   2.1378   0.4887
  Columns 17 through 24
    0.2451   0.4082  -1.9503   0.5320   1.2369   1.6296   3.8870   0.2981
```

```
            Column 25
     2.2465
```

> **注意**
> 该函数 randn 中输入的参数 m、n、和 p 应该为非负整数，如果输入的是负整数，将按 0 处理。

如果读者对创建其他矩阵非常感兴趣，可以借助 MATLAB 中"help"命令对如："vander"、"hilb"、"invhilb"、"magic"和"compan"等函数进行学习。

【**例 2.8**】 创建几个特殊矩阵。

EXAMP02008

创建几个特殊矩阵。
```
     % Generate a Vandermonde matrix following vector V=[1 2 3 4 5].
     % (创建一个 5×5 范德蒙特矩阵)
     >> vander([1 2 3 4 5])
     ans =
          1     1     1     1     1
         16     8     4     2     1
         81    27     9     3     1
        256    64    16     4     1
        625   125    25     5     1
     % Generate a 3-by-3 Hilbert matrix.
     % (生成一个 3×3 希尔伯特矩阵)
     >> hilb(3)
     ans =
         1.0000    0.5000    0.3333
         0.5000    0.3333    0.2500
         0.3333    0.2500    0.2000
     % Generate a 3-by-3 inverse Hilbert matrix.
     % (生成一个 3×3 逆希尔伯特矩阵)
     >> invhilb(3)
     ans =
           9    -36     30
         -36    192   -180
          30   -180    180
     % Generate a 5-by-5 magic matrix.
     % (生成一个 5×5 魔术矩阵)
     >> magic(5)
     ans =
         17    24     1     8    15
         23     5     7    14    16
          4     6    13    20    22
         10    12    19    21     3
         11    18    25     2     9
     % Generate a company matrix of the polynomial x^5+ 2*x^4+ 3*x^3+
     %  4*x^2+ 5*x+6.
     % (生成多项式 x^5+2*x^4+3*x^3+4*x^2+5*x+6 的伴随矩阵)
```

```
>> compan([1 2 3 4 5 6])
ans =
    -2   -3   -4   -5   -6
     1    0    0    0    0
     0    1    0    0    0
     0    0    1    0    0
     0    0    0    1    0
```

2.2 构建数组方法

2.2.1 数组生成命令

当矩阵离开线性代数就变成了二维数字数组。对矩阵进行四则运算是将整个矩阵看成一个操作对象，而数组操作是对单个元素进行的。仅仅从外表看，数组与矩阵具有完全相同的外在形式，所以我们可以借助所有生成矩阵的命令来创建数组。矩阵和数组有一个明显的不同之处是，矩阵是二维的，数组可以超过二维，比如在"EXAMP02005"和"EXAMP02006"中创建的所谓数组"Array"。在此我们将不再对这一内容进行深入讨论。如果读者对此仍有不解之处或仍有疑问，请参照 MATLAB 软件中的帮助命令"help"进行深入学习。

2.2.2 矢量生成命令

1. 冒号":"符号

冒号符号在 MATLAB 中是最有用的符号之一，它不仅能用来引用、添加、删除数组的元素，而且可以用它来创建矢量数组。

【例 2.9】 通过冒号":"创建几个数组。

EXAMP02009

借助冒号":"符号创建几个数组。
```
% Generate a vector A=[1 2 3 4 5 6 7 8 9 10] by colon operator.
% (借助冒号创建一个矢量A=[1 2 3 4 5 6 7 8 9 10])
>> A=1:10
A = 1   2   3   4   5   6   7   8   9   10
% Generate a vector ranging from 1 to 20 with step-length 2.
% (生成一个矢量，元素从1到20，步长为2)
>> B=1:2:20
B = 1   3   5   7   9   11   13   15   17   19
% Generate a vector ranging from 100 to 20 with step-length 9.
% (生成一个矢量，元素从100到20，步长为-9).
>> C=100:-9:20
C = 100   91   82   73   64   55   46   37   28
```

2. linspace 函数

linspace 函数是用来生成等差数列的命令，其语法与相应描述参见表 2.6。

表 2.6 linspace 函数

语法	描述
linspace(x1, x2)	在从 x1 到 x2 区间内创建含有 100 个元素的等差数列
linspace(x1, x2, n)	在从 x1 到 x2 区间内创建含有 n 个元素的等差数列

【例 2.10】 通过 linsapce 函数创建几个等差数列。

EXAMP02010

```
通过 linspace 函数创建几个等差数列。
   % Generate a vector of 100 linearly equally spaced points
   % in the region [5,9].
   % (生成一个含 100 个元素的矢量，元素为 5 到 9 区间的等差数列)
>> linspace(5,9)
ans =
  Columns 1 through 8 ……
  5.0000 5.0404 5.0808 5.1212 5.1616 5.2020 5.2424  5.2828 ……
  Columns 89 through 96
  8.5556 8.5960 8.6364 8.6768 8.7172 8.7576 8.7980 8.8384
  Columns 97 through 100
  8.8788 8.9192 8.9596 9.0000
   % Generate a vector of 9 linearly equally spaced points
   % in closed internal [11,99].
   % (生成一个含 9 个元素矢量，元素为 11 到 99 区间的等差数列)
>> linspace(11,99,9)
   ans = 11    22    33    44    55    66    77    88    99
```

3. logspace 函数

logspace 函数用来创建指数等差数列，其语法与描述参见表 2.7。

表 2.7 logspace 函数

语法	描述
logspace(x1, x2)	在从 10^x1 到 10^x2 区间内生成 50 个指数为等差数列的元素
logspace(x1, x2, n)	在从 10^x1 到 10^x2 区间内生成 n 个指数为等差数列的元素

【例 2.11】 通过函数 logspace 创建几个数组。

EXAMP02011

```
通过函数 logspace 创建几个数组。
   % Set all numbers with long format.
   % (设定所有数为 long 格式)
>> format long
   % Generate a vector of 50 logarithmically equally spaced points
   % between 10^2 and 10^9.
```

```
% (在10^2 到10^9 区间生成一个含 50 个元素的对数等差数组)
>> logspace(2,9)
ans =
  1.0e+09 *
  Columns 1 through 3.......
  0.000000100000000   0.000000138949549   0.000000193069773......
  Columns 48 through 50
   0.517947467923122   0.719685673001153   1.000000000000000
% Generate a vector of 8 logarithmically equally spaced points
% between 10 and 10^9.
% (在10 到10^9 区间,生成含 8 个元素的对数等差数组)
>> logspace(1,9,8)
ans =
  1.0e+009 *
  Columns 1 through 4
 0.00000001000000,0.00000013894955,0.00000193069773,0.00002682695795
  Columns 5 through 8
 0.00037275937203,0.00517947467923,0.07196856730012,1.00000000000000
% Generate a vector of 4 logarithmically equally spaced points
% between 10^0.01 and pi.
% (在 10^0.01 到 pi 区间,生成含 4 个元素的对数等差数组)
>> logspace(0.01,pi,4)
ans =
  1.02329299228075, 1.48724764838776, 2.16155644993225, 3.14159265358979
% Reset the format of number as short.
% (重新将输入数设置为 short 格式)
>> format short
```

4. 基于已有矩阵或数组创建数组

在某些情况下,已经定义一些矩阵或数组。如果待创建数组的所有元素或部分元素在已有数组或矩阵已经存在,我们可以借助这些已有矩阵或数组来创建需要的新数组。下面提供了一些示例。

【例 2.12】 基于已有矩阵或数组构建新数组。

EXAMP02012

基于已有矩阵或数组构建新数组。

```
% Clear Command Window.
% (清除命令窗口)
>>clc
% Clear variables in workspace.
% (清除工作空间中的变量)
>> clear
% Generate a 5-by-5 Hilbert matrix.
% (生成一个 5×5 希尔伯特矩阵)
>> A=hilb(5)
A =
```

```
    1.0000    0.5000    0.3333    0.2500    0.2000
    0.5000    0.3333    0.2500    0.2000    0.1667
    0.3333    0.2500    0.2000    0.1667    0.1429
    0.2500    0.2000    0.1667    0.1429    0.1250
    0.2000    0.1667    0.1429    0.1250    0.1111
% Define an array (or vector) B with all elements of
% the predefined matrix A.
% (通过已定义矩阵A的元素,定义一个新数组或矢量,记为B)
>> B=A(:)
B =
    1.0000
    0.5000
    0.3333
    0.2500
    0.2000
    0.5000
    0.3333
    0.2500
    0.2000
    0.1667
    0.3333
    0.2500
    0.2000
    0.1667
    0.1429
    0.2500
    0.2000
    0.1667
    0.1429
    0.1250
    0.2000
    0.1667
    0.1429
    0.1250
    0.1111
% Define an array (or vector) C with elements in the predefined
% matrix A from the first to the last with step-length 4.
% (借助已定义矩阵A,定义一个新数组或矢量,记为C)
>> C=A([1:4:25])
C =
    1.0000    0.2000    0.2000    0.2000    0.2000    0.2000    0.1111
% Define an array (or vector) D with all elements in the first
% column of predefined matrix A.
% (借助已定义矩阵A,定义一个新数组或矢量,记为D,只含有矩阵第一列)
>> D=A(:,1)
D =
    1.0000
    0.5000
    0.3333
    0.2500
    0.2000
```

```
% Define an array (or matrix) E with all elements in the first
% and forth columns in the predefined matrix A.
% (借助已定义矩阵A，定义一个新数组或矢量，记为E，只含有矩阵的第一列与第二列)
>> E=A(:,[1:3:5])
E =
    1.0000    0.2500
    0.5000    0.2000
    0.3333    0.1667
    0.2500    0.1429
    0.2000    0.1250
% Define an array (or matrix) F with some special elements in
% the matrix A.
% (基于已定义矩阵A，定义一个新数组或矩阵，记为F，只含有部分特定元素)
>> F=A([1,3,5],[2,4])
F =
    0.5000    0.2500
    0.2500    0.1667
    0.1667    0.1250
% Defined an array (or scalar) G with the special element in
% the matrix A.
% (选定已定义矩阵A中特定元素，记为一个新数组或标量，记为G)
>> G=A(1,5)
G = 0.2000
% Create a matrix H with all elements of C and 99.
% (创建一个矩阵H，含有矩阵C和99)
>>H=[C,99]
H=
    1.0000    0.2000    0.2000    0.2000    0.2000    0.2000    0.1111    99
```

2.3 矩阵数组的运算操作

本节将对矩阵、数组数学运算进行介绍，有初等的四则运算，也有相对复杂的高等运算。为了根据相应情况进行相应运算，本节将对逻辑运算符号和关系运算符号进行阐述。在对运算函数与符号进行阐述前，首先介绍一些常数变量参见表2.8。

表 2.8 常数变量

变量	描述	变量	描述
pi	圆周率π的近似值	date	用来显示日期，格式为日-月-年（例如17-sep-15）
realmin	最小正实数	eps	机器零阈值
realmax	最大正实数	ans	计算结果的默认数值变量
bitmax	最大正整数	nargin	函数输入参数个数
i, j	虚数单位	nargout	函数输出参数个数
inf, Inf	无穷大	lasterr	存放最新的错误信息
nan, NaN	非数	lastwarn	存放最新的警告信息
clock	一般显示6个元素，分别为年、月、日、时、分、秒		

常用符号描述参见表 2.9。

表 2.9 常用符号

符号	描述
{ }	单元素数组构建符号
[ ]	数组或矩阵构建符号
( )	标注角标参数
' '	标注字符串
,	用来区分下标或矩阵元素
;	(1) 可以压缩命令窗口的运算结果，让其不显示； (2) 断开矩阵行； (3) 断开 MATLAB 中的执行语句，表示上一指令到此为止
%	标注解释内容
:	用于创建等差数列
+	加法
-	减法
.*	点乘，数组乘法
*	乘法
./	数组右除
.\	数组左除
/	右除
\	左除
.^	数组乘方
^	数乘方，矩阵乘方
'	矩阵和数组转置，复数取共轭
…	悬挂，代表上一行语句没有输入完

常见的操作命令描述参见表 2.10。

表 2.10 常见的操作命令

变量	描述	变量	描述
cd	显示或改变工作目录	pack	整理内存碎片
clc	清空命令窗口	path	显示搜索目录
clear	清空工作空间的变量	quit	退出 MATLAB
clf	清空图形窗口	save	保存内存变量
diary	日志文件命名	type	显示文件内容
dir	显示当前目录下的文件	what	列出所在目录 MATLAB 文件
disp	显示变量或文字的内容	which	定位 MATLAB 文件的路径
echo	命令窗口信息显示开关	who	列出工作空间的变量
hold	图形保持命令	whos	详细列出工作空间的变量
load	加载制订文件中的变量		

输出数据显示格式命令描述参见表 2.11。

表 2.11 输出数据显示格式命令

变 量	描 述
format short	小数点后包括 4 位有效数字,最多不超过 7 位有效数字;如果数值大于 1000,那么按照科学计数法表示
format long	用 15 位数字表示
format short e	5 位科学计数法表示
format long e	15 位科学计数法表示
format short g	从 format short 和 format short e 中自动选择最佳的数值表示方法
format long g	从 format long 和 format long e 中自动选择最佳的数值表示方法
format rat	采用近似有理数来表示
format hex	十六进制表示
format +	用+、-和空格来分别表示正数、负数和零,复数中的虚部不表示
format bank	金融表示方法,元、角、分等
format compact	显示变量之间没有空格
format loose	显示变量之间有空格

注意:

在 MATLAB 函数语法中,只有小括号,不能使用中括号和大括号,并且函数一般都是小写,不能出现大写字母,只有在函数所带的参数中有时有大写字母出现,比如后面绘图中将要讲述 plot 命令可带的参数 "LineWidth" 等。

所有带点的运算命令一般只能作用到数组,不能作用到矩阵,如果作用到矩阵时,已经把矩阵当作二维数组来处理了。带点的运算是对矩阵或数组里的元素单个操作的,不能把矩阵看成一个整体进行运算操作。

2.3.1 四则运算

1. 矩阵四则运算

众所周知,标量、矢量和复数都可以进行四则运算。为简单起见,我们将对矩阵四则运算进行简单介绍,参见表 2.12。

表 2.12 矩阵四则运算符号

+	加	\	左除
-	减	^	乘方
*	乘	'	转置
/	右除	( )	优先操作运算顺序

如果读者对矩阵四则运算不清楚,请阅读线性代数和矩阵相关参考书。

【例 2.13】 矩阵四则运算示例。

EXAMP02013

矩阵四则运算实例。

```
% Clear variable in workspace.
% (清除工作空间中的变量)
>>clear
% Clear Command Window.
% (清除命令窗口)
>>clc
% Generate a 3-by-3 magic matrix as A.
% (生成3×3魔术矩阵,记为A)
>> A=magic(3)
A =
     8     1     6
     3     5     7
     4     9     2
% Generate a 3-by-3 Hilbert matrix as B.
% (生成3×3希尔伯特矩阵,记为B)
>> B=hilb(3)
B =
    1.0000    0.5000    0.3333
    0.5000    0.3333    0.2500
    0.3333    0.2500    0.2000
% Solve the sum of A and B.
% (求解矩阵A和B的和)
>> A+B
ans =
    9.0000    1.5000    6.3333
    3.5000    5.3333    7.2500
    4.3333    9.2500    2.2000
% Solve the subtraction of A and B.
% (求解矩阵A和B之差)
>> A-B
ans =
    7.0000    0.5000    5.6667
    2.5000    4.6667    6.7500
    3.6667    8.7500    1.8000
% Solve the multiplication of A and B.
% (求解矩阵A和B之乘积)
>> A*B
ans =
   10.5000    5.8333    4.1167
    7.8333    4.9167    3.6500
    9.1667    5.5000    3.9833
% Solve the result of A divided by B from left side.
% (计算矩阵A右除矩阵B的结果)
>> A/B
ans =
  1.0e+003 *
    0.2160   -1.1760    1.1400
    0.0570   -0.4080    0.4500
   -0.2280    1.2240   -1.1400
% Solve the result of B divided by A from right side.
```

```
% (求解矩阵 B 左除矩阵 A 的结果)
>> A\B
ans =
    0.0963    0.0414    0.0257
   -0.0148    0.0032    0.0063
    0.0407    0.0275    0.0202
% Note the operation rules of matrix power.
% (注意矩阵幂的计算规则)
>> A^B
??? Error using ==> mpower
At least one operand must be scalar.
% Solve 2 powers of matrix A.
% (计算矩阵 A 的 2 次幂)
>> A^2
ans =
    91    67    67
    67    91    67
    67    67    91
% Solve 3 powers of matrix B.
% (计算矩阵 B 的 3 次幂)
>> B^3
ans =
    1.9111    1.0618    0.7462
    1.0618    0.5912    0.4159
    0.7462    0.4159    0.2927
% Solve the sum of matrix A and 100.
% (计算矩阵 A 和 100 的和)
>> A+100
ans =
   108   101   106
   103   105   107
   104   109   102
% Solve the subtraction of matrix A and 100.
% (计算矩阵 A 减 100 的结果)
>> A-100
ans =
   -92   -99   -94
   -97   -95   -93
   -96   -91   -98
% Solve the subtraction of 100 and matrix A.
% (计算 100 减矩阵 A 的结果)
>> 100-A
ans =
    92    99    94
    97    95    93
    96    91    98
% Solve the multiplication of 12 and matrix A.
% (计算 12 乘矩阵 A 之积)
>> 12*A
ans =
```

```
           96    12    72
           36    60    84
           48   108    24
% Note the operation rules of matrix division.
% (注意矩阵除法的运算规则)
>> (1+1/5)/A
??? Error using ==> mrdivide
Matrix dimensions must agree.
% Solve the result of matrix A divided by (1+1/5) from right side.
% (计算(1+1/5)左除矩阵A的结果)
>> (1+1/5)\A
ans =
    6.6667    0.8333    5.0000
    2.5000    4.1667    5.8333
    3.3333    7.5000    1.6667
% Solve the transpose of matrix A.
% (计算矩阵A的转置)
>> A'
ans =
     8     3     4
     1     5     9
     6     7     2
% Define a complex 2-by-2 matrix as D.
% (定义一个2×2复数矩阵,记为D)
>> D=[1+2*i,3-i;22-15*i,1/5-1/3*i]
D =
    1.0000 + 2.0000i   3.0000 - 1.0000i
   22.0000 -15.0000i   0.2000 - 0.3333i
% Solve the transpose of matrix D.
% (计算矩阵D的转置)
>> D'
ans =
    1.0000 - 2.0000i  22.0000 +15.0000i
    3.0000 + 1.0000i   0.2000 + 0.3333i
```

> **注意:**
> 转置符号"'"作用到矩阵上,是整个矩阵进行转置,如果作用到复数上,是对复数进行求共轭。

2. 数组

对数组进行四则运算操作,实际上是对数组中每个元素分别进行操作。数组四则运算符号参见表2.13。

表2.13 数组四则运算

+	加	./	右除	( )	限定运算顺序
-	减	.\	左除	dot	矢量点乘
.*	乘	.^	乘方	cross	矢量叉乘

【例 2.14】 数组四则运算实例。

EXAMP02014

数组四则运算实例。

```
%   Clear variable in workspace.
%   (清除工作空间中的变量)
>>clear
%   Clear Command Window.
%   (清除命令窗口)
>>clc
%   Generate an array with integer ranging from 1 to 9 as A.
%   (生成整数数组 A, 含有 9 个元素, 从 1 到 9)
>> A=1:9
A =
     1    2    3    4    5    6    7    8    9
%   Generate an array with integer ranging from 2 to 10 as B.
%   (生成整数数组 B, 含有 9 个元素, 从 2 到 10)
>> B=2:10
B =
     2    3    4    5    6    7    8    9   10
%   Solve the sum of A and B.
%   (求解数组 A 和 B 之和)
>> A+B
ans =
     3    5    7    9   11   13   15   17   19
%   Solve the subtraction of A and B.
%   (求解数组 A 和 B 之差)
>> A-B
ans =
    -1   -1   -1   -1   -1   -1   -1   -1   -1
%   Solve the multiplication of A and B.
%   (求解数组 A 和 B 之积)
>> A.*B
ans =
     2    6   12   20   30   42   56   72   90
%   Solve the result of B divided by A from left side.
%   (计算数组 B 左除数组 A 的结果)
>> A.\B
ans =
  Columns 1 through 8
   2.0000   1.5000   1.3333   1.2500   1.2000   1.1667   1.1429   1.1250
  Column 9
   1.1111
%   Solve the result of A divided by B from right side.
%   (计算数组 A 右除数组 B 的结果)
>> A./B
ans =
  Columns 1 through 8
```

```
         0.5000    0.6667    0.7500    0.8000    0.8333    0.8571    0.8750    0.8889
  Column 9
         0.9000
% Solve the power of A with B.
% （计算数组 A 的 B 次幂）
>> A.^B
ans =
  1.0e+009 *
  Columns 1 through 8
         0.0000    0.0000    0.0000    0.0000    0.0000    0.0003    0.0058    0.1342
  Column 9
         3.4868
```

> **注意**：在数组四则运算时，如果两个数组都含有多个元素，那么两个数组元素个数必须相等。

2.3.2 初等数学运算

本节相关知识，对于一般读者都非常熟悉，所以这里我们只是对一些初等函数进行罗列，没有进行举例讲解，参见表 2.14。如果读者想了解更加详细的知识，请借助 MATLAB 帮助系统进行学习，或阅读相关资料。

表 2.14 初等数学函数

函　数	描　述
abs(x)	计算绝对值|x|
angle	复数的相角
conj	复数共轭函数
imag	复数的虚部
real	复数的实部
cos(x)	x 的余弦值，x 的单位为弧度
acos(x)	x 的反余弦函数，结果单位为弧度
sin(x)	x 的正弦值，x 的单位为弧度
asin(x)	x 的反正弦函数，结果单位为弧度
tan(x)	x 的正切值，x 的单位为弧度
atan(x)	x 的反正切函数，结果单位为弧度
atan2(y,x)	计算圆上 4 个象限的所有点反正切函数，结果取值范围为[-pi, pi]
sqrt(x)	x 的平方根
log(x)	x 的自然对数
log10(x)	x 的对数
logm(A)	A 的矩阵对数，A 为单精度或双精度浮点数
exp(x)	x 的指数
expm(x)	x 的矩阵指数，expm(x)=V*diag(exp(diag(d)))/V，其中 V 为 x 完备本征矢
mod(x,y)	x 除 y 的余数，符号与 y 相同（又称对数法取余数）
rem(x,y)	x 除 y 的余数，符号与 x 相同（又称对数法取余数）
sign(x)	符号函数

在 MATLAB 中，我们经常需要取四舍五入，下面给出一些常用的四舍五入函数，参见表 2.15。

表 2.15 四舍五入函数

函 数	描 述
ceil(x)	向正无穷舍入，如 ceil(3.1)=4 and ceil(-3.1)=-3
fix(x)	向 0 舍入，如 fix(3.1)=3 and fix(-3.1)=-3
floor(x)	向负无穷舍入，如 floor(3.1) = 3 and floor(-3.1)=-4
round(x)	向最近整数舍入

2.3.3 矩阵运算操作函数

在本节，我们将简要介绍关于矩阵的一些基本运算操作命令，比如：det、eig、inv、rank、trace、diag，等等。矩阵运算函数描述参见表 2.16。

表 2.16 矩阵运算函数

函 数	描 述
det	求矩阵行列式
eig	求矩阵特征值与相应特征向量，如[D,V]=eig(B),B*D=D*V
eigs	一般是用来求解大矩阵的特征值与特征向量的，但是显示结果上有些区别，需要注意的是，该命令只给出 6 个最大的特征值，采用了 Rice 大学所编写的求解大型矩阵的子程序包 ARPACK
inv	求矩阵的逆矩阵
rank	求矩阵的秩
trace	求矩阵的迹，对角线元素之和
svd	[U,S,V] = SVD(X)，S 为对角矩阵，满足 X = U*S*V'关系
diag	取矩阵的对角线元素，如 diag(A,k)，其中 k 可取大于 0 或小于 0 的数，但默认值为 0

【例 2.15】 首先构建矩阵：

$$B = \begin{bmatrix} 17 & 24 & 1 & 8 & 15 \\ 23 & 5 & 7 & 14 & 16 \\ 4 & 6 & 13 & 20 & 22 \\ 10 & 12 & 19 & 21 & 3 \\ 11 & 18 & 25 & 2 & 9 \end{bmatrix}$$

然后求矩阵 **B** 的行列式、特征值与特征向量、逆、秩、迹、矩阵对角化等。

EXAMP02015

为矩阵操作函数提供示例。

```
%   Clear variable in workspace.
%   (清除工作空间中的变量)
>>clear
%   Clear Command Window.
%   (清除命令窗口)
```

```
>>clc
% Generate a 5-by-5 magic matrix as B.
% （生成一个5×5魔术矩阵，记为B）
>> B=magic(5)
B =   17    24     1     8    15
      23     5     7    14    16
       4     6    13    20    22
      10    12    19    21     3
      11    18    25     2     9
% Solve the determinant of matrix B.
% （计算矩阵B的行列式）
>> det(B)
ans = 5070000
% Solve the eigenvectors and eigenvalues of B as D and V.
% （求解矩阵B的本征矢和本征值分别记为D和V）.
>> [D,V]=eig(B)
D =  -0.4472    0.0976   -0.6330    0.6780   -0.2619
     -0.4472    0.3525    0.5895    0.3223   -0.1732
     -0.4472    0.5501   -0.3915   -0.5501    0.3915
     -0.4472   -0.3223    0.1732   -0.3525   -0.5895
     -0.4472   -0.6780    0.2619   -0.0976    0.6330
V =  65.0000         0         0         0         0
           0  -21.2768         0         0         0
           0         0  -13.1263         0         0
           0         0         0   21.2768         0
           0         0         0         0   13.1263
% Solve the inverse of matrix B.
% （求解矩阵B的逆）
>> inv(B)
ans = -0.0049    0.0512   -0.0354    0.0012    0.0034
       0.0431   -0.0373   -0.0046    0.0127    0.0015
      -0.0303    0.0031    0.0031    0.0031    0.0364
       0.0047   -0.0065    0.0108    0.0435   -0.0370
       0.0028    0.0050    0.0415   -0.0450    0.0111
% Solve the rank of matrix B.
% （求解矩阵B的秩）
>> rank(B)
ans =    5
% Solve the trace of matrix B.
% （求解矩阵B的迹）
>> trace(B)
ans =   65
% Solve singular value decomposition for B.
% （求解矩阵B的奇异值分解）
>> [U,S,V]=svd(B)
U =  -0.4472   -0.5456    0.5117    0.1954   -0.4498
     -0.4472   -0.4498   -0.1954   -0.5117    0.5456
     -0.4472   -0.0000   -0.6325    0.6325   -0.0000
     -0.4472    0.4498   -0.1954   -0.5117   -0.5456
     -0.4472    0.5456    0.5117    0.1954    0.4498
S =  65.0000         0         0         0         0
```

```
                0    22.5471          0          0          0
                0          0    21.6874          0          0
                0          0          0    13.4036          0
                0          0          0          0    11.9008
    V =   -0.4472   -0.4045     0.2466    -0.6627     0.3693
          -0.4472   -0.0056     0.6627     0.2466    -0.5477
          -0.4472    0.8202    -0.0000    -0.0000     0.3568
          -0.4472   -0.0056    -0.6627    -0.2466    -0.5477
          -0.4472   -0.4045    -0.2466     0.6627     0.3693
    % Solve the diagonal of matrix B.
    % (求解矩阵 B 的对角矩阵)
>>  diag(B)'
    ans =    17    5    13    21    9
```

在一些特定条件下，可能需要重新构建已有矩阵，MATLAB 提供了丰富的操作矩阵函数，参见表 2.17。

表 2.17 操作矩阵函数

函数	描述
rot90	逆时针方向旋转矩阵 90°，如果带有参数 K，就是逆时针方向旋转 $K×90°$，$K=+1,-1,+2,-2,\cdots$
tril	取矩阵下三角部分，省略参数 K，取默认值为 0，一般取相对对角线向上 K 行的下三角矩阵部分（如果 K 为负值，取向下的下三角矩阵部分）
triu	取矩阵上三角部分，省略参数 K，取默认值为 0，一般取相对对角线向上 K 行的上三角矩阵部分（如果 K 为负值，取向下的上三角矩阵部分）
flipud	上下翻转
fliplr	左右翻转
[ ]	删除矩阵的任意行或列

【例 2.16】 首先创建矩阵：

$$C = \begin{bmatrix} 1.0000 & 0.5000 & 0.3333 & 0.2500 \\ 0.5000 & 0.3333 & 0.2500 & 0.2000 \\ 0.3333 & 0.2500 & 0.2000 & 0.1667 \\ 0.2500 & 0.2000 & 0.1667 & 0.1429 \end{bmatrix}$$

然后通过函数"chol"构建一个新矩阵 D，再对矩阵 D 进行旋转，对矩阵 C 取上三角、下三角等各种操作。

EXAMP02016

创建一个 4×4 希尔伯特矩阵为 C，然后再通过函数"chol"创建一个新矩阵 D，再借助上面讲述的函数对这两个矩阵进行各种操作。

```
    % Clear Command Window.
    % (清除命令窗口)
>>clc
    % Clear variables in workspace.
    % (清除工作空间中的变量)
>>clear
```

```
% Generate a 4-by-4 Hilbert matrix as C.
% (生成一个 4×4 希尔伯特矩阵,记为 C)
>> C=hilb(4)
C =
    1.0000    0.5000    0.3333    0.2500
    0.5000    0.3333    0.2500    0.2000
    0.3333    0.2500    0.2000    0.1667
    0.2500    0.2000    0.1667    0.1429
% Solve the Cholesky factorization of C as D, C=D'*D.
% (求解矩阵 C 的丘拉斯基分解,记为 D,有 C=D'*D)
>> D=chol(C)
D =  1.0000    0.5000    0.3333    0.2500
         0    0.2887    0.2887    0.2598
         0         0    0.0745    0.1118
         0         0         0    0.0189
% Rotate D 90 degrees counterclockwise.
% (矩阵 D 逆时针旋转 90°)
>> rot90(D)
ans =   0.2500    0.2598    0.1118    0.0189
        0.3333    0.2887    0.0745         0
        0.5000    0.2887         0         0
        1.0000         0         0         0
% Rotate D 2*90 degrees counterclockwise.
% (矩阵 D 逆时针旋转 2*90°)
>> rot90(D,2)
ans =
        0.0189         0         0         0
        0.1118    0.0745         0         0
        0.2598    0.2887    0.2887         0
        0.2500    0.3333    0.5000    1.0000
% Rotate D 90 degrees clockwise.
% (矩阵 D 顺时针旋转 90°)
>> rot90(D,-1)
ans =
             0         0         0    1.0000
             0         0    0.2887    0.5000
             0    0.0745    0.2887    0.3333
        0.0189    0.1118    0.2598    0.2500
% Select the lower triangle part of C from the upper 1 line.
% (从主对角线的上一行,取矩阵 C 的左下角)
>> tril(C,1)
ans =1.0000    0.5000         0         0
     0.5000    0.3333    0.2500         0
     0.3333    0.2500    0.2000    0.1667
     0.2500    0.2000    0.1667    0.1429
% Select the lower triangle part of C from the lower 1 line.
% (从主对角线的下一行,取矩阵 C 的左下角)
>> tril(C,-1)
ans =        0         0         0         0
        0.5000         0         0         0
```

```
         0.3333     0.2500          0          0
         0.2500     0.2000     0.1667          0
% Select the upper triangle part of C from the upper 1 line.
% (从主对角线的上一行,取矩阵C的右上角)
>> triu(C,1)
ans =      0     0.5000     0.3333     0.2500
           0          0     0.2500     0.2000
           0          0          0     0.1667
           0          0          0          0
% Select the upper triangle part of C from the lower 1 line.
% (从主对角线的下一行,取矩阵C的右上角)
>> triu(C,-1)
ans =1.0000     0.5000     0.3333     0.2500
     0.5000     0.3333     0.2500     0.2000
          0     0.2500     0.2000     0.1667
          0          0     0.1667     0.1429
% Flip the matrix triu(C,1) up/down direction.
% (对矩阵triu(C,1)上下翻转)
>> flipud(triu(C,1))
ans =      0          0          0          0
           0          0          0     0.1667
           0          0     0.2500     0.2000
           0     0.5000     0.3333     0.2500
% Flip the matrix triu(C,-1) left/right direction as T.
% (对矩阵triu(C,-1)左右翻转,记为T)
>>T= fliplr(triu(C,-1))
T =
     0.2500     0.3333     0.5000     1.0000
     0.2000     0.2500     0.3333     0.5000
     0.1667     0.2000     0.2500          0
     0.1429     0.1667          0          0
% Exchange the second column and the third in T as S.
% (互换矩阵T中的第二列与第三列,记为S)
>> S=T(:,[1 3 2 4])
S =  0.2500     0.5000     0.3333     1.0000
     0.2000     0.3333     0.2500     0.5000
     0.1667     0.2500     0.2000          0
     0.1429          0     0.1667          0
```

2.4 多项式

2.4.1 多项式表述

MATLAB 中,通常可以借助含有所有系数的数值矢量来表示符号多项式。对于矢量、数组和矩阵可以通过键盘直接输入,也可以通过 MATLAB 中已定义函数来实现。但是如果需要把数值矢量表示的多项式转化为习惯意义上的多项式,我们可以通过函数"poly2str"来完成。

【例 2.17】 首先创建一个矢量数组,然后将其所表示的多项式转化为常见形式。

EXAMP02017

首先创建一个矢量数组，然后将其所表示的多项式转化为常见形式。

```
% Clear Command Window.
% (清除命令窗口)
>>clc
% Clear variables in workspace.
% (清除工作空间中的变量)
>>clear
% Generate a vector.
% (生成一个矢量)
>> A=[2 5 9 6 3 8]
A =    2    5    9    6    3    8
% Change the predefined vector A into a symbolic polynomial.
% (将已定义矢量A转换为一符号多项式)
>> poly2str(A,'x')
ans = 2 x^5 + 5 x^4 + 9 x^3 + 6 x^2 + 3 x + 8
```

注意：
在函数"poly2str"的语法中，需要明确注明变量是什么，否则将给出错误提示。

2.4.2 多项式操作

MATLAB 为符号多项式也提供了很多操作函数，参见表 2.18。

表 2.18 多项式函数

函数	描述	函数	描述
poly2str	将多项式的数值矢量表示形式转化为常见形式	polyfit	多项式曲线拟合
roots	求解多项式的根	interp	插值
poly	根据特定根给出多项式	polyval	多项式求值
conv	多项式相乘	polyvalm	多项式矩阵求值
deconv	多项式相除	fminsearch	寻找最小值所对应的自变量值
polyder	多项式微分	fzero	寻找因变量为0对应的自变量值

表里所列函数有限，如对该类函数感兴趣，可以借助 MATLAB 帮助系统进一步学习，或查阅相关书籍/资料。

【例 2.18】 首先创建两个多项式矢量数组 *A* 和 *B*，然后求解这两个多项式 *A* 和 *B* 的积和商。在程序编写调试窗口，输入如下命令，然后直接存为 EXAMP02018.m 即可。（当然本例题完全可以在命令窗口输入命令完成）

EXAMP02018

```
% Clear Command Window.
% (清除命令窗口)
clc
% Clear variables in workspaces.
```

```
% (清除工作空间中的变量)
clear
% Define two row vectors A and B.
% (定义两行矢量,记为A和B)
A=[1,2,3,4,5]
B=[9,8,6,4,0]
% Solve the multiplication of the two polynomial A and B.
% (计算由A和B对应的多项式之积)
C=conv(A,B)
% Return polynomial C as string with a variable x.
% (返回多项式C的字符表达式,变量为x)
poly2str(C,'x')
% Solve the division of the two polynomials A and B as D.
% (求解多项式A和B相除,记为D)
D=deconv(A,B)
% Return polynomial D as string with a variable x.
% (返回多项式D的字符表达式,变量为x)
poly2str(D,'x')
% Solve the differential of polynomial C, and save as
% a function depending of t.
% (对多项式C求微商,保存为自变量为t的函数)
Ct_1=poly2str(polyder(C),'t')
```

运行结果如下:

```
A = 1    2    3    4    5
B = 9    8    6    4    0
C = 9   26   49   76  103   76   46   20   0
ans = 9 x^8 + 26 x^7 + 49 x^6 + 76 x^5 + 103 x^4 + 76 x^3 + 46 x^2 + 20 x
D = 0.1111
ans = 0.11111
Ct_1 = 72 t^7 + 182 t^6 + 294 t^5 + 380 t^4 + 412 t^3 + 228 t^2 + 92 t + 20
```

【**例 2.19**】 首先给出两个数值矢量式数组,然后基于这两数组拟合出最高次幂为 5 的关于 t 的多项式,最后在 $t=3.1$ 处进行插值。

EXAMP02019

```
% Generate two row vectors S and T. Then based on the two row vectors,
% a 5-order t polynomial of SE is given by polyfit command, and
% a new value is interpolated at point t=3.1.
% (生成两行矢量记为S和T。然后基于这两个矢量,拟合出最高阶为5次幂的多项式,
%  并且基于已知数据,在t=3.1处进行插值)
%
% Clear Command Window.
% (清除命令窗口)
clc
% Clear variables in workspaces.
% (清除工作空间中的变量)
clear
% Define two row vectors S and T.
```

```
%   (生成两行矢量,记为 S 和 T)
S=[12.3 21.5 34.2 46.2 39.5];
T=[1 2 3 4 5];
%   Solve a polynomial of 5-order t from the previously defined
%   two vectors S and T as SE.
%   (基于前面两个矢量,拟合 5 次幂变量为 t 的多项式,记为 SE).
SE=polyfit(T,S,5)
%   Return polynomial SE as string with a variable t.
%   (返回多项式 SE 的字符表达式,变量为 t).
se=poly2str(SE,'t')
%   Interpolate a new value of S at t=3.1 point.
%   (在 t=3.1 处,插入一个数值).
v=interp1(T,S,3.1)
```

运行结果如下:

```
Warning: Polynomial is not unique; degree >= number of data points.
In polyfit at 72    (警告提示:多项式不唯一)
In EXAMP02019 at 9
SE =-0.0912    0.7936    -2.7055    6.3542    0    7.9489
se =-0.091241 t^5 + 0.79361 t^4 - 2.7055 t^3 + 6.3542 t^2 + 7.9489
v = 35.4000
```

【例 2.20】 首先定义一个函数 $f = x^2 \sin x$,然后在 $x = 3$ 附近和范围为 $[-2,2]$ 内寻找 x 对应函数最小值的取值,再在 $x = -2$ 附近寻找函数为 0 的 x 对应的值。

EXAMP02020

```
%   Define a function f=sin(x)*x^2, and then find the x-position for
%   the minimum value of f around x=3, and the x-position for
%   the minimum value of f with x ranging from -2 to 2,
%   the x-position for the zero-value of f around x=-2.
%   (先定义一个函数 f=sin(x)*x^2,然后在 x=3 附近寻找极小值的 x 坐标,在[-2,2]范围内
%   寻找极小值的 x 坐标,在 x=-2 附近寻找函数为 0 的 x 坐标)
%
%   Clear Command Window.
%   (清除命令窗口)
clc
%   Clear variables in workspaces.
%   (清除工作空间中的变量)
clear
%   Define a function depending on x as f=sin(x)*x^2.
%   (定义一个 x 的函数为 f=sin(x)*x^2)
f='sin(x)*x^2';
%   Find the x-position for the minimum value of f.
%   (寻找函数 f 极值对应的 x 坐标)
xmin=fminsearch(f,3)
xmin=fminbnd(f,-2,2)
%   Plot the line of the function f to demonstrate that the above
%   results are right.
```

```
% (对函数 f 进行绘图, 以证明上面结果是正确的)
fplot(f,[-2*pi,2*pi])
% Hold on the present figure.
% (保留在当前图形窗口)
hold on
% Define a new independent variable x1.
% (定义一个新自变量 x1)
x1=linspace(-2*pi,2*pi,100);
% Define a new dependent variable y1.
% (定义一个新因变量 y1)
y1=zeros(1,length(x1));
% Plot the line defined by x1 and y1.
% (绘制 x1 和 y1 的曲线)
plot(x1,y1,'r:')
% Hold off the present figure.
% (释放当前图形)
hold off
% Find the x-position for the zero value of f.
% (寻找函数 f 为 0 对应的 x 坐标)
xzero=fzero(f,-2)
```

运行结果如下：

```
xmin = 5.0870
xmin = -1.9999
xzero = -3.1416
```

例 2.20 运行结果示意图如图 2.1 所示。

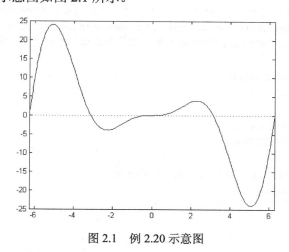

图 2.1 例 2.20 示意图

2.5 符号表达式的生成

如果要对符号函数、符号多项式、符号方程中的符号变量进行操作，那么在定义这些函数、多项式、方程前就要定义其中所含有的所有符号变量。本节将介绍如何定义符号数、符号变量、符号对象，如何操作这些符号对象，如何简化符号多项式。

2.5.1 创建符号对象

MATLAB 语言中，提供了两个函数"sym"和"syms"，可以定义符号数、符号变量、符号对象等。"syms"是一次定义多个符号对象的简捷办法，下面介绍它们的语法格式。

1. sym 函数

sym 函数可以用来创建符号数、符号变量和符号对象等。

语法格式：S=sym(A)基于已知 A，创建对象 S 为符号。如果针对 A 输入的是一段字符，那么结果就是由这段字符为名字的符号变量；如果输入的是一个数和数值矩阵，那么结果符号 S 代表这符号数或符号矩阵。

x=sym('x')创建符号变量 x，将存储给 x。其他语法形式，如：x=sym('x', 'real')是将 x 符号变量定义为实数符号变量，并存储给 x，x=sym('x', 'positive')为正数符号变量，x=sym('x', 'unreal')为非实数符号变量，等等。

S=sym(A, flag) 将数值标量或矩阵变为符号形式。第二个参数选项可以是"f"、"r"、"e"、"d"。其中，"f"代表浮点数；"r"代表有理数；"e"代表区别错误精度，一般与"eps"相当；"d"代表十进制。

2. syms 函数

syms 是一次创建多个符号对象的快捷命令。为简单起见，syms 使用率比 sym 高。

【例 2.21】 借助函数 sym 和 syms 可创建符号对象，比如：符号数、符号变量、符号表达式等。

EXAMP02021

```
% Use two functions "sym" and "syms" to create symbolic numbers,
% symbolic variables and symbolic expressions and so on.
% (用量函数"sym"和"syms"创建符号数、符号变量和符号表达式，等等)
%
% Clear Command Window.
% (清除命令窗口)
clc
% Clear variables in workspace.
% (清除工作空间中的变量)
clear
% Define symbolic numbers a, b and c.
% (定义符号数 a、b、c)
a=sym('1/3')
b=sym('2/5')
c=sym('1+4/5')
% Calculate the expression depending on the defined a, b and c.
% (计算关于 a、b、c 的表达式)
myvalue1=a*b-a^2/c
% Define symbolic variables x, y and z.
% (定义符号变量 x、y、z)
x=sym('x')
y=sym('y')
```

```
z=sym('z')
% Solve a function funxyz depending on x, y and z.
%  (计算关于已定义变量 x、y、z 的函数)
funxyz=x^2+2*y*z
pretty(funxyz)
```

运行结果如下:

```
a =1/3
b =2/5
c =9/5
myvalue1 =29/405
x =x
y =y
z =z
funxyz =x^2 + 2*y*z
   2
  x  + 2 y z
```

> **注意:**
> 符号数与数值数是不同的,符号数严格准确,数值数通常都是近似的。

通过函数"sym"可以直接创建符号表达式和符号方程,借助已经定义的符号变量也可以创建符号表达式和符号方程。如果从操作上看,通过函数"sym"直接完成很简单,但是这两种方法有些不同,就是借助函数"sym"直接创建的符号表达式不能与其中所含有的符号变量进行运算,而借助已定义的符号变量创建的符号表达式或符号方程可以与其中符号变量进行计算,如例 2.22。

【**例 2.22**】 举例说明:直接通过函数"sym"直接创建的符号表达式与先创建符号变量再创建符号表达式之间的区别。

EXAMP02022

```
% Distinguish the difference between the symbolic expression defined
% by "sym" and that expressed by the defined symbolic variables
% by "sym" or "syms".
% (说明:直接通过"sym"定义的表达式与先定义符号变量然后通过它们定义的表达式之
% 间的区别)
%
% Clear variables in workspace.
%  (清除工作空间中的变量)
clear
% Clear Command Window.
%  (清除命令窗口)
clc
% Define a, b, c, x, y and z as symbols before the definition a symbolic
% function fxyz.
%  (定义符号函数 fxyz 之前,先定义符号变量 a、b、c、x、y、z)
syms a b c x y z
% Define a symbolic function fxyz.
```

```
%  (定义一个符号函数 fxyz)
fxyz=a*x^2+b*y*z^3-c*z
%  Calculate the subtraction between the function fxyz and
%  an expression depending on the defiend symbolic varables.
%  (计算函数 fxyz 和一个表达式的差)
fxyz-a*x^2
%  Define a symbolic function fxyzw, and then to be subtracted
%  by "a*x^2".
%  (清除所有已定义符号变量,定义一个符号函数 fxyz,然后减去上面出现的一个表达式)
%  Notice to clear variables in workspace.
%  (注意清除工作空间中的变量)
clear
fxyzw=sym('a*x^2+b*y*z^3-c*z')
fxyzw-a*x^2
```

运行结果如下:

```
fxyz =a*x^2 + b*y*z^3 - c*z
ans =b*y*z^3 - c*z
fxyzw =a*x^2 + b*y*z^3 - c*z
Undefined function or variable 'a'.  (错误提示,说明没有定义符号变量"a")
Error in EXAMP02022 (line 31)
fxyzw-a*x^2
```

如果欲将符号数转化为数值数,"digits"和"vpa"两个函数会助你一臂之力。这两个函数我们将对其进行介绍。

3. 函数 digits 和 vpa

1) 函数 digits

digits 决定 Maple 数值计算的有效位数。

语法格式:digits 显示当前有效位数设定值。

digits(D)将计算有效位数值设定为 D,D 是一个整数,或者是一个符号表达式数值为整数。

D=digits 与只输入命令 digits 一样,给出当前数值计算有效位数,并存储给 D。

2) 函数 vpa

语法格式为:R=vpa(S)对符号矩阵 S 进行数值估算,有效位数为当前默认有效位数。

R=vpa(S,D)对符号矩阵 S 进行数值估算,有效位数设定为 D。

【例 2.23】 通过"digits"和"vpa"两个函数定义需要有效位数。

EXAMP02023

```
%  The examples for the usage of the two functions"digits" and "vpa".
%  (两个函数"digits"和"vpa"的用法示例)
%
%  Clear Command Window.
%  (清除命令窗口)
clc
```

```
%  Clear variables in workspace.
%  (清除工作空间中的变量)
clear
%  Define symbolic numbers a, b and c.
%  (定义符号数a、b、c)
a=sym('1/3')
b=sym('2/5')
c=sym('1+4/5')
%  Evaluate the numeric value of the symbolic expression "a*c- b/c ".
%  (估算表达式"a*c- b/c"的数值)
vpa(a*c-b/c)
%  Define the accurate number is 5.
%  (定义计算精度为5位有效数字)
digits(5)
%  Evaluate the numeric value of "a*c- b/c ".
%  (估算表达式"a*c- b/c"的数值)
vpa(a*c- b/c)
%  Define the accurate number is 20.
%  (定义计算精度为20位有效数字)
digits(20)
%  Evaluate the numeric value of "a*c- b/c ".
vpa(a*c- b/c)
%  Evaluate the numeric value of "a*c- b/c " with the accurate
%  numver as 100.
%  (估算表达式"a*c- b/c"的数值,保留有效数字为100位)
vpa(a*c-b/c,100)
```

运行结果如下：

```
a =1/3
b =2/5
c =9/5
ans =0.37777777777777777777777777777777777777777777777778
(有效位数为50位)
ans =0.37778  (有效位数为5位)
ans =0.37777777777777777778  (有效位数为20位)
ans
=0.3777777777777777777777777777777777777777777777777777777777777777777777777777777777777777777777777778 (有效位数为100位)
```

符号对象与数值对象是有一定区别的，从例2.24很容易看出。

【例2.24】 符号数与数值数之间的区别。

EXAMP02024

```
%  The examples for the distinguishment of the difference
%  between symbolic numbers and numerical numbers.
%  (区别符号数与数值数间不同的示例)
%
%  Clear Command Window.
%  (清除命令窗口)
```

```
clc
% Clear variables in workspace.
% (清除工作空间中的变量)
clear
% Define symbolic numbers as as, bs and cs.
% (定义符号数 as、bs、cs)
as=sym('pi')
bs=sym('9')
cs=sym('sqrt(3)')
% Calculate sin(as/4).
% (计算 sin(as/4))
sin(as/4)
% Express the above result in usual mathematical form.
% (将上面结果表示为通常数学形式)
pretty(sin(as/4))
% Calculate sqrt(as*bs-cs).
% (计算 sqrt(as*bs-cs))
sqrt(as*bs-cs)
% Express the above result in usual mathematical form.
% (将上面结果表示为通常数学形式)
pretty(sqrt(as*bs-cs))
% Define and numerical numbers a, b and c.
% (定义数值数 a、b、c)
a=pi
b=9
c=sqrt(3)
% Calculate sin(a/4).
% (计算 sin(a/4))
sin(a/4)
% Calculate sqrt(a*b-c).
% (计算 sqrt(a*b-c))
sqrt(a*b-c)
```

运行结果如下：

```
as =pi
bs =9
cs =3^(1/2)
ans =2^(1/2)/2
    1/2
   2
   ----
    2
ans =(9*pi - 3^(1/2))^(1/2)
         1/2  1/2
   (9 pi - 3   )
a = 3.1416
b = 9
c = 1.7321
ans = 0.7071
ans = 5.1519
```

4. 创建符号矩阵

矩阵是 MATLAB 操作的元素，所以在讨论符号对象时，首先要知道符号矩阵，然后可以通过操作矩阵命令完成一些基本运算操作，如 diag、triu、tril、inv、det、rank、eig、transpose，等等。

【例 2.25】 创建一个符号矩阵，然后对该矩阵进行运算操作。

EXAMP02025

```
% Create symbolic matrices, and operate them by the functions
% of matrix.
% (创建符号矩阵，并通过矩阵函数进行操作)
%
% Clear Command Window.
% (清除命令窗口)
clc
% Clear variables in workspace.
% (清除工作空间中的变量)
clear
% Create a symbolic matrix A.
% (创建一个符号矩阵 A)
A=sym('[sin(x),cos(x);acos(x),asin(x)]')
% Manipulate matrix A through transpose, det, inv, pretty.
% (通过函数 transpose、det、inv、pretty 对矩阵 A 进行操作)
AT=transpose(A)
Adet=det(A)
Ainv=inv(A)
pretty(Ainv)
```

运行结果如下：

```
A =[ sin(x),  cos(x)]
   [ acos(x), asin(x)]
AT =[ sin(x), acos(x)]
    [ cos(x), asin(x)]
Adet = asin(x)*sin(x)-acos(x)*cos(x)
Ainv =[-asin(x)/(acos(x)*cos(x)-asin(x)*sin(x)),
       cos(x)/(acos(x)*cos(x) - asin(x)*sin(x))]
[acos(x)/(acos(x)*cos(x)-asin(x)*sin(x)),
       -sin(x)/(acos(x)*cos(x)-asin(x)*sin(x))]
 +-                         -+
 |     asin(x)      cos(x)   |
 |    - -------,    ------   |
 |       #1           #1     |
 |                           |
 |     acos(x)      sin(x)   |
 |     -------,   - ------   |
 |       #1           #1     |
 +-                         -+
where  #1 == acos(x)cos(x)-asin(x)sin(x).
```

2.5.2 操作符号对象

1. 符号对象的四则运算

对于符号对象,尤其是符号多项式能够进行四则运算,下面将给出一些示例。

【例 2.26】 符号对象四则运算示例。

EXAMP02026

```
% Create two symbolic polynomials, and operate them.
% (创建两个符号多项式,并对其进行四则运算)
%
% Clear Command Window.
% (清除命令窗口)
clc
% Clear variables in workspace.
% (清除工作空间中的变量)
clear
% Define symbolic variables a, b, c, x, y, z.
% (定义符号变量a、b、c、x、y、z)
syms a b c x y z
% Define two functions f and g depending on a, b, c, x, y, and z.
% (定义两个函数 f 和 g)
f=a*x^2*y^3+5*b*x*y*z-2*c*z
g=b*x*y*z-3*a*c*z
% Manipulate f and g.
% (操作两个函数 f 和 g)
Sumfg=f+g
pretty(Sumfg)
Subfg=f-g
pretty(Subfg)
Timesfg=f*g
pretty(Timesfg)
Divfg=f/g
pretty(Divfg)
P2f=f^2
pretty(P2f)
```

运行结果如下:

```
f =a*x^2*y^3+5*b*z*x*y-2*c*z
g =b*x*y*z-3*a*c*z
Sumfg =a*x^2*y^3+6*b*z*x*y-2*c*z-3*a*c*z
     2  3
  ax  y  +6bzxy-2cz-3acz
Subfg =a*x^2*y^3+4*b*z*x*y-2*c*z+3*a*c*z
     2  3
  ax  y  +4bzxy-2cz+3acz
Timesfg =-(3*a*c*z-b*x*y*z)*(a*x^2*y^3+5*b*z*x*y-2*c*z)
          2  3
```

```
       -(3acz-bxyz)(ax y +5bzxy-2cz)
Divfg =-(a*x^2*y^3+5*b*z*x*y-2*c*z)/(3*a*c*z-b*x*y*z)
         2 3
       ax y +5bzxy-2cz
     - --------------------
            3acz-bxyz
P2f =(a*x^2*y^3+5*b*z*x*y-2*c*z)^2
      2 3              2
    (ax y +5bzxy-2cz)
```

2. 简化多项式函数

MATLAB 提供了大量简化多项式的函数，参见表 2.19。

表 2.19 简化多项式函数

函数	描述	函数	描述
collect	合并同类项	numden	提取有理多项式的分子、分母无法消除公因子
expand	多项式展开	subexpr	替换重复字符串
horner	嵌套式层乘表达式	subs	代换符号字符串
factor	因式分解	pretty	将符号表达式转化为常见数学形式
simplify	简化不确定符号对象	finverse	求反函数
simple	给出最简单模式符号表达式	compose	合成函数

【例 2.27】 简化多项式示例。

EXAMP02027

```
% Examples for the simplifying polynomials.
% (简化多项式的示例)
%
% Clear Command Window.
% (清除命令窗口)
clc
% Clear variables in workspace.
% (清除工作空间中的变量)
clear
% Define two symbolic variables x and y.
% (定义两个符号变量 x 和 y)
syms x y
% In terms of x and y, one define a function fxy.
% (基于 x 和 y，定义一个函数 fxy)
fxy=x^2*y+3*x*y^2-4*x*y-x^2+2*y^2-5*x+6*y
% Collect the function fxy.
% (对函数 fxy 合并同类项)
fxy1=collect(fxy)
% Collect the function fxy with respect to x.
% (对函数 fxy 中 x 为变量合并同类项)
fxyx=collect(fxy,x)
% Collect the function fxy with respect to y.
% (对函数 fxy 中 y 为变量合并同类项)
```

```
fxyy=collect(fxy,y)
% Expand a symbolic expression (x+2)^5.
% (对符号表达式(x+2)^5展开)
Resul_1=expand((x+2)^5)
% Express the above result in the usual mathematical form.
% (将上面结果表示为通常数学形式)
pretty(Resul_1)
% Expand a symbolic expression sin(2*x+3*y).
% (对符号表达式 sin(2*x+3*y)展开)
Resul_2=expand(sin(2*x+3*y))
% Express the above result in usual mathematical form.
% (将上面结果表示为通常数学形式)
pretty(Resul_2)
% Horner a symbolic expression x^7+ 2*x^6+ 3*x^5+ 4*x^4+ 5*x^3 + 6*x^2+
% 7*x+ 8.
% (分解多项式 x^7+2*x^6+3*x^5+4*x^4+5*x^3+6*x^2+7*x+8 为嵌套形式)
horner(x^7+2*x^6+3*x^5+4*x^4+5*x^3+6*x^2+7*x+8)
% Define a symbolic function f=x^3-7*x+6.
% (定义一个符号函数 f=x^3-7*x+6)
f=x^3-7*x+6;
% Factor the symbolic function.
% (对符号函数进行因式分解)
facf=factor(f)
% Define a symbolic function g=sin(x)*cos(x)+sin(2*x)*cos(2*x).
% (定义一个符号函数 g=sin(x)*cos(x)+sin(2*x)*cos(2*x))
g=sin(x)*cos(x)+sin(2*x)*cos(2*x);
% Symplify the symbolic function g.
% (简化符号函数 g)
facg=simple(g)
% Obtain the numerator and denominator of a symbolic expression.
% (通过一个符号表达式,给出分子和分母表达式)
[n,d]=numden(x/y+y/x)
% Solve a symbolic equation.
% (解一个符号方程)
r=solve('a*x^3+b*x^2+c*x+d=0')
% Simplify the above results.
% (简化上面结果)
[roots,sg]=subexpr(r,'sg')
% Define two symbolic variables a and b.
% (定义两个符号变量 a 和 b)
syms a b
% Replace a symbolic variable in an symbolic expression with
% a specific value.
% (将一个符号表达式中符号特定值代入)
s1=subs(a+b,a,4)
s2=subs(a+b,4,a)
% Notice the following statement.
% (注意下面表述)
s3=subs(a+b,4,a,0)
% Replace two symbolic variables in the symbolic expression with
```

```matlab
%   two specific values.
%   (将符号表达式中两个符号变量特定值代入)
s4=subs(a+b,{a,b},{2,4})
%   Define a symbolic function fxy_1=3*x^2+6*y.
%   (定义一个符号函数 fxy_1=3*x^2+6*y)
fxy_1=3*x^2+6*y
%   Solve the inverse function of fxy_1 with respect to x.
%   (求函数 fxy_1 的 x 为变量的反函数)
gx_1=finverse(fxy_1,x)
%   Solve the inverse function of fxy_1 with respect to y.
%   (求函数 fxy_1 的 y 为变量的反函数)
gy=finverse(fxy_1,y)
%   Define three symbolic variables x, y and z.
%   (定义三个符号变量 x, y 和 z)
syms x y z
%   Define a symbolic function f_2=3*x^2+2*x-5.
%   (定义一个符号函数 f_2=3*x^2+2*x-5)
f_2=3*x^2+2*x-5
%   Define a symbolic function g_2=2*y-3.
%   (定义一个符号函数 g_2=2*y-3)
g_2=2*y-3
%   Illustrate the usage of the function "compose".
%   (说明函数"compose"的用法)
comf1=compose(f_2,g_2)
comf2=compose(f_2,g_2,z)
comf3=compose(f_2,g_2,x,z)
```

运行结果如下:
```
fxy =x^2*y - x^2 + 3*x*y^2 - 4*x*y - 5*x + 2*y^2 + 6*y
fxy1 =(y - 1)*x^2 + (3*y^2 - 4*y - 5)*x + 2*y^2 + 6*y
fxyx =(y - 1)*x^2 + (3*y^2 - 4*y - 5)*x + 2*y^2 + 6*y
fxyy =(3*x + 2)*y^2 + (x^2 - 4*x + 6)*y - x^2 - 5*x
Resul_1 =x^5 + 10*x^4 + 40*x^3 + 80*x^2 + 80*x + 32
    5      4       3       2
   x + 10 x + 40 x + 80 x + 80 x + 32
Resul_2 =sin(y)-2*cos(x)^2*sin(y)-4*cos(y)^2*sin(y)
-6*cos(x)*cos(y)*sin(x)+8*cos(x)*cos(y)^3*sin(x)
+8*cos(x)^2*cos(y)^2*sin(y)
                    2            2                   2
  8 sin(y) cos(x)  cos(y)  - 2 sin(y) cos(x)   +
                        3
      8 sin(x) cos(x) cos(y)  -
                                                    2
      6 sin(x) cos(x) cos(y) - 4 sin(y) cos(y) + sin(y)
ans=x*(x*(x*(x*(x*(x*(x + 2) + 3) + 4) + 5) + 6) + 7) + 8
facf =(x + 3)*(x - 1)*(x - 2)
facg =sin(2*x)/2 + sin(4*x)/2
n =x^2 + y^2
d =x*y
r =
```

```
             (((d/(2*a) + b^3/(27*a^3) - (b*c)/(6*a^2))^2 + (- b^2/(9*a^2) +
c/(3*a))^3)^(1/2) - b^3/(27*a^3) - d/(2*a) + (b*c)/(6*a^2))^(1/3) - b/(3*a) - (-
b^2/(9*a^2) + c/(3*a))/(((d/(2*a) + b^3/(27*a^3) - (b*c)/(6*a^2))^2 + (c/(3*a)
- b^2/(9*a^2))^3)^(1/2) - b^3/(27*a^3) - d/(2*a) + (b*c)/(6*a^2))^(1/3)
             (-b^2/(9*a^2) + c/(3*a))/(2*(((d/(2*a) + b^3/(27*a^3) - (b*c)/(6*a^2))^2
+ (c/(3*a) - b^2/(9*a^2))^3)^(1/2) - b^3/(27*a^3) - d/(2*a) + (b*c)/(6*a^2))^(1/3))
+ (3^(1/2)*((-b^2/(9*a^2) + c/(3*a))/(((d/(2*a) + b^3/(27*a^3) - (b*c)/(6*a^2))^2
+ (c/(3*a) - b^2/(9*a^2))^3)^(1/2) - b^3/(27*a^3) - d/(2*a) + (b*c)/(6*a^2))^(1/3)
+ (((d/(2*a) + b^3/(27*a^3) - (b*c)/(6*a^2))^2 + (-b^2/(9*a^2) + c/(3*a))^3)^(1/2)
- b^3/(27*a^3) - d/(2*a) + (b*c)/(6*a^2))^(1/3))*i)/2 - b/(3*a) - (((d/(2*a) +
b^3/(27*a^3) - (b*c)/(6*a^2))^2 + (-b^2/(9*a^2) + c/(3*a))^3)^(1/2) - b^3/(27*a^3)
- d/(2*a) + (b*c)/(6*a^2))^(1/3)/2
             (-b^2/(9*a^2) + c/(3*a))/(2*(((d/(2*a) + b^3/(27*a^3) - (b*c)/(6*a^2))^2
+ (c/(3*a) - b^2/(9*a^2))^3)^(1/2) - b^3/(27*a^3) - d/(2*a) + (b*c)/(6*a^2))^(1/3))
- (3^(1/2)*((-b^2/(9*a^2) + c/(3*a))/(((d/(2*a) + b^3/(27*a^3) - (b*c)/(6*a^2))^2
+ (c/(3*a) - b^2/(9*a^2))^3)^(1/2) - b^3/(27*a^3) - d/(2*a) + (b*c)/(6*a^2))^(1/3)
+ (((d/(2*a) + b^3/(27*a^3) - (b*c)/(6*a^2))^2 + (-b^2/(9*a^2) + c/(3*a))^3)^(1/2)
- b^3/(27*a^3) - d/(2*a) + (b*c)/(6*a^2))^(1/3))*i)/2 - b/(3*a) - (((d/(2*a) +
b^3/(27*a^3) - (b*c)/(6*a^2))^2 + (-b^2/(9*a^2) + c/(3*a))^3)^(1/2) - b^3/(27*a^3)
- d/(2*a) + (b*c)/(6*a^2))^(1/3)/2
      roots =sg^(1/3)-b/(3*a)-(-b^2/(9*a^2)+c/(3*a))/sg^(1/3)
             (-b^2/(9*a^2)+c/(3*a))/(2*sg^(1/3))-sg^(1/3)/2 + (3^(1/2)*(sg^(1/3) +(-
b^2/(9*a^2)+c/(3*a))/sg^(1/3))*i)/2-b/(3*a)
             (-b^2/(9*a^2)+c/(3*a))/(2*sg^(1/3))-sg^(1/3)/2 - (3^(1/2)*(sg^(1/3) +(-
b^2/(9*a^2)+c/(3*a))/sg^(1/3))*i)/2-b/(3*a)
      sg =((d/(2*a) +b^3/(27*a^3) - (b*c)/(6*a^2))^2 +(-b^2/(9*a^2) +c/(3*a))^3)
^(1/2)-b^3/(27*a^3)-d/(2*a)+(b*c)/(6*a^2)
      s1=b + 4
      s2=a + b
      Warning: SUBS(S,OLD,NEW,0) will not accept 0 in a
      future release. Use SUBS(S,OLD,NEW) instead.
      > In sym.subs at 62
        In EXAMP02027 at 72
      s3 =a + b
      s4 =6
      fxy_1 =3*x^2 + 6*y
      gx_1 =(3^(1/2)*(x - 6*y)^(1/2))/3
      gy =- x^2/2 + y/6
      f_2 =3*x^2 + 2*x - 5
      g_2 =2*y - 3
      comf1 =4*y + 3*(2*y - 3)^2 - 11
      comf2 =4*z + 3*(2*z - 3)^2 - 11
      comf3 =4*z + 3*(2*z - 3)^2 - 11
```

2.6 微 积 分

  微积分在科学工程技术方面应用占有非常重要的地位。在一些特殊情况下,微积分的求解显得非常困难。在 MATLAB 软件面前,一切变得简单起来。本节将对微积分相关函数进行较详细的应用分析,比如:微分、积分、求和、求极限以及泰勒展开式等,参见表 2.20。

表 2.20 微积分函数

函 数	描 述	函 数	描 述
findsym	在符号表达式和符号矩阵中寻找符号变量	sum	元素求和
diff	微分和近似求导	limit	求表达式极限
int	积分	taylor	泰勒展开

【例 2.28】 微积分相关示例。

EXAMP02028

```
% Display some examples for the functions listed in table 2.20.
% (对表 2.20 中函数给出一些示例)
%
% Clear Command Window.
% (清除命令窗口)
clc
% Clear variables in workspace.
% (清除工作空间中的变量)
clear
% Define symbolic variables a, b, c, d, e, f, x, y and z.
% (定义符号变量 a、b、c、d、e、f、x、y 和 z)
syms a b c d e f x y z
% Define a symbolic funtion f_1= a*x^3+b*y-c*z+d.
% (定义一个符号函数 f_1= a*x^3+b*y-c*z+d)
f_1=a*x^3+b*y-c*z+d
% Find all symbolic variables in the symbolic function f_1.
% (在符号函数 f_1 中寻找所有符号变量)
findsym(f_1)
% Find the first symbolic variable in f_1.
% (给出 f_1 中被 MATLAB 默认为第一个的符号变量)
findsym(f_1,1)
% Find the first and second symbolic variables in f_1.
% (给出 f_1 中被 MATLAB 默认为第一个和第二个的符号变量)
findsym(f_1,2)
% Find the front four symbolic variables in f_1.
% (给出 f_1 中前四个符号变量)
findsym(f_1,4)
% Derivative of f_1 with respect to x.
% (函数 f_1 对于 x 的一阶微商)
f1x=diff(f_1)
% Derivative of f_1 with respect to y.
% (函数 f_1 对于 y 的一阶微商)
f1y=diff(f_1,y)
% Two order derivative of f_1 with respect to x.
% (函数 f_1 对于 x 的二阶微商)
f1x2=diff(f_1,2)
f1xx=diff(f_1,x,2)
% Two order derivative of f_1 with respect to x and y.
% (对 f_1 求关于 x 和 y 的二阶微商)
f1xy=diff(diff(f_1,x),y)
```

```matlab
% Define a new symbolic function f_2=2*x^2*y+3*y-5*z+8.
% (定义一个新符号函数)
f_2=2*x^2*y+3*y-5*z+8
% Solve the integral of f_2.
% (求解函数 f_2 的积分)
f2intx=int(f_2)
% Solve the integration of x for f_2.
% (求解函数 f_2 对于变量 x 的积分)
f2intx1=int(f_2,x)
% Solve the integration of y for f_2.
% (求解函数 f_2 对于变量 y 的积分)
f2inty=int(f_2,y)
% Solve the double integration of x and y for f_2.
% (求解函数 f_2 对于 x 和 y 的积分)
f2intxy=int(int(f_2,y),x)
% Integrate two variable x and y for f_2.
% (积分函数 f_2 中两变量 x 和 y)
f2intxy2=int(f_2,'x','y')
% Integrate x ranging from -5 to 5 for f_2.
% (函数 f_2 对变量 x 从-5 到 5 的积分)
f2intxval=int(f_2,'x',-5,5)
% Integrate f_2 with x ranging from 0,5, y from 1 to 3*x, z from
% -x*y to x*y.
% (对 f_2 进行积分,x 为从 0 到 5,y 从 1 到 3*x,z 从-x*y 到 x*y)
f2val=int(int(int(f_2,'z','-x*y','x*y'),'y','1','3*x'),'x',0,5)
% Define the accurate number as 9.
% (定义有效数字为 9 位)
vpa(f2val,9)
% Define a matrix A.
% (定义一个矩阵 A)
A=[1 2 3; 5 6 7; 9 10 11]
% Change the matrix A into a symbolic matrix as B.
% (将矩阵 A 转化为符号矩阵,记为 B)
B=sym(A)
% Sum the matrix B.
% (对矩阵 B 求和)
sum(B)
% Sum the columns of B.
% (对矩阵 B 的各列进行求和)
sum(B,1)
% Sum the rows of B.
% (对矩阵 B 的各行进行求和)
sum(B,2)
% Defien a symbolic function f_3.
% (定义一个符号函数 f_3)
f_3=3*x^2*y+4*y^2-x*y+6*x-y+11
% Limit x->0 for the function f_3.
% (函数 f_3 对 x→0 求极限)
limitx=limit(f_3)
% Limit x->5 for f_3.
% (函数 f_3 对 x→5 求极限)
```

第2章 矩阵、数组、符号运算

```
limitx2=limit(f_3,'x',5)
%  Limit f_3 two times.
%  (对 f_3 求两重极限)
limitxy=limit(limit(f_3))
%  Limit x->11 and y->-12 for f_3.
%  (函数 f_3 对 x→11 和 y→-12 求极限)
limitxyab=limit(limit(f_3,'x',11),'y',-12)
%  Solve the right limit of x->0 and the left limit y->1 for f_3.
%  (函数 f_3 对 x→0 求右极限,对 y→1 求左极限)
limitxryl=limit(limit(f_3,'x',0,'right'),'y',1,'left')
%  Define two symbolic functions f_4 and g_4.
%  (定义两个符号函数 f_4 和 g_4)
f_4=sin(x)*cos(y)
g_4=sin(x)*cos(x)
%  Play the taylor expansion of f_4.
%  (对函数 f_4 泰勒展开)
taylorx=taylor(f_4,'x',3)
%  Play the taylor expansion of g_4.
%  (对函数 g_4 泰勒展开)
taylorxg=taylor(g_4)
%  Play the taylor expansion of f_4 for y.
%  (函数 f_4 对 y 变量泰勒展开)
taylorxf=taylor(f_4,y)
```

运行结果如下：

```
f_1=a*x^3 + d + b*y - c*z
ans=a,b,c,d,x,y,z
ans=x
ans=x,y
ans=x,y,z,d
f1x=3*a*x^2
f1y=b
f1x2=6*a*x
f1xx=6*a*x
f1xy=0
f_2=2*y*x^2 + 3*y - 5*z + 8
f2intx=(2*y*x^3)/3 + (3*y - 5*z + 8)*x
f2intx1=(2*y*x^3)/3 + (3*y - 5*z + 8)*x
f2inty=y^2*(x^2 + 3/2) - y*(5*z - 8)
f2intxy=(x^3*y^2)/3 - x*(y*(5*z - 8) - (3*y^2)/2)
f2intxy2=y*((2*y^3)/3+3*y-5*z+8)-x*((2*y*x^2)/3+3*y-5*z+8)
f2intxval=(590*y)/3-50*z+80
f2val=9375500/21
ans=446452.381
A=    1    2    3
      5    6    7
      9   10   11
B =[ 1,  2,  3]
   [ 5,  6,  7]
   [ 9, 10, 11]
```

```
ans=[ 15, 18, 21]
ans=[ 15, 18, 21]
ans=   6
      18
      30
f_3=3*x^2*y - x*y + 6*x + 4*y^2 - y + 11
limitx=4*y^2 - y + 11
limitx2=4*y^2 + 69*y + 41
limitxy=11
limitxyab=-3559
limitxryl=14
f_4=cos(y)*sin(x)
g_4=cos(x)*sin(y)
taylorx=sin(3)*cos(y) + cos(3)*cos(y)*(x - 3) - (cos(3)*cos(y)*(x - 3)^3)/6
+ (cos(3)*cos(y)*(x - 3)^5)/120 - (sin(3)*cos(y)*(x - 3)^2)/2 + (sin(3)*cos(y)*(x
- 3)^4)/24
taylorxg=(2*x^5)/15 - (2*x^3)/3 + x
taylorxf=(sin(x)*y^4)/24 - (sin(x)*y^2)/2 + sin(x)
```

2.7 求解符号方程

符号方程包含一般方程、一般微分方程和偏微分方程3类。本节只对求解一般方程、一般微分方程相关函数进行介绍。

2.7.1 解代数方程

代数方程在数学中占有重要位置，在科学与工程技术方面有较广泛应用。本节将对求解代数方程的常用函数"solve"进行讨论，给出示例。

solve 是一求解代数方程的函数。

语法格式：solve('eqn1','eqn2',…,'eqnN')
　　　　　solve('eqn1','eqn2',…,'eqnN','var1','var2',…,'varN')
　　　　　solve('eqn1','eqn2',…,'eqnN','var1','var2',…,'varN')

其中，"eqn1"，"eqn2"，…，"eqnN"表示待求的方程；"var1"，"var2"，…，"varN"表示待求的变量。如果运行语句中没有指定待求变量，MATLAB将根据findsym函数找到变量次序为待求变量次序。如果待求方程没有解析解，MATLAB将尝试数值解，如果常用方法也很难给出数值解，将给出错误或警告提示。

【例2.29】 线性代数方程。

EXAMP02029

```
% Display some examples for the usage of the function "solve".
%  (函数"solve"用法的示例)
%
% Clear Command Window.
%  (清除命令窗口)
clc
```

```
% Clear variables in workspaces.
% (清除工作空间中的变量)
clear
% Define symbolic variables a, b, c and x.
% (定义符号变量a、b、c和x)
syms a b c x
% Define a symbolic expression as S.
% (定义一个符号表达式为S)
S=a*x^2+b*x+c
% Solve an algebraic equation a*x^2+b*x+c=0.
% (求解代数方程a*x^2+b*x+c=0)
Roots=solve(S)
% Express Roots as the usual mathematical form.
% (将Roots表示为通常数学形式)
pretty(Roots)
% Solve an algebraic equation a*x^2+b*x+c=0 with taking b as variable.
% (将b作为变量,求解代数方程a*x^2+b*x+c=0)
solve(S,b)
% Solve an algebraic equation sin(x)*cos(2*x)=x-5.
% (求解代数方程sin(x)*cos(2*x)=x-5)
solution1=solve('sin(x)*cos(2*x)=x-5')
% Solve an algebraic equation x^3-2*x^2=x-1.
% (求解代数方程x^3-2*x^2=x-1)
Results=solve('x^3-2*x^2=x-1')
% Evaluate the specific values for the Results.
% (给出Results的具体数值)
double(Results)
```

运行结果如下：

```
S=a*x^2 + b*x + c
Roots= -(b + (b^2 - 4*a*c)^(1/2))/(2*a)
       -(b - (b^2 - 4*a*c)^(1/2))/(2*a)
 +-                    -+
 |           2    1/2   |
 |    b + (b - 4 a c)   |
 |  - ----------------- |
 |          2 a         |
 |                      |
 |           2    1/2   |
 |    b - (b - 4 a c)   |
 |  - ----------------- |
 |          2 a         |
 +-                    -+
ans=-(a*x^2 + c)/x
solution1=5.3075348633325381451
Results=
7/(9*(108^(1/2)*((7*i)/108) + 7/54)^(1/3)) + ((108^(1/2)*7*i)/108 + 7/54)^(1/3)
 + 2/3
        2/3 - ((108^(1/2)*7*i)/108 + 7/54)^(1/3)/2 + (3^(1/2)*(7/(9*(108^(1/2)*
((7*i)/108) + 7/54)^(1/3)) - ((108^(1/2)*7*i)/108 + 7/54)^(1/3))*i)/2 -
```

```
7/(18*(108^(1/2)*((7*i)/108) + 7/54)^(1/3))
      2/3 - ((108^(1/2)*7*i)/108 + 7/54)^(1/3)/2 - (3^(1/2)*(7/(9*(108^(1/2)*
((7*i)/108) + 7/54)^(1/3)) - ((108^(1/2)*7*i)/108 + 7/54)^(1/3))*i)/2 -
7/(18*(108^(1/2)*((7*i)/108) + 7/54)^(1/3))
      ans=2.2470 + 0.0000i
          0.5550 + 0.0000i
         -0.8019 - 0.0000i
```

【例 2.30】 非线性代数方程。

EXAMP02030

```
% Displays some examples of linear algebraic equation for the
% function "solve".
% (用函数"solve"求解线性代数方程的几个示例)
%
% Clear Command Window.
% (清除命令窗口)
clc
% Clear variables in workspaces.
% (清除工作空间中的变量)
clear
% Define symbolic variables u, v, x and y.
% (定义符号变量u、v、x 和y)
syms u v x y
% Solve linear algebraic equations x+y=u and x-y=v.
% (求线性方程组 x+y=u 和 x-y=v 的解)
S=solve('x+y=u','x-y=v')
% Display the x part of the result S.
% (显示结果 S 的 x 部分)
S.x
% Display the y part of the result S.
% (显示结果 S 的 y 部分)
S.y
% Solve a nonlinear algebraic equation x^2+y^2=5 and x+a*y^2=b.
% (解线性方程组 x^2+y^2=5 和 x+a*y^2=b)
S1=solve('x^2+y^2=5','x+a*y^2=b')
% Display the x part of the result S1.
% (显示结果 S1 的 x 部分)
S1.x
% Display the y part of the result S1.
% (显示结果 S1 的 y 部分)
S1.y
```

运行结果如下:

```
S = x: [1x1 sym]
    y: [1x1 sym]
ans =u/2 + v/2
ans =u/2 - v/2
S1 = x: [4x1 sym]
     y: [4x1 sym]
```

```
ans = b + ((20*a^2 - 4*b*a + 1)^(1/2) - 2*a*b + 1)/(2*a)
      b - ((20*a^2 - 4*b*a + 1)^(1/2) + 2*a*b - 1)/(2*a)
      b - ((20*a^2 - 4*b*a + 1)^(1/2) + 2*a*b - 1)/(2*a)
      b + ((20*a^2 - 4*b*a + 1)^(1/2) - 2*a*b + 1)/(2*a)
ans = (-((20*a^2 - 4*b*a + 1)^(1/2) - 2*a*b + 1)/(2*a^2))^(1/2)
      (((20*a^2 - 4*b*a + 1)^(1/2) + 2*a*b - 1)/(2*a^2))^(1/2)
     -(((20*a^2 - 4*b*a + 1)^(1/2) + 2*a*b - 1)/(2*a^2))^(1/2)
     -(-((20*a^2 - 4*b*a + 1)^(1/2) - 2*a*b + 1)/(2*a^2))^(1/2)
```

2.7.2 解微分方程

代数微分方程在现代科学与工程技术中扮演着比代数方程更重要的角色。在 MATLAB 中，dsolve 是一专门用来求解代数微分方程的函数。

其语法格式：dsolve('eqn1','eqn2',…)。一般"eqn1"为通常意义上的微分方程，"eqn2"等为初始条件。注意，描述微分求导一般用"D"来表述，用"D2"来表示"d^2/dt^2"。通常默认的自变量为"t"。

【例 2.31】 解微分方程示例。

EXAMP02031

```
%   Some examples for the usage of the function "dsolve".
%   (函数"dsolve"用法的示例)
%
%   Clear Command Window.
%   (清除命令窗口)
clc
%   Clear variables in workspaces.
%   (清除工作空间中的变量)
clear
%   Solve a first-order differential equation Dx=1+2*x+x^2
%   without initial condition.
%   (解无初始条件的一阶微商方程 Dx=1+2*x+x^2)
dsolve('Dx=1+2*x+x^2')
%   Solve a first-order differential equation Dx=1+2*x+x^2
%   with initial condition x(0)=2.
%   (解初始条件为 x(0)=2 的一阶微商方程 Dx=1+2*x+x^2)
Result=dsolve('Dx=1+2*x+x^2','x(0)=2')
%   Express the above result in the usual mathematical form.
%   (将上面结果表示为通常的数学形式)
pretty(Result)
%   Solve a first-order differential equation Dx=1+2*x+x^2
%   with initial condition x(0)=9.
%   (解初始条件为 x(0)=9 的一阶微商方程 Dx=1+2*x+x^2)
Result1=dsolve('Dx=1+2*x+x^2','x(0)=9')
%   Solve a first-order differential equation 3*(Dx)^2+2*x=2
%   with initial condition x(0)=9.
%   (解初始条件为 x(0)=9 的一阶微商方程 3*(Dx)^2+2*x=2)
X=dsolve('3*(Dx)^2+2*x=2','x(0)=9')
%   Solve a second-order differential equation D2y=tan(x)-y+x
%   with initial conditions y(0)=0 and Dy(0)=5, independent
```

```
%   variable x.
%  (x 作为自变量，解初始条件为 y(0)=0 和 Dy(0)=5 的二阶微商方程
%   D2y=tan(x)-y+x)
Y2=dsolve('D2y=tan(x)-y+x','y(0)=0','Dy(0)=5','x')
%  Solve a third-order differential equation D3w=x*w+2*x
%  with initial conditions w(0)=0, Dw(0)=2 and D2w(0)=2*pi,
%   independent variable y.
%  (y 为自变量，解初始条件为 w(0)=0、Dw(0)=2 和 D2w(0)=2*pi 的三阶微商
%   方程 D3w=x*w+2*x)
W3=dsolve('D3w=x*w+2*x','w(0)=0','Dw(0)=2','D2w(0)=2*pi','y')
%  Solve two first-order differential equations Df=f+g and Dg=f-g
%  with initial conditions f(0)=2 and g(0)=5, independent variable x.
%  (自变量为 x，解初始条件为 f(0)=2 和 g(0)=5 的两个一阶微商方程)
S=dsolve('Df=f+g','Dg=f-g','f(0)=2','g(0)=5','x')
%  Display the f part of S.
%  (显示结果 S 的 f 部分)
sf=S.f
%  Display the g part of S.
%  (显示结果 S 的 g 部分)
sg=S.g
```

运行结果如下：

```
ans =            -1
      - 1/(C3 + t) - 1
Result =- 1/(t - 1/3) - 1
        1
-  ------- - 1
    t - 1/3
Result1 =- 1/(t - 1/10) - 1
X = 1-(2^(1/2)*4*i+(2^(1/2)*3^(1/2)*t)/3)^2/4
    1-(2^(1/2)*4*i-(2^(1/2)*3^(1/2)*t)/3)^2/4
Y2=x/2+5*sin(x)-(x*cos(2*x))/2-cos(x)*(2*atanh(tan(x/2))
   -x*cos(x))
W3 =(2*exp(x^(1/3)*y)*(pi+x^(1/3)+x^(2/3)))/(3*x^(2/3)) + (3^(1/2)
    *exp(y*((3^(1/2)*x^(1/3)*i)/2-x^(1/3)/2))*(3*pi*x^(1/3)
    -3^(1/2)*x*2*i+3^(1/2)*x^(2/3)*i-3*x^(2/3)+ pi*3^(1/2)
    *x^(1/3)*i)*i)/(9*x)+(3^(1/2)*exp(-y*((3^(1/2)*x^(1/3)*i)/2
    +x^(1/3)/2))*(3*x^(2/3)-3*pi*x^(1/3)+3^(1/2)*x^(2/3)*i
    -3^(1/2)*x*2*i + pi*3^(1/2)*x^(1/3)*i)*i)/(9*x) - 2
S =  g: [1x1 sym]
     f: [1x1 sym]
sf =-exp(-2^(1/2)*x)*(exp(2*2^(1/2)*x)*((3*2^(1/2))/4 - 5/2)
    +(5*2^(1/2))/2-(2^(1/2) *(5*2^(1/2) + 3))/4 + 2^(1/2)
    *exp(2*2^(1/2)*x)*((3*2^(1/2))/4-5/2)+3/2
sg =-exp(-2^(1/2)*x)*(exp(2*2^(1/2)*x) *((3*2^(1/2))/4-5/2)
    -(2^(1/2)*(5*2^(1/2) + 3))/4)
```

2.8 积分变换

积分变换在信号处理以及其他工程领域有着广泛的应用。MATLAB 为积分变换提供了大

量函数,可以将在校学生以及工程技术人员从烦琐的积分变换计算中解脱出来。MATLAB 中提供的积分变换函数有傅里叶变换、傅里叶逆变换、拉布拉斯变换、拉布拉斯逆变换、Z 变换以及 Z 逆变换等。积分变换函数语法参见表 2.21。

表 2.21 积分变换函数

函 数	语 法
fourier	傅里叶变换: F = fourier(f)　　求: $F(w) = \int_{-\infty}^{\infty} f(x)\mathrm{e}^{-iwx}\mathrm{d}x$ F = fourier(f,v)　　求: $F(v) = \int_{-\infty}^{\infty} f(x)\mathrm{e}^{-ivx}\mathrm{d}x$ F=fourier(f,u,v)　　求: $F(v) = \int_{-\infty}^{\infty} f(u)\mathrm{e}^{-ivu}\mathrm{d}u$
ifourier	傅里叶逆变换: f = ifourier(F)　　求: $f(x) = \dfrac{1}{2\pi}\int_{-\infty}^{\infty} F(w)\mathrm{e}^{iwx}\mathrm{d}w$ f = ifourier(F,u)　　求: $f(u) = \dfrac{1}{2\pi}\int_{-\infty}^{\infty} F(w)\mathrm{e}^{iwu}\mathrm{d}w$ f = ifourier(F,v,u)　　求: $f(u) = \dfrac{1}{2\pi}\int_{-\infty}^{\infty} F(v)\mathrm{e}^{ivu}\mathrm{d}v$
laplace	拉布拉斯变换: L = laplace(F)　　求: $L(s) = \int_{0}^{\infty} F(t)\mathrm{e}^{-st}\mathrm{d}t$ L = laplace(F,t)　　求: $L(t) = \int_{0}^{\infty} F(x)\mathrm{e}^{-tx}\mathrm{d}x$ L = laplace(F,w,z)　　求: $L(z) = \int_{0}^{\infty} F(w)\mathrm{e}^{-zw}\mathrm{d}w$
ilaplace	拉布拉斯逆变换: F = ilaplace(L)　　求: $F(t) = \int_{c-i\infty}^{c+i\infty} L(s)\mathrm{e}^{st}\mathrm{d}s$ F = ilaplace(L,y)　　求: $F(y) = \int_{c-i\infty}^{c+i\infty} L(y)\mathrm{e}^{sy}\mathrm{d}s$ F = ilaplace(L,y,x)　　求: $F(x) = \int_{c-i\infty}^{c+i\infty} L(y)\mathrm{e}^{xy}\mathrm{d}y$
ztrans	Z 变换: F = ztrans(f)　　求: $F(z) = \sum_{n=0}^{\infty} \dfrac{f(n)}{z^n}$ F = ztrans(f,w)　　求: $F(w) = \sum_{n=0}^{\infty} \dfrac{f(n)}{w^n}$ F = ztrans(f,k,w)　　求: $F(w) = \sum_{k=0}^{\infty} \dfrac{f(k)}{w^k}$
iztrans	Z 逆变换: f = iztrans(F) f = iztrans(F,k) f = iztrans(F,w,k)

为了节约篇幅,对积分变换就不再举例,如有兴趣,可以阅读与信号处理 MATLAB 相关的专业书籍。

2.9 实 例 应 用

2.9.1 解多项式

在代数中,众所周知多项式:

$$f(x) = C_n x^n + C_{n-1} x^{n-1} + \cdots + C_1 x + C_0$$

有伴随矩阵：

$$A = \begin{bmatrix} -\dfrac{C_{n-1}}{C_n} & -\dfrac{C_{n-2}}{C_n} & -\dfrac{C_{n-3}}{C_n} & \cdots & -\dfrac{C_1}{C_n} & -\dfrac{C_0}{C_n} \\ 1 & 0 & 0 & \cdots & 0 & 0 \\ 0 & 1 & 0 & \cdots & 0 & 0 \\ \vdots & \vdots & \vdots & \ddots & \vdots & \vdots \\ 0 & 0 & 0 & 1 & 0 & 0 \\ 0 & 0 & 0 & 0 & 1 & 0 \end{bmatrix}$$

$f(x)$称为矩阵 A 的本征多项式。方程 $f(x)=0$ 的根就是矩阵 A 的特征值。所以可以借用MATLAB求解矩阵特征值来求解多项式的根。

【例 2.32】 求解方程 $5x^8 - 6x^4 + 3x^3 - 21x^2 + 15 = 0$。

EXAMP02032

```
% Clear Command Window.
% (清除命令窗口)
clc
% Clear variables in workspace.
% (清除工作空间中的变量)
clear
% Define a polynomial expressed by a vector P.
% (定义一个多项式表示矢量形式，记为P)
P=[5 0 0 0 -6 3 -21 0 15]
% Solve the campany matrix of the polynomial P as A.
% (求得多项式P的伴随矩阵，记为A)
A=compan(P)
% Solve the eigenvalues of the matrix A.
% (解矩阵A的本征值)
x=eig(A)
```

运行结果如下：

```
P =  5     0      0      0     -6     3    -21     0     15
A =  0     0      0    1.2000 -0.6000 4.2000  0   -3.0000
  1.0000  0      0      0      0      0      0      0
     0  1.0000   0      0      0      0      0      0
     0    0   1.0000    0      0      0      0      0
     0    0      0   1.0000    0      0      0      0
     0    0      0      0   1.0000    0      0      0
     0    0      0      0      0   1.0000    0      0
     0    0      0      0      0      0   1.0000    0
x = -0.5272 + 1.2194i
    -0.5272 - 1.2194i
     0.5400 + 1.1600i
     0.5400 - 1.1600i
     1.2143 + 0.0000i
     0.8468 + 0.0000i
    -1.3240 + 0.0000i
    -0.7626 + 0.0000i
```

> **注意：**
> MATLAB 语言也提供了求解多项式根的函数 "roots"。

2.9.2 解线性方程组

基于线性代数知识，可以借助矩阵逆矩阵来解线性方程组 $A \times X = B$，解为 $X = A^{-1} \times B$。当然也可以通过函数 "solve" 求解。

【例 2.33】 求解线性方程组：

$$\begin{cases} 5x + 6y - 12z = 1 \\ 6x + 5y - 2z = 129 \\ 4x - 9y + 2z = 17 \end{cases}$$

EXAMP02033

```matlab
% Clear Command Window.
% (清除命令窗口)
clc
% Clear variables in workspace.
% (清除工作空间中的变量)
clear
% Define the coefficient matrix for the linear equations as A.
% (定义线性方程组的系数矩阵，记为 A)
A=[5 6 -12;6 5 -2; 4 -9 2]
% Define the column vector as B.
% (定义等式列矢量，记为 B)
B=[1 129 17]'
% Solve the inverse matrix of A as invA.
% (求矩阵 A 的逆矩阵，记为 invA)
invA=inv(A)
% Solve the results of the linear equations
% (解线性方程组的解)
X1=invA*B
% Solve the problem by left division.
% (通过左除求解结果)
X2=A\B
% Solve the problem through right division.
% (通过右除求解结果)
AT=A'
BT=B'
X3=(BT/AT)'
```

运行结果如下：

```
A =    5     6    -12
       6     5    -2
       4    -9     2
B =    1
     129
      17
invA =  -0.0110    0.1319    0.0659
```

```
                -0.0275    0.0797   -0.0852
                -0.1016    0.0948   -0.0151
    X1 =18.1209
         8.8022
        11.8681
    X2 =18.1209
         8.8022
        11.8681
    AT =   5        6       4
           6        5      -9
         -12       -2       2
    BT =   1      129      17
    X3 = 18.1209
          8.8022
         11.8681
```

很明显三种方法得到的结果"X1"、"X2"、"X3"完全一致。

2.9.3 求平行六面体体积

在立体几何中，求解平行六面体体积公式为 $V = (\boldsymbol{a} \times \boldsymbol{b}) \cdot \boldsymbol{c}$，其中 $\boldsymbol{a}$、$\boldsymbol{b}$ 和 $\boldsymbol{c}$ 为平行六面体的三个边。

【例2.34】 求三边分别为 $\boldsymbol{a}$=(12,0,0)、$\boldsymbol{b}$=(2,15,0) 和 $\boldsymbol{c}$=(2,5,30) 的平行六面体体积。

EXAMP02034

```
% Clear Command Window.
%  (清除命令窗口)
clc
% Clear variables in workspace.
%  (清除工作空间中的变量)
clear
% Solve the volume of the predefined parallelepiped with three
% edges a, b, and c.
% Define three vectors a, b and c.
%  (定义三个矢量，记为a、b和c)
a=[12 0 0];
b=[2 15 0];
c=[2 5 30];
% Solve the volume of the predefined parallelepiped with three
% edges a, b, and c.
%  (计算以a、b、c为三个基础边的平行六面体体积)
V=dot(c,cross(a,b))
```

运行结果如下：
```
V = 5400
```

2.9.4 特征值与特征向量

许多问题都是特征值问题，如：

第 2 章 矩阵、数组、符号运算

其中，A 为 $n \times n$ 矩阵；v 为 $n \times 1$ 向量；λ 为标量，λ 取值为矩阵 A 的特征值时，方程有解。v 的解是对应特征值的特征向量。特征值问题可以表述如下：

$$A \cdot v - \lambda \cdot v = 0$$
$$A \cdot v - \lambda \cdot I \cdot v = 0$$
$$(A - \lambda \cdot I) \cdot v = 0$$

如果 v 有非零解，那么必有

$$|A - \lambda \cdot I| = 0$$

该方程成为矩阵 A 的特征方程。关于 λ 的 n 阶多项式的解为矩阵 A 的特征值。我们将解决不同根的情况，对于有相同根的情况不予介绍。有相同根的情况，在量子力学中称为简并态。

【例 2.35】 现有系数矩阵 A：

$$A = \begin{bmatrix} 0 & 1 \\ -2 & -3 \end{bmatrix}$$

相应特征方程为

$$|A - \lambda \cdot I| = \begin{vmatrix} -\lambda & 1 \\ -2 & -3-\lambda \end{vmatrix} = 0$$

可求得特征值为

$$\lambda_1 = -1, \lambda_2 = -2$$

对第 1 个解，有特征矢量为

$$\begin{bmatrix} 1 & 1 \\ -2 & -2 \end{bmatrix} \cdot \begin{bmatrix} v_{1,1} \\ v_{1,2} \end{bmatrix} = 0$$

可得

$$v_1 = k_1 \cdot \begin{bmatrix} +1 \\ -1 \end{bmatrix}$$

其中，k_1 为任意常数。

类似，可求另一个解为

$$v_2 = k_2 \cdot \begin{bmatrix} +1 \\ -2 \end{bmatrix}$$

将上面问题分析过程转化为 MATLAB 可执行语言。

EXAMP02035

```
%   Clear Command Window.
%   (清除命令窗口)
clc
%   Clear variables in workspace.
%   (清除工作空间中的变量)
```

```
clear
% Define a matrix A=[0 1; -2 -3].
% (定义一个矩阵A=[0 1; -2 -3])
A=[0 1; -2 -3];
% Solve the eigenvectors and eigenvalues of A.
% (解矩阵A的特征向量与特征值)
[V,D]=eig(A)
```

运行结果如下：

```
V =   0.7071   -0.4472
     -0.7071    0.8944
D =  -1    0
      0   -2
```

2.9.5 多元数据

MATLAB 提供一些函数可以对多元数据进行分析。下面示例就是对多元数据进行分析，并给出平均值和标准差。

【例 2.36】 求下面数据的平均值和标准差：

$$\begin{bmatrix} 72 & 81 & 69 & 82 & 75 \\ 134 & 201 & 156 & 148 & 170 \\ 3.2 & 3.5 & 7.1 & 2.4 & 1.2 \end{bmatrix}$$

EXAMP02036

```
% Clear Command Window.
% (清除命令窗口)
clc
% Clear variables in workspace.
% (清除工作空间中的变量)
clear
% Type a matrix D=[72 134 3.2; 81 201 3.5; 69 156 7.1;
% 82 148 2.4; 75 170 1.2].
% (输入矩阵D)
D=[72 134 3.2; 81 201 3.5; 69 156 7.1; 82 148 2.4; 75 170 1.2]'
% Solve the mean of the rows of D.
% (解矩阵的行平均值)
mu=mean(D')
% Solve the standard deviation of the columns of D.
% (计算矩阵的行标准差)
sigma=std(D)
```

运行结果如下：

```
D =  72.0000   81.0000   69.0000   82.0000   75.0000
    134.0000  201.0000  156.0000  148.0000  170.0000
      3.2000    3.5000    7.1000    2.4000    1.2000
mu = 75.8000  161.8000   3.4800
sigma = 65.4295   99.5092   74.8018   72.9058   84.6216
```

2.9.6 电路问题

【例2.37】 电路如图2.2所示，$R_1 = 2\Omega$，$R_2 = 5\Omega$，$R_3 = 3\Omega$，$R_4 = 2\Omega$，$R_5 = 7\Omega$，$R_6 = 2\Omega$，$R_7 = 1\Omega$。求解下面2个问题：

（1）如果$U_s = 36\text{V}$，请求解I_s和I_0；

（2）如果$I_0 = 1.5\text{A}$，请求解U_s和I_s。

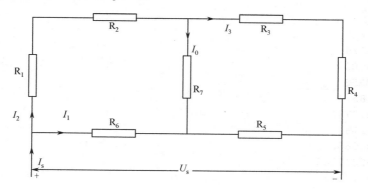

图 2.2 电路图

分析与求解：根据电路如图 2.2 所示可得方程：

$$\begin{cases} (R_6 + R_5)I_1 - R_6 I_2 - R_5 I_3 = U_s \\ -R_6 I_1 + (R_1 + R_2 + R_7 + R_6)I_2 - R_7 I_3 = 0 \\ -R_5 I_1 - R_7 I_2 + (R_3 + R_4 + R_5 + R_7)I_3 = 0 \end{cases}$$

借助 MATLAB 命令可以求解。

EXAMP02037

```
% Clear Command Window.
%  (清除命令窗口)
clc
% Clear variables in workspace.
%  (清除工作空间中的变量)
clear
% Create a coefficient matrix as A and a column matrix as U.
A=[9 -2 -7;-2 10 -1;-7 -1 13]
U=[36 0 0]'
% Solve the result of currents by left division.
%  (通过左除求电流结果)
I=A\U
% The solutions of (1) are given as the following
%  (第一问的解如下)
Is=I(1)
I0=I(2)-I(3)
% The solutions of (2) are given as the following
%  (第二问的解如下)
k1=Is/U(1)
k2=I0/U(1)
```

```
Us2=1.5/k2
Is2=Us2*k1
```

运行结果如下：

```
A =    9    -2    -7
      -2    10    -1
      -7    -1    13
U = 36
     0
     0
I = 7.8579
    2.0102
    4.3858
Is = 7.8579
I0 = -2.3756
k1 = 0.2183
k2 = -0.0660
Us2 = -22.7308
Is2 = -4.9615
```

2.9.7　稀疏矩阵绘图

【例 2.38】　绘制碳-60 图。

EXAMP02038

```
% Clear Command Window.
% (清除命令窗口)
clc
% Clear variables in workspace.
% (清除工作空间中的变量)
clear
% Define the variables by the function "bucky".
% (借助函数"bucky"定义两个变量 B 和 V)
[B,V]=bucky;
% Define a sparse matrix as H.
% (定义一个稀疏矩阵，记为 H)
H=sparse(60,60);
% Define a row vector ranging from 31 to 60 with a step-length
% 1 as k.
% (定义一行矢量从 31 到 60，记为 k)
k=31:60;
% Define the diagonal elements form 31 to 60 H as
% the same values of B.
% (定义矩阵 H 从 31 到 60，对角元为矩阵 B 对应值)
H(k,k)=B(k,k);
% Visualize the variables.
% (可视化这些变量).
gplot(B-H,V,'.b-');
hold on
```

```
gplot(H,V,'.r-');
hold off
axis off equal
```

运行结果如图 2.3 所示。

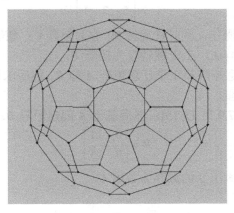

图 2.3 碳-60

小　　结

本章首先介绍了如何创建和操作矩阵、数组并举例说明；其次对多项式的构建、求解、简化分别进行阐述并举例说明；再次对符号对象的创建、操作、微积分运算、积分变化进行阐述并举例说明；最后给出 7 个实例为所讲知识的应用给予提示和启发。

习　　题

2.1　至少采用两种方法创建数组 A=[1,2,3,4,5,6,7,8,9]。

2.2　首先创建两个数组 A=[1,3,9,15] 和 B=[28,96,21,55]，然后计算 $A+B$，$A-B$，$A \times B$，$A \div B$。

2.3　首先创建 2 个矩阵：

$$A = \begin{bmatrix} 1 & 2 & 3 & 4 \\ 5 & 6 & 7 & 8 \\ 9 & 11 & 12 & 13 \end{bmatrix}, \quad B = \begin{bmatrix} 10 & 11 & 12 & 13 \\ 14 & 15 & 16 & 17 \\ 18 & 19 & 20 & 21 \end{bmatrix}$$

然后求解 $A \times B$，$A \div B$。

2.4　这里有三个数组 A=[1,2,3]，B=[3,1,4] 和 C=[9,-1,4]。求解下列问题：

（1）数组 A 和 B 的点乘；

（2）数组 B 和 C 的叉乘。

2.5　已知矩阵：

$$A = \begin{bmatrix} 16 & 2 & 3 & 13 \\ 5 & 11 & 10 & 8 \\ 9 & 7 & 6 & 12 \\ 4 & 14 & 15 & 1 \end{bmatrix}$$

求该矩阵的逆矩阵、行列式、迹和秩。

2.6 已知矩阵：
$$B = \begin{bmatrix} 1 & 3 & 5 \\ 2 & 4 & 9 \\ 5 & 7 & 1 \end{bmatrix}$$

请对该矩阵进行如下操作：顺时针旋转 90° 存为矩阵 C；逆时针旋转 180° 存为矩阵 D；然后左右翻转存为矩阵 E，上下翻转存为矩阵 F。

2.7 创建 1 个 4×4 的准任意矩阵和 1 个任意矩阵 A 和 B，然后求这 2 个矩阵的上三角、下三角、特征值与特征向量。

2.8 首先输入数组 A=[3,5,7,9,11,13,15]，然后基于 A 创建矩阵 B：
$$B = \begin{bmatrix} 5 & 11 \\ 9 & 15 \end{bmatrix}$$

再从数组 A 中挑选出大于 9 的数存为数组 C。

2.9 有矩阵为
$$A = \begin{bmatrix} 16 & 2 & 3 & 13 \\ 5 & 11 & 10 & 8 \\ 9 & 7 & 6 & 12 \\ 4 & 14 & 15 & 1 \end{bmatrix}$$

基于矩阵 A，创建数组 C=[16 5 9 4] 和 D=[5 6 13 15]，还有矩阵 E：
$$E = \begin{bmatrix} 16 & 13 \\ 9 & 12 \end{bmatrix}$$

2.10 求解下面线性方程组。

（1） $\begin{cases} 2x + 3y - 4z = 1 \\ 3x + 3z = 10 \\ 43x + 22y + 10z = 198 \end{cases}$

（2） $\begin{cases} x + 2y - z = 3 \\ 4x - 7y + z = 10 \\ 3x + 2y - 100z = 1 \end{cases}$

（3） $\begin{cases} 122x + 231y - 1888z = 1292 \\ 394x - 983y + 23z = -1000 \\ 39x + 19y - 243z = 23 \end{cases}$

2.11 请将多项式 $13x^5 + 45x^4 - 98x^3 + 108x^2 - 35x - 12$ 和 $11x^5 + 8x^3 - 35x$ 表示为数值矢量形式，然后对这 2 个多项式进行加、减、乘、除运算。

2.12 基于两个数组 T=[1,2,3,4,5] 和 S=[12.1,32.4,45.3,36.7,78.9]，请拟合出最高次幂为 5 的多项式，并在 4.6 处插值。

2.13 解方程 $12x^5 + 13x^4 + 14x^3 + 15x^2 + 16x + 17 = 0$。

2.14 这里有数组 A=[12.3,13/3,24/7,29,97] 和矩阵 $\begin{bmatrix} 1 & 2 & 3 \\ 1/2 & 1/3 & 1/4 \\ 9 & 8 & 7 \end{bmatrix}$，请计算它们的最大值、

最小值、平均值、中值、标准差，并借助函数 sort 进行升序排列。

2.15 对下列函数在-2、2、5 附近进行寻找最小值、零值对应的 x 值，并绘图。

（1） $x^3 + 8x^{1/5} - 1$

（2） $\sin x \cos(\frac{1}{2}x) + \tan x^2 \sin(2x)$

（3） $e^{x^2-3} \sin x - e^{\frac{1}{2}x} \cos x$

（4） $\sin hx + \cos h(\frac{1}{7}x^{\frac{1}{3}})x$

2.16 创建两个符号表达式 $f(x) = \sin x + x\cos x$ 和 $f(x, y) = \sin x \cos y - \tan x \sin y$。

2.17 创建一个符号表达式 $f(x, y, z) = ax^3yz + bxyz - cyz + c$，并计算 $f(x, y, z) - cyz$。

2.18 对有效位数为 1、5、29 和 90 情况下，当 $x = \pi - 1$ 时，计算多项式 $f(x) = x + \cos x \sin x - \tan x$。

2.19 基于符号表达式 $f(x) = x^4 + 2x^2 + 1$ 和 $g(x) = x^3 + 6x^2 + 3x + 5$。求解下列问题：

（1） $f(x) + g(x)$

（2） $f(x) \times g(x)$

（3） 因式分解 $f(x)$

（4） $g(x)$ 的反函数

2.20 解下列表达式：

（1） $\lim\limits_{x \to 0} \dfrac{\tan x}{x}$

（2） $\lim\limits_{x \to \pi} \dfrac{\sin x}{x - \pi}$

2.21 对以下两个函数 $f(x) = \dfrac{1}{x-1} + \arcsin x^2$ 和 $g(x) = \log x^2 - x\sin x$ 求不定积分。

2.22 在 $[-2\pi, 2\pi]$ 范围内求解以下两个函数 $f(x) = \dfrac{1}{x-1} + \arcsin x^2$ 和 $g(x) = \log x^2 - x\sin x$ 的定积分。

2.23 求解下列表达式。

（1） $\int_{-1}^{1} dx \int_{x}^{\sqrt{x}} (\sin x \cos(\frac{1}{2}xy^2) - x^2y + 3x - 7) dy$

（2） $\int_{0}^{3} dx \int_{-x}^{x} dy \int_{-\sqrt{xy}}^{\sqrt{xy}} (e^{x^2-1} + \dfrac{yz}{x^2-1} - y^2z - 6\cos z) dz$

2.24 求导：

（1） $\dfrac{d^2}{dx}(xe^{-x^2-y^2})$

（2） $\dfrac{d^2 f}{dxdy} (f = e^{x^2-y^2} + \sin x \cos y + 9)$

（3） $\dfrac{d^2 f}{dxdy}$、$\dfrac{d^2 f}{dydz}$、$\dfrac{d^2 f}{dzdx}$、$\dfrac{d^2 f}{dydx}$、$\dfrac{d^2 f}{dzdy}$、$\dfrac{d^2 f}{dxdz}$ 和 $\dfrac{d^3 f}{dxdydz}$

（$f = 3x^3y^2z - 5x^2yz^2 - 11yz + \cos x \sin y \tan z$）

2.25 对下列函数进行傅里叶变换：

（1） $f(x) = \sin x e^{-x^2}$

（2） $g(w) = e^{-|w|} + \cos w$

（3） $f(x) = e^{-|x|}$

（4） $f(x,v) = e^{-x^2|v|} \dfrac{\cos v}{v}$（$x$ 实数）

2.26 对下列函数进行傅里叶逆变换：

（1） $f(w) = e^{w/(4a^2)}$

（2） $g(x) = xe^{-|x|}$

（3） $f(w) = 2e^{-|w|} + w - 1$

（4） $f(w,v) = e^{-w^2|v|} \dfrac{\sin v}{v+1}$（$w$ 实数）

2.27 对下列函数进行拉布拉斯变换：

（1） $f(t) = t^4 + t - 1$

（2） $g(s) = \dfrac{s}{\sqrt{s+2}}$

（3） $f(t) = e^{-st}$

2.28 对下列函数进行拉布拉斯逆变换：

（1） $f(s) = \dfrac{s}{s^2+1}$

（2） $g(t) = \dfrac{1}{(t-a)^2}$

（3） $f(u) = \dfrac{1}{u^2+a^2}$

2.29 对下列函数进行 Z 变换：

（1） $f(n) = n^3$

（2） $g(z) = (a+1)^z$

（3） $f(n) = \sin an + \cos an$

2.30 对下列函数进行 Z 逆变换：

（1） $g(n) = \dfrac{n(n+1)(n+2)}{n^3+3n^2+3n+1}$

（2） $f(z) = \dfrac{z-1}{(z-a)^2}$

第 3 章 编 程

本章将对 MATLAB 语言编程进行简要阐述。MATLAB 编程有两种形式：一类是通常意义上的编程为脚本程序；另一类是函数程序。程序结构有三种：顺序结构、分支结构和循环结构。

顺序结构是最简单的结构，该结构中语句严格按照顺序一句接一句执行。顺序结构程序一般首先要读取数据，对这些数据进行运算处理，然后再给出预期答案。该类程序所有语句能且只能运行一次，不允许有语句运行次数超过一次，也不允许有语句可以选择运行或者选择不运行。

而分支结构（Branches 结构），可根据用户的选择使得部分语句参加程序运算，部分语句不参加程序运算。

循环结构（Loops 结构），可以让特定语句重复运行，也就是可以根据用户需要重复运行任意次。

三类结构的嵌套组合使得程序变得复杂，错误出现的概率大大增加。为了减少编程错误，本书将基于大家熟悉的"top-down 设计"给出常用程序设计模式，同时引入一种通用的运算方法"伪代码法"进行讲解。

3.1 Top-Down 设计模式

假如你在一家公司工作，需要编写程序来解决一些问题，通常你会怎么做呢？

接到新任务时，通常情况下我们会立马坐到电脑前开始编写程序，不会去仔细思考待解决的问题。如果问题简单，这种"运指如飞"的办法一般可以解决，就如本书中的一些示例。但是，实际中问题往往很复杂，编程者在使用这种方法时将会陷入难于自拔的困境。对于相对庞大的程序，在开始编写程序之前，建议编程者花费足够的时间思考清楚需要解决的问题和选择解决问题的具体方法和步骤。

本章将逐步给出一种程序设计规范，并将它应用到后面示例。虽然对于简单程序，该规范显得冗余，但是，当问题复杂时，这种规范就变得非常重要。正如 Chapman 教授的大学老师讲的那样，编写程序很容易，知道该编写什么比较困难。这是因为只有一个人完全理解需要解决的问题后，才能给出解决问题的正确方法。无论这问题多么复杂总是可以分解为很多个子问题，子问题就很简单、很容易解决，分别把子程序转化为可执行程序，最后把子程序整合为一个程序，就完成了。

Top-down 设计就是一个分解大任务的过程，首先将一个大任务分解为几个相对的小任务，如果这些小任务仍然不容易解决可以再将其分解为更小的任务，直至所分解的小任务都很容易编程完成为止；然后完成每个小任务的编程和测试，直到确保每一部分都没有错误为止；最后完成程序整合和整体测试。

Top-down 设计是程序设计规范的基础。所谓程序设计规范实际上就是自己的编程习惯，如图 3.1 所示。

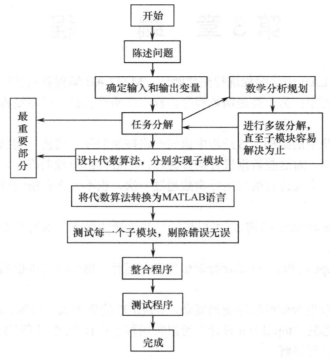

图 3.1 程序设计规范示意图

程序设计规范操作步骤如下所述。

1．陈述问题

程序通常是针对某类问题，清楚的阐述对读者来讲是必要的。简明扼要的阐述能够避免使用者的错误理解，并且对设计者理清思路也很有帮助。

2．确定输入和输出变量

理解问题后的第一件事就是确定输入变量和输出变量，统一变量可以避免后面子模块（子任务）中出现变量混乱的情况。

3．任务分解

首先，要清楚待解决问题需要分为几个部分，每个部分设计需要完成什么任务，是否能够继续分解为更小的独立问题；其次，对所有子问题进行分析，给出解决方案；最后，搜寻解决方法的相关知识。

4．设计代数算法，分别实现子模块

对所有子程序分别设计出可行的数学运算，并用数学语言对其进行阐述。

5. 将代数算法转换为 MATLAB 语言

借助 MATLAB 语言，对所有子程序的数学运算进行"翻译"。

6. 测试每一个子模块，剔除错误确保无误

分别对所有编译的子程序进行测试，尽量全面考虑各种可能找出子程序缺陷并进行修正，为最终的程序调试节省大量时间，尽量避免一些不必要的错误。

7. 整合程序

根据待求问题的需要，将所有子程序有机结合，完成最终程序的组装。

8. 测试程序

根据待求问题的各个方面需求进行测试，尽量考虑到各种合理的输入数据，考虑所有分支都要得到运行，以使得所编写程序能够满足用户尽量多的用途和需求。这需要相当长的时间对其进行检测，以避免错误。

> **好习惯：**
> 认真贯彻 Top-down 设计规范步骤编程，为编译出可用无误的程序提供保障，同时提高程序的可移植性。

编写大程序过程中，真正用在编写程序上的时间是非常少的。正如 Frederick P. Brooks Jr. 说的那样，在典型的软件开发工程中，大概 1/3 的时间用来计划做什么（1 到 4 步），1/6 的时间用来写程序（5 和 7 步），1/2 的时间用来测试和调试程序。

3.2 伪 代 码

伪代码是程序编写过程必要的和重要的组成部分，它对程序中将要执行的代数运算进行了详细阐述。最重要的是这些阐述易于他人理解，并能够为程序编写人员提供帮助。但伪代码中阐述的步骤设计是以编写者数学能力为基础的，数学水平高的设定运算步骤简捷快速，数学水平稍差的设定的程序运算步骤烦琐，这方面我们很难给出什么有价值的意见。

伪代码是借助数学表述和文字语言对所编写程序运算过程的阐述。也就是说，按照伪代码中的步骤和方法，通过数学计算就能够给出所要解决问题的正确答案，即使用人工计算也可以（前提是人工能够完成）。伪代码也就是所要解决问题的解决过程的详细陈述。也可以说，伪代码是待编写程序的另一种表述形式，将伪代码中的数学语言和文字表述转化为 MATLAB 可执行语句，就是要编写的程序。

伪代码编写时，一般将每一个运算和操作步骤作为一行，并对每一行都附以浅显易懂的文字解释。因为伪代码对程序员修改程序非常有帮助，便于对每一个语句的清楚理解；对想了解 MATLAB 内置程序思想的程序学习者和使用者很有用；对想了解 MATLAB 程序编写过程的不同用户也很有用。当程序非常简单时，也就是读者或使用者较容易看懂程序的情况下，伪代码往往被省略。

3.3 顺 序 结 构

顺序结构是所有程序结构中最简单的一种。顺序结构程序是由一系列 MATLAB 语句组成,执行是一行紧接一行进行的。提高编程能力的最有效途径就是做大量练习,为此本书将提供大量示例。

【例 3.1】 输入学生信息。

请编写新生登录程序,提示学生输入姓名、家庭住址、院系、年龄,并在窗口显示所输入信息。

分析:根据题意知该程序的用户为入学新生,包含内容有 4 个部分,即姓名、家庭住址、院系、年龄;功能有二,一是新生输入自己的相关信息,二是显示输入的相关信息。根据待编程序,伪代码设计如下所述。

(1)该程序需要 4 个变量分别为 name、address、department 和 age。

MATLAB 表述:syms name address department age

(2)在光标提示处显示"请输入您的姓名:"或"Please type your name:",然后按回车键。

MATLAB 表述:name=input('Please type your name: ', 's')

或者:name=input('请输入您的姓名:',' s')

(3)在光标提示处显示"请输入您的家庭住址:"或"Please type your address: ",然后按回车键。

MATLAB 表述:address=input('Please type your address: ', 's')

或者:address=input('请输入您的家庭住址', 's')

(4)在光标提示处显示"请输入您的院系专业:"或"Please type your department: ",然后按回车键。

MATLAB 表述:department=input('Please type your department: ', 's')

或者:department=input('请输入您的院系专业', 's')

(5)在光标提示处显示"请输入您的年龄:"或"Please type your age: ",然后按回车键。

MATLAB 表述:age=input('Please type your age: ', 's')

或者:age=input('请输入您的年龄', 's')

(6)显示您所输入的信息。

MATLAB 表述:disp('Your name is '), disp(name); disp('Your address is'), disp(address); disp('Your department is'), disp(department); disp('Your age is'), disp(age)

或者:disp('姓名'), disp(name); disp('家庭住址'), disp(address); disp('院系'), disp(department); disp('年龄'), disp(age)

在本程序中,没有考虑新生输入错误的修改功能。如果要为新生提供修改机会,这里将会出现两种情况:一种情况是保存输入信息并显示;另一种情况是选择修改,重新输入信息,直到确定输入正确再保存。为实现这一个功能,程序出现分支结构,即两种选择,这将在 3.4 节中进行介绍。

EXAMP03001

```matlab
%  Purpose:
%  This program is to supply an access for fresh students to enter
%  their information including name, address, department and age.
%  And then the entered contents appear on the window.
%  (本程序为入学新生提供输入端口，新生可以输入包含姓名、家庭地址、院系专业和年龄
%  的相关信息，并且这些信息显示在桌面上)
%
%  Records of Revision: (修改记录)
%  Date: 2015-07-12
%  Programmer: Wang Yonglong
%  Description of change: Second revision
%
%  Clear command window.
%  (清除命令窗口)
clc
%  Clear variables in workspace.
%  (清除工作空间中的变量)
clear
%  Define four variables Name, Address, Department, Age.
%  (定义4个变量：name、address、department 和 age)
syms Name Address Department Age
%  Prompt user to enter whose name.
%  (提示用户输入自己的姓名)
Name=input('请输入您的姓名：','s');
%  Prompt user to enter whose address.
%  (提示用户输入家庭住址)
Address=input('请输入您的家庭住址：','s');
%  Prompt user to enter whose department.
%  (提示用户输入院系专业)
Department=input('请输入您的院系：','s');
%  Prompt user to enter whose age.
%  (提示用户输入年龄)
Age=input('请输入您的年龄：','s');
%  Display the contents entered by user.
%  (显示用户输入的信息).
disp('姓名：'), disp(Name);
disp('家庭住址：'), disp(Address);
disp('院系：'), disp(Department);
disp('年龄：'), disp(Age);
```

运行结果如下：

```
请输入您的姓名：王永龙
请输入您的家庭住址：山东临沂
请输入您的院系：理学院
请输入您的年龄：18
姓名：王永龙
家庭住址：山东临沂
```

```
院系：理学院
年龄：18
```

> **注意：**
> 输入数据通常都是数字，一旦输入的是字符时需要带上参数's'。在 MATLAB 环境下，所有命令中引字符串的单引号"'"和"'"必须为半角，否则会变为红色，运行时提示错误。

为了与 MATLAB 自带函数命令相一致，所有编程示例的程序帮助都采用英汉双语注释，方便学习者学习。需要注意的是，每行帮助注释前面都有"%"符号，标示该行为 MATLAB 非执行行。

【例 3.2】 任意两个数之和。

请编写程序，计算并显示用户输入的任意两个数之和。

分析： 本问题需要 3 个变量，其中 2 个变量用来存放用户输入的任意两个数，余下的 1 个变量用来存放输入两个数之和。为容易理解起见，3 个变量就好比 3 个人需要 3 把椅子一样，1 把椅子只能坐 1 个人。无论我们在什么时候叫某人，也可以直接用相对应椅子的编号来代替。本程序的伪代码设计如下所述。

（1）本程序需要 3 个变量，分别为 number1、number2、sum2numbers。（变量命名需注意，比如上面命名就很方便，它们的中文意思为"数1"，"数2"和"两个数之和"，这样容易理解与记忆。）

MATLAB 表述：syms number1 number2 sum2numbers

（2）提示用户输入第一个数 number1。

MATLAB 表述：number1=input('Please enter number1 as：')

或者：number1=input('请输入第一个数 number1 为：')

（3）提示用户输入第二个数 number2。

MATLAB 表述：number2=input('Please enter number2 as：')

或者：number2=input('请输入第二个数 number2 为：')

（4）求解 number1 和 number2 两个数之和 sum2numbers。

MATLAB 表述：sum2numbers=number1+number2;

（5）显示所求结果。

MATLAB 表述：fprintf('The sum of %f and %f is %f.\n', number1, number2, sum2numbers)

EXAMP03002

```
% Purpose:
%  This program solves the sum of two arbitrary numbers entered by
%  the user and shows the result in the command window.
%  (本程序可以求解用户输入的任意两个数之和，并将该和显示在命令窗口)
%
% Records of revision:
% Date: 2015-07-12
% Programmer: Wang Yonglong
% Description of changer: original code
%
```

```
% Clear variables in workspace.
% (清除工作空间中的变量)
clear
%
% Define three variables number1, number2, sum2numbers.
% (定义3个变量：number1、number2 和 sum2numbers)
syms number1 number2 sum2numbers
% Prompt user to enter the first number number1.
% (提示用户输入第一个数 number1)
number1=input('Please enter number1: ');
% Prompt user to enter the second number number2.
% (提示用户输入第二个数 number2)
number2=input('Please enter number2: ');
% Solve the sum of the two numbers number1 and number2.
% (求解 number1 和 number2 两个数之和)
sum2numbers=number1+number2;
% Display the sum of number1 and number2.
% (显示 number1 和 number2 之和)
fprintf('The sum of %f\n and %f\n is %f\n', number1, number2, sum2numbers);
% Save number1, number2 and sum2numbers as two symbolic numbers.
% (将 number1、number2 和 sum2numbers 保存为符号数)
nub_1=num2str(number1);
nub_2=num2str(number2);
sum2nub=num2str(sum2numbers);
% Display the tree symbolic numbers number1, number2 and sum2numbers
% as string.
% (将3个符号数 number1、number2 和 sum2numbers 显示为文本)
fprintf('The sum of %s\n and %s\n is %s\n', nub_1, nub_2, sum2nub)
```

运行结果如下：

```
Please enter number1: 23
Please enter number2: 52
The sum of 23.000000
 and 52.000000
 is 75.000000
The sum of 23
 and 52
 is 75
```

> **注意：**
> 在函数 fprintf 语句中，要注意有 "%f" 符号表示在此处插入一类数据为 f 型（除 f 型数据外，在 MATLAB 中还有 d 型、i 型、o 型、u 型、x 型、X 型、e 型、E 型、g 型、G 型、c 型、s 型数据）；另外还有 "\n" 表示为换行。

本程序不能求两个复数之和，如果想增加该功能，需要由分支结构来完成，这一结构将在 3.4 节讲述。另外，如果本程序只是需要给出结果就可以，用户输入的两个数都不需要保存，需要参量为 2 个，也就把 2 个椅子上的 2 个人捆绑起来放在其中 1 个椅子上即可。需要注意的是，被放了 2 个人的椅子，在放 2 个人的同时也丢失了原先那个人的所有信息，没有被占用的椅子却依然保留着前 1 个人的信息。

【例 3.3】 三重定积分。

请为求解三重定积分编写程序，要求能够显示每一步的积分结果。

分析：尽管 MATLAB 为三重积分提供了积分函数"int"，但是没有给出实际积分步骤。为给出大家熟悉的积分步骤，在此我们给出求解三重积分的程序。首先，对输入变量、输出变量、需要显示的数学结果进行分析。

本程序主题是被积函数，需要输入。积分过程分为 3 步，对 3 个变量一个接着一个进行积分。这里需要 6 个变量和表达式，因为每个积分变量都需要上积分限和下积分限。所以该程序至少需要 7 个输入通道。同时，为了提供每一步积分结果，这里还至少需要 3 个输出通道。本题中积分变量的积分顺序为 MATLAB 程序中默认顺序，z, y，随后是 x。伪代码设计如下所述。

（1）需要 3 个变量作为积分变量，记为 x、y、z。还需要 10 个变量，分别为 1 个被积函数、6 个积分限、3 个计算结果。

MATLAB 表述：syms x, y, z, f, x1, x2, y1, y2, z1, z2, f1, f2, f3

（2）提示用户输入被积函数，该函数是变量 x、y、z 的函数。

MATLAB 表述：f=sym(input('Please enter integral function as f=', 's'))

（3）提示用户输入 x 变量的下积分限 x1 和上积分限 x2。

MATLAB 表述：x1=input('Please enter the integral bottom-limit of x, x1= ');
　　　　　　　x2=input('Please enter the integral up-limit of x, x2= ');

（4）提示用户输入 y 变量的下积分限 y1 和上积分限 y2。

MATLAB 表述：y1=input('Please enter the integral bottom-limit of y, y1= ');
　　　　　　　y2=input('Please enter the integral up-limit of y, y2= ');

（5）提示用户输入 z 变量的下积分限 z1 和上积分限 z2。

MATLAB 表述：z1=input('Please enter the integral bottom-limit of z, z1= ');
　　　　　　　z2=input('Please enter the integral up-limit of z, z2= ');

（6）首先被积函数对变量 z 积分，下积分限为 z1，上积分限为 z2，积分结果存为 f1，为变量 x 和 y 的函数。

MATLAB 表述：f1=int('f','z','z1','z2');
　　　　　　　pretty(f1)

（7）积分结果 f1 再对积分变量 y 进行积分，积分下限为 y1，积分上限为 y2，积分结果存为 f2，为变量 x 的函数。

MATLAB 表述：f2=int('f1','y','y1','y2');
　　　　　　　pretty(f2)

（8）积分结果 f2 再对积分变量 x 进行积分，积分下限为 x1，积分上限为 x2，积分结果存为 f3，为一个数值。

MATLAB 表述：f3=int('f2','x','x1','x2');

（9）提示用户输入输出结果有效位数，并输出结果。

MATLAB 表述：n=input('The effective number of the result of f3 is');
　　　　　　　vpa(f3, n)

EXAMP03003

```
% Purpose:
```

```
% This program solves a three-folds integration based on the integral
% function and the integral down- and up-limits of x, y, and z entered
% by user.
% (本程序可根据用户针对变量x、y和z输入的积分上下限求解三重积分)
%
% Records of revision:
% Date: 2015-07-16
% Programmer: Wang Yonglong
% Description of change: Original code
%
% x: a independent variable.
% y: a independent variable.
% z: a independent variable.
% f: a integrated function f depending on x, y and z.
% x1: the integral down-limit of x.
% x2: the integral up-limit of x.
% y1: the integral down-limit of y.
% y2: the integral up-limit of y.
% z1: the integral down-limit of z.
% z2: the integral up-limit of z.
% f1: the result of f integrated by z with z1 and z2.
% f2: the result of f1 integrated by y with y1 and y2.
% f3: the final result of f2 integrated by x with x1 and x2.
%
% Clear command window.
% (清除命令窗口)
clc
% Clear variables in workspace.
% (清除工作空间中的变量)
clear
% Define input and output variables f, f1, f2, f3, x, y, z,
% x1, x2, y1, y2, z1, z2.
% (定义输入输出变量f、f1、f2、f3、x、y、z、x1、x2、y1、y2、z1、z2).
syms x y z f x1 x2 y1 y2 z1 z2 f1 f2 f3
% Prompt user to enter the integrated function f.
% (提示用户输入被积函数f)
f=sym(input('Enter the integrated function ','s'));
% Prompt user to enter the integral bottom-limit of x, x1.
% (提示用户输入x的积分下限x1)
x1=input('Enter the integral bottom-limit of x, a constant, x1= ');
% Prompt user to enter the integral up-limit of x, x2.
% (提示用户输入x的积分上限x2)
x2=input('Enter the integral up-limit of x, a constant, x2= ');
% Prompt user to enter the integral bottom-limit of y, y1.
% (提示用户输入y的积分下限y1)
y1=sym(input('Enter the integral bottom-limit of y depending only on x, y1= ','s'));
% Prompt user to enter the integral up-limit of y, y2.
% (提示用户输入y的积分上限y2)
```

```
        y2=sym(input('Enter the integral up-limit of y depending only on x, y2=
','s'));
        % Prompt user to enter the integral bottom-limit of z, z1.
        % (提示用户输入 z 的积分下限 z1)
        z1=sym(input('Enter the integral bottom-limit of z depending only on x and
y, z1= ','s'));
        % Prompt user to enter the integral up-limit of z, z2.
        % (提示用户输入 z 的积分上限 z2)
        z2=sym(input('Enter the integral up-limit of z depending only on x and y,
z2= ','s'));
        % Integrate f by z from z1 to z2 as f1.
        % (被积函数 f 对 z 从 z1 到 z2 积分,存为 f1)
        f1=int(f,'z',z1,z2);
        % Express f1 in the usual mathematical form.
        % (将 f1 表示为通常的数学形式)
        pretty(f1)
        % Integrate f1 by y from y1 to y2 as f2.
        % (结果 f1 对 y 从 y1 到 y2 积分,存为 f2)
        f2=int(f1,'y',y1,y2);
        % Express f2 in the usual mathematical form.
        % (将 f2 表示为通常的数学形式)
        pretty(f2)
        % Integrate f2 by x from x1 to x2 as f3.
        % (结果 f2 对 x 从 x1 到 x2 积分,记为 f3)
        f3=int(f2,'x',x1,x2);
        % Express f3 in the usual mathematical form.
        % (将 f3 表示为通常的数学形式)
        pretty(f3)
        % Prompt user to enter the precision of f3, n.
        % (提示用户输入 f3 的有效位数)
        n=input('The precision of the result of f3 is n= ')
        % Show the final result in the command window.
        % (在命令窗口输出最终结果)
        vpa(f3,n)
```

运行结果如下:

```
        Enter the integrated function x^2+y*z
        Enter the integral bottom-limit of x, a constant, x1= -1
        Enter the integral up-limit of x, a constant, x2= 3
        Enter the integral bottom-limit of y depending only on x, y1= -x
        Enter the integral up-limit of y depending only on x, y2= x
        Enter the integral bottom-limit of z depending only on x and y, z1= -x+y
        Enter the integral up-limit of z depending only on x and y, z2= x+y
              2   2
          2x(x + y )

                4
          16 x
          -----
            3
```

```
    3904/15
    The precision of the result of f3 is n= 6
    n = 6
    ans =260.267
```

> **注意:**
> 在编写程序过程，如果需要输入符号表达式，需要在"input"命令输入语句的外围加上"sym"函数。

3.4 分支结构

3.4.1 关系算符和逻辑算符

1. 关系算符

分支结构程序，运行程序过程中需要对运行模块进行选择。编程人员会经常遇到这类问题，不同情况会有不同结果。为了解决这类问题，程序中首先必须将其分为多个子任务，然后再将这些子任务通过语句整合在一起，借助特定命令语句进行选择。在做出判定选择过程中，会用到关系算符和逻辑算符。常用关系算符描述参见表 3.1，关系算符能够比较两个标量值、两个数组，也能用来比较标量与数组，返回正确结果只有两种情况，即"真"和"假"。否则，将给出错误提示。基于算符比较结果，我们能够一个一个进行解决。换句话说，能借助关系算符将复杂问题变得简单。

表 3.1 关系算符

算 符	描 述	算 符	描 述
==	等于	>=	大于或等于
~=	不等于	<	小于
>	大于	<=	小于或等于

> **注意:**
> 注意关系算符"=="与赋值符号"="的不同。另外，需注意在判定两个数值是否相等时，可能由于四舍五入导致本该相等的两个数却给出不等的结果。一般比较两个数相等，通过比较两个数值差是否达到要求精度来判定。例如，x2-x1<0.00001，粗略情况下即可认为相等。

2. 逻辑算符和逻辑函数

在编写程序过程中，关系算符一般用来处理包含多种情况的问题，逻辑算符通常用来描述相对比较复杂的问题。常用逻辑算符描述参见表 3.2。

表 3.2 逻辑算符

算 符	描 述	算 符	描 述
&	与	xor	非或
|	或	~	非

逻辑算符示例结果参见表 3.3。

表 3.3 逻辑算符示例结果表

输入		与	或	非或	非
L1	L2	L1&L2	L1|L2	xor(L1,L2)	~L1
0	0	0	0	0	1
0	1	0	1	1	1
1	0	0	1	1	0
1	1	1	1	0	0

MATLAB 逻辑函数描述参见表 3.4。

表 3.4 MATLAB 逻辑函数

函数	描述
ischar(A)	如果 A 是字符数组,结果为 1,否则为 0
isempty(A)	如果 A 为空数组,结果为 1,否则为 0
isinf(A)	如果 A 的值是无穷,结果为 1,否则为 0
isnan(A)	如果 A 是一个非数,结果为 1,否则为 0
isnumeric(A)	如果 A 是数值数组,结果为 1,否则为 0

3. 算符运算次序

四则运算顺序和数学规定相同,即

(1) 首先,关系算符(==、~=、>、>=、<、<=)运算顺序为从左到右依次执行;
(2) 其次,运算"xor"或非算符和"~"非算符;
(3) 再次,"&"与算符;
(4) 最后,所有或算符"|"。

补充:如果上面计算顺序不好区分,可以通过"()"强行限定计算顺序。

3.4.2 if 结构

If 结构一般具有如图 3.2 至图 3.4 所示基本结构形式。在 if 结构中,最简单结构是含有两个分支的结构,其一当 if 判定条件正确时执行,另外一个是在 if 判定条件错误时执行。一旦待解决问题非常复杂时,if 结构将含有的分支远大于两个。本书为复杂问题提供两种解决办法,一种是同级别情况下,借助 elseif 增加分支;另一种结构是多个"if—else—end"结构嵌套。后者结构也就是,在 if 紧随的分支再次含有"if—else—end"结构,else 分支中也可含有"if—else—end"结构,如果需要可以继续向下细分。对于前一种方法,可以借助 elseif 命令来任意增加需要的分支选择。

好习惯:
　　总是在每个 if 结构开始缩进两个字符或更多个字符是个好习惯,这可增强程序的可读性。

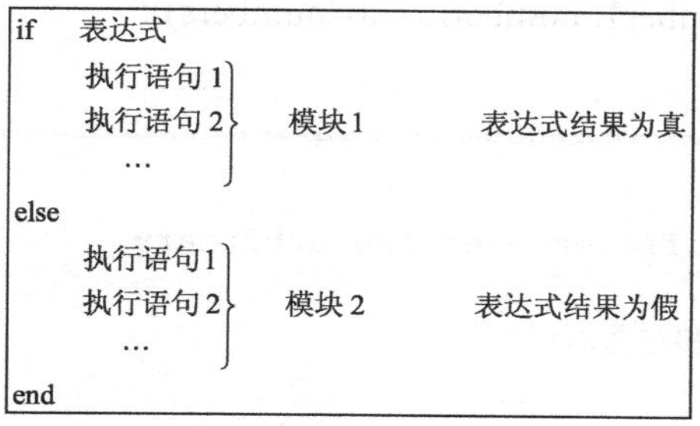

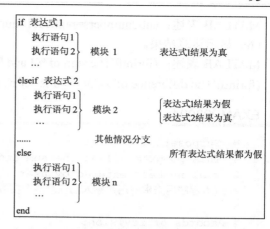

图 3.2 最简单 if 结构形式　　　　图 3.3 含有大于两个分支的 if 结构形式

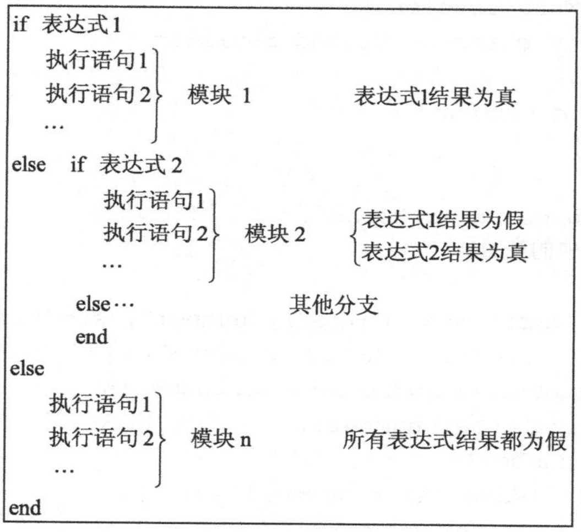

图 3.4 含有多层 if 结构形式

【例 3.4】 任意两个数的和与差。

编写程序能够根据用户输入的任意两个实数,给出两个数之和与之差。

分析:该问题可分为两个部分,第一个部分求和,第二个部分求差,所以该问题可以用一个"if—else—end"结构来完成。本问题需要 4 个变量,2 个用来存储用户输入的 2 个任意数,2 个用来存储这两个数的和与差。该程序的伪代码设计如下所述。

(1)本程序需要 4 个数,记为 number1, number2, sum2numbers 和 sub2numbers。

MATLAB 表述:syms number1, number2, sum2numbers, sub2numbers

(2)提示用户输入第一个数 number1。

MATLAB 表述:number1=input('Please enter number1= ');

(3)提示用户输入第二个数 number2。

MATLAB 表述:number2=input('Please enter number2= ');

(4)求 number1 和 number2 之和。

MATLAB 表述:sum2numbers=number1+number2;

(5)求 number1 和 number2 之差。

MATLAB 表述：sub2numbers=number1-number2;

（6）显示运算结果。

MATLAB 表述：(fprintf('The sum of %f and %f is %f.\n', number1, number2, sum2numbers));
(fprintf('The difference of %f and %f is %f\n', number1, number2, sub2numbers));

EXAMP03004

```
%   Purpose:
%   This program solves the sum and difference of two arbitrary
%   real numbers entered by user.
%   (本程序用来求解用户输入的任意两个实数之和与之差)
%
%   Records of revision:
%   Date: 2015-07-12
%   Programmer: Wang Yonglong
%   Description of changer: Second revision
%
%   Clear command window.
%   (清除命令窗口)
clc
%   Clear variables in workspace.
%   (清除工作空间中的变量)
clear
%   Define four variables: number1, number2, sum2numbers, sub2numbers.
%   (定义4个变量：number1、number2、sum2numbers 和 sub2numbers)
syms number1 number2 sum2numbers sub2numbers
%   Prompt user to enter number1.
%   (提示用户输入 number1)
number1=input('Please enter number1 as: ');
%   Prompt user to enter number2.
%   (提示用户输入 number2)
number2=input('Please enter number2 as: ');
%   Prompt user to choose the operation of the entered two numbers.
%   (提示用户对两个数的操作进行选择)
n=input('如果求和，请您输入 "1", 如果求差，请您输入 "-1" ');
%   By using the structure "if-else-end", one can accomplish this work.
%   (借助"if-else-end"结构完成该任务)
if n==1
%   number1 pluses number2 as string, and display.
%   (number1+number2 存为字符并显示)
sum2numbers=num2str(number1+number2);
numb_1=num2str(number1);
numb_2=num2str(number2);
fprintf('The sum of %s and %s is %s.\n', numb_1, numb_2, sum2numbers);
elseif n==-1
%   number1 subtracts number2 as string, and display.
%   (number1-number2 存为字符并显示)
    sub2numbers=num2str(number1-number2);
    numb_1=num2str(number1);
    numb_2=num2str(number2);
```

```
fprintf('The sum of %s and %s is %s.\n', number1, number2, sub2numbers);
end
```

运行结果如下：

```
Please enter number1 as: 99
Please enter number2 as: 12
如果求和，请您输入 "1"，如果求差，请您输入 "-1": 1
The sum of 99 and 12 is 111.
```

本程序不能计算两个复数的和与差。如果想增加这项功能，需要增加分支结构来完成。该扩展程序留给读者独立完成。

【例 3.5】 解一元二次方程。

请编写程序能够求解一元二次方程。

分析：本程序将用来求解任意常系数一元二次方程 $ax^2+bx+c=0$。首先，该程序具有判定输入方程具有怎么样的根，比如：两个实根、两个相等实根、两个复数根。其次，需要变量包含 3 个系数 a、b、c，1 个判别式 s，2 个根 x_1、x_2，根的实部与虚部 real 和 image。最后，该程序能够显示所求的解。伪代码设计如下所述。

（1）需要 3 个输入变量 a, b, c；2 个输出变量 x1, x2；3 个中间变量判别式 s，解的实部和虚部 real 和 image。

MATLAB 表述：syms a b c s x1 x2 real image

（2）提示用户输入一元二次方程的 3 个系数。

MATLAB 表述：a=input('请输入系数 a: ');
　　　　　　　b=input('请输入系数 b: ');
　　　　　　　c=input('请输入系数 c: ');

（3）根据输入参数，计算判别式 $s = b^2 - 4ac$。

MATLAB 表述：s=b^2+4*a*c;

（4）根据判别式给出方程的解，两个实根（$s>0$），解为

$$x_1 = \frac{-b+\sqrt{b^2-4ac}}{2a} = \frac{-b+\sqrt{s}}{2a}$$

$$x_2 = \frac{-b-\sqrt{b^2-4ac}}{2a} = \frac{-b-\sqrt{s}}{2a}$$

两个相等实根（$s=0$），解为

$$x_1 = x_2 = \frac{-b}{2a}$$

两个复数根（$s<0$），解为

$$x_1 = \frac{-b+\mathrm{i}\sqrt{|b^2-4ac|}}{2a} = \frac{-b+\mathrm{i}\sqrt{|s|}}{2a} = \text{real} + \mathrm{i} \times \text{image}$$

$$x_2 = \frac{-b-\sqrt{|b^2-4ac|}}{2a} \frac{-b-\mathrm{i}\sqrt{|s|}}{2a} = \text{real} - \mathrm{i} \times \text{image}$$

MATLAB 表述：if s>0
　　　　　　　　x1=(-b-sqrt(s))/(2*a);
　　　　　　　　x2=(-b+sqrt(s))/(2*a);

```
            disp('方程有两个实根：');
            fprintf('x1=%f\n',x1);
            fprintf('x2=%f\n',x2);
        elseif s==0
            x1=x2=-b/(2*a);
            disp('方程有两个相等实根：');
            fprintf('x1=x2=%f\n', x1);
        else
            real=-b/(2*a);
            image=(sqrt(abs(s)))/(2*a);
            disp('方程有两个复数根：');
            fprintf('x1=%f-i*%f\n', real, image);
            fprintf('x2=%f+i*%f\n', real, image);
        end
```

EXAMP03005

```
%   Script file: EXAMP03005.m  (可省略)
%
%   Purpose:
%   This program solves the roots of an arbitrary quadratic equation
%   entered by user as a*x^2+b*x+c=0, irrespective of the roots being
%   two distinct real roots, or a single repeated real root, or two
%   complex roots.
%   (本程序可求解用户输入的任意一元二次方程，无论其解为两个实根、两个相等实根
%   还是两个复数根)
%
%   Records of revision:
%   Date: 2007-11-12
%   Programmer: Wang Yonglong
%   Description of changer: original code
%
%   Descriptions of variables:
%   a: the coefficient of x^2 term.
%   b: the coefficient of x term.
%   c: the constant coefficient.
%   s: the discriminant of the equation entered by user.
%   image: the image part of roots.
%   real: the real part of roots.
%   x1: the first root of the equation.
%   x2: the Second root of the equation.
%
%   Clear command window.
%   (清除命令窗口)
clc
%   Clear variables in workspace.
```

```matlab
%   (清除工作空间中的变量)
clear
%   Define eight variables a b c s image real x1 x2.
%   (定义8个变量为a、b、c、s、image、real、x1、x2)
syms a b c s real image x1 x2
%   Prompt user to enter the coefficients of the solved equation.
%   (提示用户输入待求解方程的系数)
disp('This program solves for the roots of the quadratic');
disp('equation in the form a*x^2+b*x+c=0');
a=input('Please enter a: ');
b=input('Please enter b: ');
c=input('Please enter c: ');
%   Calculate the discriminant s.
%   (计算判别式s)
s=b^2-4*a*c;
%   In terms of the discriminant s, one solves the roots.
%   (根据判别式s,求解方程的根)
%   There are two real roots. (有两个实根情况)
if s>0
    x1=(-b-sqrt(s))/(2*a);
    x2=(-b+sqrt(s))/(2*a);
    disp('This equation has two real roots:')
    fprintf('x1=%f\n',x1);
    fprintf('x2=%f\n',x2);
%   There are only single repeated real root.
%   (方程有两个相等实根)
elseif s==0
    x1=(-b)/(2*a);
    disp('This equation has a single repeated root (two identical real roots):');
    fprintf('x1=x2=%f\n',x1);
%   There are two complex roots.
%   (方程有两个复数根)
else
    real=-b/(2*a);
    image=(sqrt(abs(s)))/(2*a);
    disp('This equation has two complex roots:');
    fprintf('x1=%f-i*%f\n', real, image);
    fprintf('x2=%f+i*%f\n', real, image)
end
```

运行结果如下:

```
This program solves for the roots of the quadratic
equation in the form a*x^2+b*x+c=0.
Please enter a: 3
Please enter b: 5
Please enter c: -1
This equation has two real roots:
x1=-1.847127
```

```
x2=0.180460
```

> **注意：**
> 上面程序中第一行"% Script file: EXAMP03005.m"标出本程序为"脚本程序"，文件名为"EXAMP03005.m"，该行可以省略。需要注意的是，如果省略该行，文件名按要求就可以，由字母、数字和下画线构成，开头必须为字母即可，但是如果未省略，保存文件名必须和该行标注的文件名完全一致。

3.4.3 switch 结构

switch 结构是另外一种分支结构。该结构允许编程人员根据整数、字符、逻辑表达式的结构选择执行语句。switch 结构一般结构形式如图 3.5 所示。

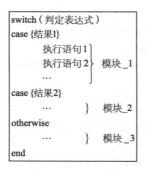

图 3.5 switch 结构形式

如果"判定表达式"的值为"结果 1"，程序将执行"模块 1"；如果"判定表达式"的值为"结果 2"，程序将执行"模块 2"，等等。需要注意的是，所陈述的"结果 1"、"结果 2"等中所包含的可能是多个结果，只要"判定表达式"等于其中一个，就可执行对应模块。

> **注意：**
> 每次只能执行一个模块，执行完一个模块，程序将直接跳至 switch 结构结尾，然后执行后面的语句。如果"判定表达式"的结果与多个"case"中的结果相一致，只执行第一个"case"模块，其他模块将直接跳过。

【例 3.6】 评述学生成绩。

编写一个程序能够根据输入的学生成绩给出恰当评述。学生成绩分为五个等级：A（或 a）、B（或 b）、C（或 c）、D（或 d）、E（或 e）。根据这些等级，给出恰当评述。等级与评述有下面关系：

A（或 a）：非常好！
B（或 b）：很好！
C（或 c）：较好！
D（或 d）：合格！
E（或 e）：非常抱歉，您需要重修！

并且如果用户输入了任何其他的字符，该程序能够显示"输入错误"。

分析： 本程序能够根据用户输入成绩级别：A（或 a）、B（或 b）、C（或 c）、D（或 d）、

E(或e)给出相应评述。对于用户输入的成绩级别需要有变量存储,然后将该变量与各个"case"后面紧跟的结果进行比较,哪个结果对就对哪个模块进行运行。这里有 5 种评述,1 种错误提示,所以共有 6 个分支。伪代码设计如下所述。

(1) 需要 1 个变量存储学生考评成绩。

MATLAB 表述:syms score

(2) 提示用户输入学生考评成绩。

MATALB 表述:score=input('请输入考评成绩:', 's');

(3) 根据输入的考评成绩,给出相应评述。

MATLAB 表述:switch(score)

```
            case{'A','a'}
                disp('非常好! ');
            case{'B','b'}
                disp('很好! ');
            case{'C','c'}
                disp('较好! ');
            case{'D','d'}
                disp('合格! ');
            case{'E','e'}
                disp('非常抱歉,您需要重修! ');
            otherwise
                disp('输入错误! ');
            end
```

EXAMP03006

```
%   Purpose:
%   This program shows the evaluation of student-courses.
%   (该程序显示学生课程的评价)
%   This program shows the corresponding evaluation of course for
%   students in terms of the score entered by user. The scores
%   of course have five classes: A, B, C, D and E.
%   (本程序根据用户输入的成绩给出课程的考评评价。课程成绩分为五级:A、B、C、
%    D和E)
%
%   A (or a): You had done an excellent job!
%   B (or b): Very Good!
%   C (or c): Good!
%   D (or d): You should do better!!
%   E (or e): I am pretty sorry to tell you that you need to study
%             it again!
%
%   Records of revision:
%   Date: 2015-07-20
%   Programmer: Wang Yonglong
```

```
% Description of changer: Second revision
%
% Clear command window.
% (清除命令窗口)
% Clear variables in workspace.
% (清除工作空间中的变量)
clear
% Define a variable as score.
% (定义一个变量score)
syms score
% Prompt user to enter the score of course.
% (提示用户输入课程成绩)
score=input('Please enter the score of course ', 's');
% Present the evaluation corresponding to the entered score.
% (显示与输入成绩的相应的评价)
switch (score)
  case {'A','a'}
      disp('非常好!');
  case{'B','b'}
      disp('很好!');
  case{'C','c'}
      disp('较好!')
  case{'D','d'}
      disp('合格!')
  case{'E','e'}
      disp('非常抱歉,您需要重修!')
  otherwise
      disp('输入错误!')
end
```

运行结果如下:

```
Please enter the score of course A
非常好!
```

【例3.7】 个人所得税。

为某单位编写一个计算个人所得税程序。月工资及个人所得税计算办法参见表3.5。

表3.5 月工资及个人所得税率

月薪（元）	个人所得税率（%）	月薪（元）	个人所得税率（%）
≥20000	20	≥3600	5
≥10000	15	<3600	0
≥5000	10		

本程序是一个常见分支结构程序,有6个分支,一个是低于3600元/月收入的不缴纳个人所得税;大于3600元/月但小于5000元/月的,高出1400元部分缴纳5%的个人所得税;当大于5000元/月小于10000元/月时,个人所得税由两个部分构成,一是超过5000元部分个人所得税为10%,另一个部分是高于3600元不足5000元部分只缴纳5%的个人所得税;当大于10000元/月而小于20000元/月时,个人所得税由三个部分构成,超过10000元部分

个人所得税为15%,高于5000元不足10000元部分个人所得税为10%,高于3600元不足5000元部分个人所得税为5%;月薪超过20000元职工的个人所得税,假设收入为 n 元,那么个税为(n-20000)×20%+(20000-10000)×15%+(10000-5000)×10%+(5000-3600)×5%元。将以上思路转化为 MATLAB 陈述即可(本例题提供两种编程结构,一种是 if—elseif—else—end 结构,一种是包含了 switch—case—end 结果)。伪代码设计如下所述。

(1) 需要3个变量:月薪、个人所得税和实发工资。
MATLAB 表述:syms int_sal pers_inc_tax fin_sal
(2) 提示用户输入职工月薪。
MATLAB 表述:int_sal=input('请输入月薪');
(3) 对个税和实发工资进行赋初始值,避免没有计算结果,显示时出错。
MATLAB 表述:pers_inc_tax=0;
　　　　　　　fin_sal=0;
(4) 根据输入的月薪和个人所得税计算办法,给出个人所得税和实际工资结果。
MATLAB 表述:if int_sal>=20000
　　　　　pers_inc_tax=(int_sal-20000)*0.2+(20000-10000)*0.15+(10000-5000)*0.1+(5000-3600)0.05;
　　　　　fin_sal=int_sal-pers_inc_tax;
　　　　　fprintf('Your personal income tax and final month-salary are %f and %f, respectively\n', pers_inc_tax, fin_sal);
　　　　elseif int_sal>=10000
　　　　　pers_inc_tax=(int_sal-10000)*0.15+(10000-5000)*0.1+(5000-3600)*0.05;
　　　　　fin_sal=int_sal-pers_inc_tax;
　　　　　fprintf('Your personal income tax and final month-salary are %f and %f, respectively\n', pers_inc_tax, fin_sal);
　　　　　elseif int_sal>=5000
　　　　　　pers_inc_tax=(int_sal-5000)*0.1+(5000-3600)*0.05;
　　　　　　fin_sal=int_sal-pers_inc_tax;
　　　　　　fprintf('Your personal income tax and final month-salary are %f and %f, respectively\n', pers_inc_tax, fin_sal);
　　　　　　elseif int>=3600
　　　　　　　pers_inc_tax=(int_sal-3600)*0.05;
　　　　　　　fin_sal=int_sal-pers_inc_tax;
　　　　　　　fprintf('Your personal income tax and final month-salary are %f and %f, respectively\n', pers_inc_tax, fin_sal);
　　　　　　　else
　　　　　　　　fin_sal=int_sal-per_inc_tax;
　　　　　　　　fprintf('Your personal income tax and final month-salary are %f and %f, respectively\n', pers_inc_tax, fin_sal);
　　　　　　　end

对于 switch—case—otherwise 结构就不再详细分析了,留给读者独立完成。

EXAMP03007_1 (if—elseif—else—end)

```
%   Purpose:
%   This program calculates personal income tax.
%   (本程序计算个人所得税)
%   This program calculates personal income tax. The personal income
%   tax is listed as following
%-------------------------------------------------------------
%       month-salary (RMB)        |     personal income tax (%)
%-------------------------------------------------------------
%       the part of MS>=20000     |            20
%-------------------------------------------------------------
%       the part of MS>=10000     |            15
%-------------------------------------------------------------
%       the part of MS>=5000      |            10
%-------------------------------------------------------------
%       the part of MS>=3600      |            5
%-------------------------------------------------------------
%       the part of MS<1600       |            0
%-------------------------------------------------------------
%   And this program can show the initial month-salary and
%   the personal income tax and the final month-salary.
%
%   Records of revision:
%   Date: 2015-07-12
%   Programmer: Wang Yonglong
%   Description of changer: Second revision
%
%   Clear command window.
%   (清除命令窗口)
%   Clear variables in workspace.
%   (清除工作空间中的变量)
clear
%
%   The definition s of all variables:
%   int_sal: the initial month-salary
%   pers_inc_tax: personal income tax
%   fin_sal: the final month-salary
%
%   Define three variables int_sal, pers_inc_tax and fin_sal.
%   (定义三个变量：int_sal、pers_inc_tax 和 fin_sal)
syms int_sal pers_inc_tax fin_sal
%   Prompt user to enter the monthly initial salary.
%   (提示用户输入月薪)
int_sal=input('Please enter your initial month-salary ');
%   Define the initial values for pers_inc_tax and fin_sal.
%   (定义 pers_inc_tax 和 fin_sal 初始值)
pers_inc_tax=0;
fin_sal=0;
%   Calculate personal income tax and final month-salary.
```

第3章 编　　程

```
    %   (计算个人所得税和实发工资)
    if int_sal>=20000
    pers_inc_tax=(int_sal-20000)*0.2+(20000-10000)*0.15+…
    (10000-5000)*0.1+(5000-3600)*0.05;
    fin_sal=int_sal-pers_inc_tax;
        fprintf('Your personal income tax and final month-salary are %f and %f,
respectively.\n', pers_inc_tax, fin_sal);
    elseif int_sal>=10000
    pers_inc_tax=(int_sal-10000)*0.15+(10000-5000)*0.1+…
    (5000-3600)*0.05;
    fin_sal=int_sal-pers_inc_tax;
        fprintf('Your personal income tax and final month-salary are %f and %f,
respectively.\n', pers_inc_tax, fin_sal);
    elseif int_sal>=5000
    pers_inc_tax=(int_sal-5000)*0.1+(5000-3600)*0.05;
    fin_sal=int_sal-pers_inc_tax;
        fprintf('Your personal income tax and final month-salary are %f and %f,
respectively.\n', pers_inc_tax, fin_sal);
    elseif int>=3600
    pers_inc_tax=(int_sal-3600)*0.05;
    fin_sal=int_sal-pers_inc_tax;
        fprintf('Your personal income tax and final month-salary are %f and %f,
respectively.\n', pers_inc_tax, fin_sal);
    else
        fin_sal=int_sal-per_inc_tax;
        fprintf('Your personal income tax and final month-salary are %f and %f,
respectively.\n', pers_inc_tax, fin_sal);
    end
```

运行结果如下：

```
    Please enter your initial month-salary 23450
    Your personal income tax and final month-salary are 2760.000000 and
20690.000000 , respectively.
```

EXAMP03007_2 (switch—case—case—otherwise—end)

```
    %   Purpose:
    %   This program calculates personal income tax.
    %   (本程序计算个人所得税)
    %
    %   Records of revision:
    %   Date: 2015-07-12
    %   Programmer: Wang Yonglong
    %   Description of changer: second revision
    %
    %   Clear command window.
    %   (清除命令窗口)
    %   Clear variables in workspace.
    %   (清除工作空间中的变量)
    clear
```

```
%
% The definitions of variables:
% int_sal: the initial month-salary
% pers_inc_tax: personal income tax
% fin_sal: the final month-salary
%
% Define three variables int_sal, pers_inc_tax and fin_sal.
% (定义三个变量: int_sal、pers_inc_tax 和 fin_sal)
syms  int_sal  pers_inc_tax  fin_sal
% Prompt user to enter the initial month-salary.
% (提示用户输入月薪工资)
int_sal=input('Please enter your initial month-salary ');
% Define the initial values for pers_inc_tax and fin_sal.
% (定义 pers_inc_tax 和 fin_sal 初始值)
pers_inc_tax=0;
fin_sal=0;
% Change the salary number into 1, 2, 3, 4 and 5.
% (将工资数字转换为 1、2、3、4、5 五个等级)
if int_sal<=3600
    int_sal_c=1;
elseif int_sal<=5000
    int_sal_c=2;
elseif int_sal<=10000
    int_sal_c=3;
elseif int_sal<=20000
    int_sal_c=4;
elseif int_sal>20000
    int_sal_c=5;
else
    display('The number entered by you is a mistake!\n');
    int_sal_c=6;
end
%
% Calculate personal income tax and final month-salary.
% (计算个人所得税和实发月薪)
switch(int_sal_c)
    case{1}
        fin_sal=int_sal-pers_inc_tax;
        fprintf('Your personal income tax and final month-salary are %f and %f , respectively\n', pers_inc_tax, fin_sal);
    case{2}
        pers_inc_tax=(int_sal-1600)*0.05;
        fin_sal=int_sal-pers_inc_tax;
        fprintf('Your personal income tax and final month-salary are %f and %f , respectively.\n', pers_inc_tax, fin_sal);
    case{3}
        pers_inc_tax=(int_sal-3000)*0.1+70;
        fin_sal=int_sal-pers_inc_tax;
        fprintf('Your personal income tax and final month-salary are %f and %f , respectively.\n', pers_inc_tax, fin_sal);
    case{4}
```

```
            pers_inc_tax=(int_sal-5000)*0.15+270;
            fin_sal=int_sal-pers_inc_tax;
            fprintf('Your personal income tax and final month-salary are %f and
%f , respectively.\n', pers_inc_tax, fin_sal);
        case{5}
      pers_inc_tax=(int_sal-10000)*0.2+1020;
            fin_sal=int_sal-pers_inc_tax;
            fprintf('Your personal income tax and final month-salary are %f and
%f , respectively.\n', pers_inc_tax, fin_sal);
        otherwise
            display('You need to restart this program and enter again!')
    end
```

运行结果如下：
```
    Please enter your initial month-salary 16710
    Your personal income tax and final month-salary are 2026.500000 and
14683.500000 , respectively.
```

3.4.4 try/catch 结构

try/catch 结构是一种特殊的分支结构，一般为锁定错误而设定。通常情况下，当运行 MATLAB 程序遇到错误时，程序会自动终止运行。而 try/catch 结构修改了 MATLAB 的这种错误执行行动，使其能够继续运行下去。这种继续运行不是运行 try 模块语句而是直接执行 catch 模块语句。这让用户能够清楚知道错误语句所处的位置，并且不会导致程序停止或者进入死循环。try/catch 结构形式如图 3.6 所示。

图 3.6 Try/catch 结构形式

【例 3.8】 两矩阵相乘。

通过 MATLAB 语言编写程序，该程序能够判断两个矩阵能否相乘，如果能相乘给出结果，如果不能相乘给出错误提示。

分析：该问题可分成 3 步完成。第一步，允许用户输入两个待乘矩阵；第二步，尝试计算所输入的两个矩阵的乘积；一旦第二步有错误执行不下去，将执行第三步部分语句；第三步又分为两步，其中第一子模块为 try 模块，将对两个矩阵进行点乘计算尝试，并给出结果，如果仍不可运算，将执行第三步中的第二个 catch 子模块给出错误提示。伪代码设计如下所述。

（1）提示用户输入两个矩阵。

MATLAB 语言表述：A=sym(input('Please enter matrix A= ', 's'));

B=sym(input('Please enter matrix B= ', 's'));

(2) 将本程序分成 3 个子任务，借助两个 try/catch 结构嵌套完成。
MATLAB 语言表述如下：

```
try % 尝试矩阵乘法运算
      disp('The product of A and B');
  C=A*B;
catch
    try % 尝试矩阵点乘运算
        disp('The dot multiplication of A and B');
        D=A.*B;
    catch % 这两个矩阵既不能进行矩阵乘积运算，也不能进行点乘运算
    disp('The two matrices A and B cannot be executed matrix product and dot product.');
      end
end
```

EXAMP03008

```
% Purpose:
% This program investigates whether the two matrices A and B entered
% by user can be calculated by multiplication or dot multiplication.
% (本程序研究用户输入的两个矩阵 A 和 B 是否能进行乘积或点乘运算)
%
% Records of revision:
% Date: 2015-07-21
% Programmer: Wang Yonglong
% Description of changer: second revision
%
% Clear command window.
% (清除命令窗口)
clc
% Clear variables in workspace.
% (清除工作空间中的变量)
clear
% Prompt user to enter two matrices A and B.
% (提示用户输入量矩阵 A 和 B)
A=str2num((input('Please enter matrix A= ','s')));
B=str2num((input('Please enter matrix B= ','s')));
try
    % Calculate the product of matrices A and B.
    % (计算矩阵 A 和 B 的乘积)
    C=A*B;
    % Display the string "The product of A and B is".
    % (显示字符串"The product of A and B is")
    disp('The product of A and B is')
    C
catch
  try
        % Calculate the dot multiplication of A and B.
        % (计算矩阵 A 和 B 的点乘)
        D=A.*B;
        % Display the string "The dot multiplication of A and B is".
```

```
            %  (显示字符串"The dot multiplication of A and B is")
            disp('The dot multiplication of A and B is')
            D
     catch
            %  Display the warning words.
            %  (显示警告提示)
            disp('The two matrices A and B can not be executed matrix and dot
multiplication.');
        end
    end
```

运行结果如下：

```
    Please enter matrix A= 1 2 3; 4 5 6
    Please enter matrix B= 7 8 9;3 4 8
    The dot multiplication of A and B is
    D =    7    16    27
          12    20    48
```

3.5 循 环 结 构

循环结构赋予某些语句执行多次的能力。循环机构有两个常用结构，分别是 while 循环和 for 循环。这两种结构的根本不同之处在于它们的循环控制条件不同。通常情况下，while 结构中循环次数没有预先确定，而 for 结构的循环次数在未执行前就已经确定。从接受容易程度上，一般先讲授 for 循环结构。

3.5.1 for 结构

for 循环是一种能够让某一个特定语句模块执行超过一次的结构语句，其基本结构形式如图 3.7 所示。

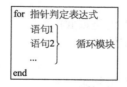

图 3.7　for 循环结构形式

这里指针是循环变量，指针判定表达式是用来判定下面循环模块是否需要停止的唯一指令。只有指针指令与指针判定表达式中某一个元素一致时，下面循环语句才会运行，当指针的指令数值在判定表达式中找不到对应数值时，循环将被停止。for 和 end 之间语句为循环语句，每循环一次，这部分语句都将从头至尾循环执行一次。一般对 for 循环结构函数有如下描述：

（1）在执行循环语句前，MATLAB 程序需要产生一个控制表达式。

（2）第一个执行循环语句，控制表达式将把它的第一个元素赋予给指针，然后将该值代入下面循环模块进行运算。

（3）循环语句设置上需注意它的连续性，也就是第一次运行结束后，能够让控制表达式的第二个元素赋予指针进行下一次运算。

（4）第三步的重复执行需要能够持续，一直到跳出控制表达式的元素范围。

控制表达式有很多结构模式，下面给出几种比较典型的模式（见图3.8）。

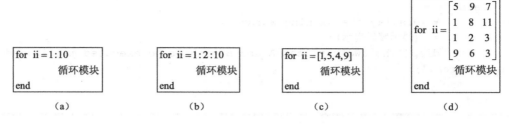

图3.8 控制表达式典型模式

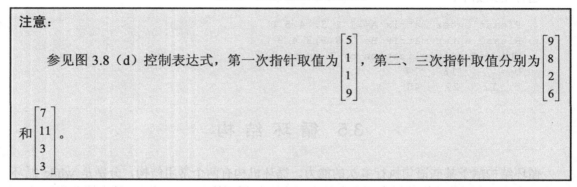

【例3.9】 阶乘函数。

阶乘函数是循环结构的典型示例。下面，我们将对该示例进行分析解答。

分析：阶乘函数计算过程中，首先实现需要用户输入要计算阶乘的数，然后是程序给出结果，所以该程序需要两个变量，循环过程中还需要指针变量。伪代码设计如下所述。

（1）需要2个变量，1个待用户输入正整数，1个为储存计算结果。

MATLAB 表述：syms n fact_n

（2）提示用户输入正整数。

MATLAB 表述：n=input('请输入正整数 n');

（3）赋予输出函数阶乘 fact_n 初始值。

MATLAB 表述：fact_n=1;

（4）执行循环运算，计算阶乘给出结果。

MATLAB 表述：for ii=1:n
　　　　　　　　fact_n=ii*fact_n;
　　　　　　end

EXAMP03009

```
%   Purpose:
%   This program calculates the factorial function of a positive integer
%   entered by user.
%   (本程序计算用户输入的正整数阶乘)
%
%   Records of revision:
```

```
%  Date: 2015-07-24
%  Programmer: Wang Yonglong
%  Description of changer: revision
%
%  Definitions of variables:
%  n:  a positive integer entered by user.
%  fact_n:  the factional function of n.
%
%  Clear variables in workspace.
%  (清除工作空间中的变量)
clear
%  Clear command window.
%  (清除命令窗口)
clc
%  Define two variables n and fact_n.
%  (定义两个变量 n 和 fact_n)
syms n fact_n
%  Define the initial value of fact_n.
%  (定义 fact_n 的初始值)
fact_n=1;
%  Prompt user to enter a positive integer.
%  (提示用户输入正整数)
n=str2num(input('Please enter a positive integer:','s'));
%
%  Calculate the factional function of n by for structure.
%  (通过 for 结构计算 n 的阶乘)
for ii=1:n
    fact_n=fact_n*ii;
end
%  Present the factional function of n.
%  (显示 n 的阶乘结果)
ns=num2str(n);
fact_ns=num2str(fact_n);
fprintf('The factional function of %s! is %s\n', ns, fact_ns);
```

运行结果如下：

```
Please enter a positive integer:9
The factional function of 9! is 362880.
```

3.5.2 while 结构

while 结构不确定循环次数只须满足循环条件，循环模块就会执行。其结构形式通常如图 3.9 所示。

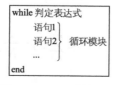

图 3.9 while 结构形式

如果判定表达式返回结果为真，循环模块将运行，然后再回到 while 循环结构前等待下一次判定。如果判定表达式在上一次循环后返回结果仍为真，循环模块将继续运行。这种处理过程直到判定表达式返回结果为假,循环终止,执行点跳到 while 循环结构的后一个语句。

如果想对测得的多组数据基本特点有所了解,可以通过平均值和标准差进行比较,而平均值和标准差 while 循环结构能够完成。

【例 3.10】 平均值和标准差。

平均值和标准差表达式分别为

$$\bar{x} = \frac{1}{N}\sum_{i=1}^{N} x_i$$

$$s = \sqrt{\frac{\sum_{i=1}^{N}(x_i - \frac{1}{N}\sum_{i=1}^{N} x_i)^2}{N}}$$

其中,x_i 是第 i 次测量获得的数据(有时也称为样品)。

分析:首先要知道该程序需要几个变量。平均值定义式显示需要 4 个变量,分别为 x_i、$\sum_{i=1}^{N} x_i$、n 和 $\bar{x}$;标准差定义式显示需要 5 个变量,分别为 x_i、$\sum_{i=1}^{N} x_i$、$\sum_{i=1}^{N}(x_i - \frac{1}{N}\sum_{i=1}^{N} x_i)^2$、$n$ 和 s。在计算过程中,数据存储需要 1 个隐含变量,该变量可为计算平均值和标准差提供必要数据支持。综上可知,需要变量总共为 7 个,分别为 A、x_i、$\sum_{i=1}^{N} x_i$、$\bar{x}$、$\sum_{i=1}^{N}(x_i - \frac{1}{N}\sum_{i=1}^{N} x_i)^2$、$n$ 和 s。伪代码设计如下所述。

(1)需要定义的变量:样品 x、取样个数 n、平均值 xbar、标准差 std_dev、样品之和 sum_x、样品均差平方之和 diff_mean_x。

MATLAB 表述:syms x sum_x n xbar std_dev diff_mean_x

(2)对变量赋初值。

MATLAB 表述:n=0; sum_x=0; sum_x2=0;

(3)提示用户输入采样数据。

MATLAB 表述:x=input('请输入第一个样品:');

(4)完成循环,计算得到计算结果需要的中间量 sum_x 和输入样品个数 n,并将 x 作为的 n 个元素存入数组 xmat 中。需要注意的是,不要忘记提示用户输入下一个样品数据。

MATLAB 表述:while isnumeric(x) && isempty(x)==0
    n=n+1;
    xmat(n)=x;
    sum_x=sum_x+x;
    x=input('请输入下一个样品:');
  end

(5)计算中间过渡变量 sum_x2,也就是 $\sum_{i=1}^{N}(x_i - \frac{1}{N}\sum_{i=1}^{N} x_i)^2$。

MATLAB 表述:x_bar=sum_x/n;
    for ii=1:n
      sum_x2=sum_x2+(xmat(ii)-x_bar)^2;
    end

（6）计算标准差。

MATLAB 表述：std_dev=sqrt(sum_x2/n);

（7）在命令窗口显示需要信息。

MATLAB 表述：fprintf('The number of data points is: %f.\n', n);
fprintf('The mean of this data set is: %f.\n', x_bar);
fprintf('The standard deviation is: %f.\n', std_dev);

EXAMP03010

```
%   Purpose:
%   This program calculates the mean and the standard deviation of
%   a series of data input or loaded by user.
%   (本程序可以计算用户输入或上传数据的平均值和标准差)
%
%   Records of revision:
%   Date: 2015-07-22
%   Programmer: Wang Yonglong
%   Description of changer: second rivision
%
%   Descriptions for variables as below:
%   x:  a simple
%   n:  the number of input samples
%   sum_x:  the sum of the input samples
%   sum_x2: the sum of the squares of the input samples
%   xbar:   the mean value of the input samples
%   std_dev:    the standard deviation of the input samples
%
%   Clear variables in workspace.
%   (清除工作空间中的变量)
clear
%   Clear command window.
%   (清除命令窗口)
clc
%   Define variables x, n, sum_x, sum_x2, xbar and std_dev.
%   (定义变量x、n、sum_x、sum_x2、xbar 和 std_dev)
syms x n sum_x sum_x2 xbar std_dev
%   Predefine the initial values of variables n, sum_x and sum_x2.
%   (预定义变量n、sum_x 和 sum_x2 初始值)
n=0;  sum_x=0;  sum_x2=0;
%   Prompt user to input first sample.
%   (提示用户输入第一个样本)
x=input('Please enter the first sample:');
%   Execute the following calculations by while structure.
%   (通过while结构执行下面计算)
%   The decision condition is that x is a number and not an empty.
%   (判定条件 x 是一个数并且不能为空)
while isnumeric(x) && isempty(x)==0
    %   Number the sample.
    %   (计数该样品)
    n=n+1;
```

```matlab
        %  Define the nth element of xmat as x.
        %  (定义 xmat 的第 n 个元素为 x)
        xmat(n)=x;
        %  Plus the sample to sum_x.
        %  (将该样品加到 sum_x 中)
        sum_x=sum_x+x;
        %  Prompt user to enter next sample.
        %  (提示用户输入下一个样品)
        x=input('Please enter next sample: ');
end
%  Calculate the mean.
%   (计算平均值)
x_bar=sum_x/n;
%  Calculate the sum_x2.
%   (计算平方和)
for ii=1:n
    sum_x2=sum_x2+(xmat(ii)-x_bar)^2;
end
%  Calculate the standard deviation.
%   (计算标准差)
std_dev=sqrt(sum_x2/n);
%
%  Present the requested results n, x_bar and std_dev.
%   (显示求解结果 x、x_bar 和 std_dev)
fprintf('The number of data points is: %f.\n', n);
fprintf('The mean of this data set is: %f.\n', x_bar);
fprintf('The standard deviation is: %f.\n', std_dev);
```

运行结果如下:

```
Please enter the first sample:12
Please enter next sample: 53
Please enter next sample: 12
Please enter next sample: 25
Please enter next sample: 43
Please enter next sample: 56
Please enter next sample: 42
Please enter next sample:
The number of data points is: 7.000000.
The mean of this data set is: 34.714286.
The standard deviation is: 17.052141.
```

3.6 函数编写

经过前面的讲述,已经了解编写一个好程序的重要性。最常用的模式为 **top-down** 模式,编程人员首先需要将待解决问题的输入、输出变量阐述清楚。紧接下来的工作就是,设计容易执行的算法,借助算法将待解决问题分解为容易完成的多个小问题。然后将这些小问题作为子模块进行编程,转化容易执行的 MATLAB 语句。最后,将这些子模块有机的系统的整合起来完成程序。

上面的示例基本采用该模式，但是没能得到所阐述的效果，主要是因为我们课堂上编写的程序都非常简单，一旦编写大程序，就能彰显这种模式的好处。因为在大程序中寻找错误犹如大海捞针，与子模块中寻找错误相比可谓有天壤之别。并且在整合完成的结果程序中，很难对子模块再进行测试。

幸运的是，MATLAB语言为我们提供了自己编写函数的环境，这使得每个子模块作为独立程序成为可能。所以，对于某个特定领域的程序员来讲，有好的编写习惯将会给工作带来难以想象的方便和快捷。普适的好处总结如下。

1．便于测试

每个子模块都是独立单元，每个独立单元都可以进行单独测试，这可大大节省测试时间。

2．重复使用

在很多情况下，某些子模块可能会被重复运用，有了独立函数，只需要调用函数无须再重复编写，这可压缩程序体积，同时可提高编程效率和质量。

3．可减少侧面效应

由于是独立的函数，所以函数内的一些参数不会影响其他函数里的变量参数，这为编写程序也带来意想不到的好处。一是可以减少出现错误的可能，二是可以降低编写程序的复杂度，三是可以为程序员提供更多的自由。

> **好习惯：**
> 无论什么时候将大程序分割为许多函数都是非常实用的，因为这些函数具有能够独立测试、重复利用、避开那些烦人的侧面（或横向）影响的优点。

3.6.1 MATLAB 函数

在 MATLAB 语言中，M 文件分为两类，一类是脚本文件，另一类是函数文件。脚本文件仅仅是将一些 MATLAB 语言中可以执行的语句放在一起，保存为一个独立文件。只要在命令窗口输入文件名并按回车键，MATLAB 软件马上对该文件进行执行，并且直接在命令窗口给出运行结果。函数文件和脚本文件有所不同，一般函数需要输入变量才能进行运行，并且会将运算结果赋值给输出变量，当然特殊情况下也可没有。任何用户如果想调用所编写函数，可以在命令窗口输入函数名字和必要的输入变量与输出变量，然后按回车键，使用方法和 MATLAB 自带函数完全一致。

比较而言，MATLAB 中编写的函数是一类特殊的 M 文件，它有自己独立的工作空间，不会影响函数外变量，函数外变量也不会影响它的内在变量。它可以接受通过输入变量输入数据，可以通过输出变量给出计算结果。MATLAB 自编函数通常具有以下编写格式：

```
function [输出变量1，输出变量2，…]=fname(输入变量1，输入变量2，…)
%    （H1 函数功能简要阐述行，一般为 lookfor 命令调用）
%    （函数功能较详细阐述，一般为 help 命令调用）
…
（运行语句）      %（运行模块）
```

...
(结束显示)

function 是函数开始的标志符号，随后是输出变量、函数名和输入变量。输入变量一般用小括号( )括起来紧跟在函数名之后，输出变量一般用方括号[ ]括起来放在函数名之前，即等号=的前面。如果只有一个输出变量，中括号可以省略。

开始陈述行，主要陈述该函数特定功能及用途。紧跟在函数名后面的第一行，称为 H1 陈述行，是该函数的最简捷表述，只有一行，一般为 lookfor 命令所调用。紧随其后的解释内容，是对该函数的主要内容及功能进行较为详尽的全面的阐述，一般为 help 命令调用。如果想让所编写的函数具有较好的可读性，请严格按照上面编写模式来编写函数，解释内容不可省略。

【例 3.11】 空间两点距离。

请编写 MATLAB 函数程序能够求解空间任意两点距离，及其与笛卡儿直角坐标系三边的夹角。

分析：这个问题是一个简单数学问题，或者说是物理问题。我们知道一个空间点的确定需要 3 个数来描述（x, y, z），所以要描述两个空间点就需要 6 个数。用户输入的两个空间点可记为（x_1, y_1, z_1）和（x_2, y_2, z_2）。需要输出的结果只有 4 个，两点间距离和与直角坐标系三边的 3 个夹角。所以，需要变量总共为 10 个。4 个输出变量的数学表达式分别为

$$s = \sqrt{(x_2 - x_1)^2 + (y_2 - y_1)^2 + (z_2 - z_1)^2}$$

$$\theta_x = \arccos(\frac{x_2 - x_1}{s})$$

$$\theta_y = \arccos(\frac{y_2 - y_1}{s})$$

$$\theta_z = \arccos(\frac{z_2 - z_1}{s})$$

定义好输入变量，然后将上面运算过程转化为可执行语句，再设定输出变量的结果传递，该程序就可以完成。伪代码设计如下所述。

（1）需要 6 个变量存储 2 个点的坐标，4 个变量来存储两点间距离和与 x、y、z 坐标轴的夹角。

MATLAB 表述：function [s, theta]=dist(x1,y1,z1,x2,y2,z2)

（2）根据用户调用函数输入的数据进行计算。

MATLAB 表述：s=sqrt((x2-x1)^2+(y2-y1)^2+(z2-z1)^2);

theta(1)=asin((x2-x1)/s)*180;

theta(2)=asin((y2-y1)/s)*180;

theta(3)=asin((z2-z1)/s)*180;

EXAMP03011

```
    function [s, theta]=EXAMP03011(x1,y1,z1,x2,y2,z2)
    %  Purpose:
    %  This function calculates the distance and coordinate azimuth angles of
two points.
    %  (本函数计算两个点间距离和坐标方位角)
```

```
% This function calculates the distance between two points (x1,y1,z1) and
% (x2,y2,z2), and
% the coordinate azimuth angles between the line defined by two points
% (x1,y1,z1)
% and (x2,y2,z2) and x axis, y axis, z axis.
% (本函数计算两个点(x1,y1,z1)和(x2,y2,z2)间距离,以及由该两点连线与 x 轴、
% y 轴和 z 轴夹角,称为坐标方位角)
% Calling sequence:
% (回调该函数格式)
%   [s, theta] = dist(x1,y1,z1,x2,y2,z2)
% Define variables:
%   x1: x-position of point 1.
%   y1: y-position of point 1.
%   z1: z-position of point 1.
%   x2: x-position of point 2.
%   y2: y-position of point 2.
%   z2: z-position of point 2.
%   s: Distance between two points.
%
% Record of revisions:
% Date: 2015-07-26
% Programmer: Wang Yonglong
% Description of change: Revision
%
% Calculate distance and coordinate azimuth angles.
% (计算距离和坐标方位角)
s=sqrt((x2-x1)^2+(y2-y1)^2+(z2-z1)^2);
theta(1)=asin((x2-x1)/s)*180;
theta(2)=asin((y2-y1)/s)*180;
theta(3)=asin((z2-z1)/s)*180;
```

运行结果如下:

```
>> [s,angle]=EXAMP03011(1,1,2,5,6,8)
s = 8.7750
angle = 85.1973   109.1279   135.5234
```

3.6.2 MATLAB 中变量传递

MATLAB 中程序与它们的函数交流是借助传递变量值来完成的。当一个函数调用已经存在的变量时,会将已知变量数值移植到调用函数中。这一移植在 MATLAB 编程中是至关重要的,函数可以对移植进来的数值进行任意操作而不影响原数值。这一特点大大降低了编程难度,并且也减少了变量相互干扰而产生的许多不必要错误。为了更好理解这一特点的重要性,请看下面示例。

【例 3.12】 计算某一个运动系统的路程和平均速度。

本例题主要体现对函数输入变量某个值进行操作,计算结果显示出操作是有效的。也就是函数里的操作会改变初始输入量,并且实现数据传输。思路简单就不再深入分析,直接给出程序。

EXAMP03012

```
function [distance,vbar]=EXAMP03012(v0,g,t)
%
%   Purpose:
%   This function calculates the distance and the average velocity.
%   (本函数计算路程和平均速度)
%   This function calculates the distance and the average velocity
%   for a dynamical system defined by initial velocity v0,
%   acceleration g and a fixed time t given by user.
%   (本函数用来计算一个运动系统的路程和平均速度,该运动系统的初始速度、加速度和
%   终止时间由用户给出)
%   Calling sequence:
%   (回调该函数格式)
%     [distance,velocity] =motion(v0,g,t)
%
%   Define variables:
%    v0: the initial velocity of the dynamical system.
%    g: the acceleration of the dynamical system.
%    distance: the distance of the dynamical system from 0 to t.
%    vbar: the average velocity of the dynamical system during [0,t].
%
%   Record of revisions:
%   Date: 2015-08-02
%   Programmer: Wang Yonglong
%   Description of change: Revision
%
%   Display the initial velocity v0.
%   (显示初始速度为v0)
v0s=num2str(v0);
fprintf('The initial velocity v0 is %s m/s.\n',v0s);
%   Display the acceleration g.
%   (显示加速度为g)
gs=num2str(g);
fprintf('The acceleration g is %s m/s^2.\n',gs);
%   Display the time t.
%   (显示时间)
ts=num2str(t);
fprintf('The time t is %s s.\n',ts);
%   Calculate the distance and the average velocity.
%   (计算路程和平均速度)
distance=v0*t+1/2*g*t^2;
vbar=distance/t;
%   Display the distance and the average velocity.
%   (显示路程和平均速度)
distances=num2str(distance) ;
fprintf('The distance of the dynamical system with initial velocity %s m/s,acceleration %s m^2/s through time t %s s is %s m.\n', v0s, gs, ts, distances);
vbars=num2str(vbar) ;
fprintf('The average velocities of the dynamical system is %s m/s.\n',
```

vbars)

运行结果如下:

```
[d,v]=EXAMP03012(15,8,3)
The initial velocity v0 is 15 m/s.
The acceleration g is 8 m/s^2.
The time t is 3 s.
The distance of the dynamical system with initial velocity 15
m/s,acceleration 8 m^2/s through time t 3 s is 81 m.
The average velocities of the dynamical system is 27 m/s.
d = 81
v = 27
```

在本例的开始，函数就调用了用户输入的输入变量值。计算过程中，对输入变量进行操作，操作后的数值将代替初始值并参加进一步的运算。注意，操作只是在函数内有效。

> 说明：
> 在 MATLAB 编程中，关于输入参量的传递实际上远比我们讨论的复杂。正如上面所指出的，伴随数值的传递也可以复制，但是这些复制能避免烦人的侧面影响。事实上，MATLAB 采取方法是，首先分析每个函数的每个输入变量，确定它们是否被修改，如果被修改就对改动后的数值进行复制并存到相应的输入变量，如果没被修改，就不进行复制，仍然使用原数值。这一操作大大提高了运算速度，同时也减少了侧面影响。

【例 3.13】 直角坐标系到极坐标系的变换。

平面上一个点的位置，可以由直角坐标系给定 (x, y)，也可以由极坐标系给定 (r, θ)。这两种表示等价，也可以相互转换，转换关系为

$$\begin{cases} x = r\cos\theta \\ y = r\sin\theta \end{cases} \quad 和 \quad \begin{cases} r = \sqrt{x^2 + y^2} \\ \theta = \tan^{-1}\dfrac{y}{x} \end{cases}$$

两个函数 rect2polar 和 polar2rect 分别具有将直角坐标转换为极坐标和将极坐标转换为直角坐标的功能。本书只给出直角坐标转换为极坐标函数 rect2polar 的参考程序，而将极坐标转换为直角坐标的函数 polar2rect 留给读者自己独立完成。本函数运行内容简单只有两个表达式，所以本例题也不深入分析，直接给出程序。

EXAMP03013

```
function [r, theta]=EXAMP03013(x,y)
% EXAMP03013 convert rectangular coordinates into polar coordinates.
% (EXAMP03013 将直角坐标转换为极坐标)
% Function EXAMP03013 accepts the rectangular coordinates (x,y) and
% converts them into the polar coordinates (r, theta), where theta
% is expressed in degrees.
% (函数EXAMP03013根据输入的直角坐标(x,y)给出极坐标(r,theta)，其中
% theta 的单位为度)
%
% Calling sequence:
```

```
%   (回调函数语法)
%   [r, theta]=rect2polar(x,y)
%
%   Define variables:
%   r: the length of polar vector.
%   theta: the ploar angle of vector in degrees.
%   x: the x-position of point.
%   y: the y-position of point.
%
%   Record of revisions:
%   Date: 2000-09-19
%   Programmer: S. J. Chapman
%   Description of change: original code
%
%   Calculate the length of the vector entered by user.
%   (计算用户输入矢量的长度)
r=sqrt(x.^2+y.^2);
%   Calculate the polar angle of the vector entered by user.
%   (计算用户输入矢量的极角)
theta=180/pi*atan2(y,x);
```

运行结果如下:

```
>> [r,theta]=EXAMP03013(8,3)
r = 8.5440
theta = 20.5560
```

3.6.3 选择变量相关函数

在 MATLAB 中,许多函数支持可选择输入和输出变量。输入变量,比如:plot 函数输入变量至少可有 2 个,多可有很多。输出变量上有 max 函数,可以有 1 个输出变量,也可以有多个输出变量。如果只有 1 个输出变量,max 函数的输出结果为一个标量;如果有 2 个输出变量,max 函数的输出结果是由 2 个数值组成的数组;以此类推。对于有多少输入和输出变量出现,MATLAB 函数会随之给出相对应结果。

MATLAB 提供了测算这些变量个数的函数,也提供了提示执行语句错误的提示函数,较常用的有以下几个,参见表 3.6。

表 3.6 常见选择变量相关函数

函数	描述
nargin	用来显示运行函数中输入变量的个数。nargin(FUN) 返回函数 FUN 里面输入变量的个数
nargout	用来显示运行函数中输出变量的个数。nargout(FUN) 返回函数 FUN 里面输出变量的个数
nargchk	有效输入变量数。nargchk(LOW, HIGH, N, 'struct')返回一个错误语句,如果 N 的数值不在 LOW 和 HIGH 之间
nargoutchk	有效输出变量数。nargoutchk(LOW, HIGH, N, 'struct')返回相应的错误信息,如果 N 的数值不在 LOW 和 HIGH 之间
error	显示错误信息并终止函数的运行。通常是遇到致命错误时,执行该命令
warning	显示警告提示,函数继续运行。通常是遇到的错误不是致命的,程序还能继续运行采用该命令
inputname	显示实际输入的输入变量名

(续表)

函　　数	描　　述
varargin	输入变量列队的长度
varargout	输出变量列队的长度
break	停止 for 或 while 循环运行
pause	暂停运行
continue	跳转到 for 或 while 循环初始行进行计算
return	回到调用函数状态或者键盘输入状态
quit	终止 MATLAB 程序

【例 3.14】　可选变量示例。

将借助一个示例对选择变量用法进行说明。这个函数含有 3 个输入变量，分别为速度、加速度和时间（v, g, t），产生 2 个输出变量，分别为距离 distance 和时间 time。如果输入的变量为 2 个时，将会自动将加速度 g 设置为 0，然后进行计算。无论输入 2 个或 3 个变量最终输出结果都是 2 个，即距离 distance 和时间 time。并且输入的时间最长数组为 3，也就是最大单位为小时。伪代码设计如下所述。

（1）函数名为 EXAMP03014，输出变量有 2 个，分别为 distance 和 time，输入变量有 3 个，分别为 v，g 和 t。

MATLAB 表述：function [distance, time]=EXAMP03014(v, g, t)

（2）输入变量应该为 2 或 3 个，并给出判定结果。

MATLAB 表述：msg=nargchk(2, 3, nargin);

　　　　　　　error(msg);

（3）对时间 t 输入的长度进行判定：如果长度超过 3 个需要给出错误信息。

MATLAB 表述：if length(t)>3

　　　　　　　　　msg='The time t entered is wrong!'

　　　　　　　　　error(msg);

　　　　　　　end

（4）对输入变量个数进行判定：如果输入变量的个数为 2 个，令 g=0。

MATLAB 表述：if nargin<3

　　　　　　　　　g=0;

　　　　　　　end

（5）挑选出特殊情况：前 2 个输入变量都为 0，给出错误提示。

MATLAB 表述：if v==0 && g==0

　　　　　　　　　msg='The investigated object is immobile.'

　　　　　　　　　error(msg);

　　　　　　　end

（6）以秒为单位换算时间。

MATLAB 表述：if length(t)==3

　　　　　　　　　time=t(1)*60*60+t(2)*60+t(3);

　　　　　　　elseif length(t)==2

```
                    time=t(1)*60+t(2);
                elseif length(t)==1
                    time=t;
                elseif
                    msg='The input arguments for the function are invalid!'
                    error(msg);
            end
```
（7）计算路程。
MATLAB 表述：distance=v*time+1/2*g*time^2;

EXAMP03014

```
function [distance, time]=EXAMP03014(v,g,t)
%  EXAMP03014 calculates distance and time.
%  (EXAMP03014 计算路程和时间)
%  Function EXAMP03014 calculates the distance and time of a motion
%  object defined by the velocity v, acceleration g and time t entered
%  by user.
%  (函数 EXAMP03014 计算运动物体的路程和时间, 运动物体的速度 v, 加速度 g,
%  时间 t 都由用户输入)
%  Calling sequence:
%     [distance,time]=EXAMP03014(v,g,[tt,tt,tt])
%
%  Record of revisions:
%  Date: 2015-07-02
%  Programmer: Wang Yonglong
%  Description of change: Revision
%
%  Define variables:
%  v:  the initial velocity of a moving object.
%  g:  the acceleration of a moving object.
%  t:  the time t.
%  distance :  the moving distance of the object by time t.
%  time:  the total time in second.
%
%  Check for whether the number of input arguments is legal.
%  (检查输入变量是否为允许个数)
msg=nargchk(2,3, nargin);
%  2 is LOW, 3 is HIGH, nargin is the number of input arguments.
%  (2 是小值, 3 为大值, nargin 为输入变量个数)
error(msg);
%  Tell whether the length of t input by user is valid.
%  (判定用户输入时间 t 的长度是否有效)
if length(t)>3
   msg='The time entered is wrong!'
   error(msg)
end
%  If the number of input arguments is 2, the first is taken as
%  the initial velocity, the second is taken as the time, the
```

```
%   acceleration would be set as 0.
%   (如果输变量的个数为2，第一个认为是初始速度，第二个认为是时间，加速度被
%   设置为0)
if nargin<3
    g=0;
end
%   If (0, 0, t) are entered as input arguments, a warning message
%   would be appeared in the command window.
%   (如果(0,0,t)为输入变量，警告信息将显示在命令窗口)
if v==0 && g==0
    msg='The investigated object is immobile.';
    warning(msg);
end
%   The total time is expressed in second.
%   (总时间换算为秒)
if length(t)==3
    time=t(1)*60*60+t(2)*60+t(3);
elseif length(t)==2
    time=t(1)*60+t(2);
elseif length(t)==1
    time=t
else
    msg='The input arguments for the function are invalid!';
    error(msg);
end
%   Calculate the distance.
%   (计算路程)
distance=v*time+1/2*g*time^2;
```

运行结果如下：

```
[d,t]=EXAMP03014(5,8,[21,2])
d = 6376886
t = 1262
```

【例3.15】 inputname 示例。

需要定义一个函数 EXAMP03015，含有 2 个输入变量 a、b 和 1 个输出变量 y。输出结果 $y=a+b$。本例题结构简单，只有 2 个输入变量，然后借助 inputname 命令调用这 2 个输入变量。如果是符号变量显示变量名，如果是数值什么也不显示。结构简单，不再进行详细分析，并省略函数表述，直接给出程序如下。

EXAMP03015

```
function y = EXAMP03015(a,b)
disp(sprintf('The first input argument is "%s".' , inputname(1)))
disp(sprintf('The second input argument is "%s".', inputname(2)))
y = a+b;
```

结果在命令窗口输入如下语句：

```
>> clear
>> x=5; y=6; z=EXAMP03015(x,y)
```

运行结果如下:

```
My first input is "x".
My second input is "y".
z = 9.
```

如果在命令窗口输入如下语句:

```
x=3; y=EXAMP03015(x,9)
```

运行结果如下:

```
My first input is "x".
My second input is " ".    %（因为第 2 个输入变量输入的是数值，不显示）
y = 12.
```

【例 3.16】 varargin 示例。

编写函数，能够判断输入变量的个数，并能根据输入变量的个数不同完成不同的计算。当变量被输入时，计算两个默认值都为 10 的乘积；当输入 1 个变量时，就计算输入变量与默认值 10 之积；当输入两个变量时，就计算这两个变量之积；如果输入变量超过两个，就显示输入不满足要求。伪代码设计如下所述。

（1）本函数名及其输入变量和输出变量定义。

MATLAB 表述：function out=　EXAMP03016(varargin)

（2）根据输入变量的个数分为 4 种情况，将分别进行处理。如果没有输入变量被输入，将计算 10 与 10 的乘积；如果输入 1 个输入变量，将计算该输入变量值与 10 的乘积；如果输入 2 个输入变量，将计算这 2 个输入变量的乘积；如果输入变量超过 2 个，将给出错误提示信息。

MATLAB 表述：switch nargin

```
            case 0
                disp('Here is no input argument entered!');
                a=10;
                b=10;
            case 1
                disp('Here is one input argument entered!');
                a=varargin{1};
                b=10;
            case 2
                disp('Here are two input arguments entered!');
                a=varargin{1};
                b=varargin{2};
            otherwise
                error('Unexpected input arguments!');
        end
```

（3）给出输出乘积结果。

MATLAB 表述：out=a*b;

EXAMP03016

```
function out=EXAMP03016(varargin)
% This function elucidates the usage of the function varargin.
% (阐述函数 varargin 的用法)
% Function EXAMP03016 elucidates the usage of the function
% varargin. If there is no input argument, EXAMP03016 will calculate
% the sum of a and b with their default values 10 and 10. If there
% appears one input argument, EXAMP03016 will calculate the sum of
% the value of a and 10. If there appears two input arguments,
% EXAMP03016 will calculate the sum of a and b. If there appears
% more than 2 input arguments, EXAMP03016 will present the error
% message.
% (函数 EXAMP03016 说明函数 varargin 的用法。当没有输入变量被输入时,
% EXAMP03016 计算 10 和 10 的乘积；当有 1 个输入变量被输入时, EXAMP03016 计算
% 该变量与 10 的乘积；当有 2 个输入变量被输入时, EXAMP03016 计算这 2 个输入变量
% 的乘积；当超过 2 个输入变量时, 显示错误提示)
%
% Record of revisions:
% Date: 2015-07-21
% Programmer: Wang Yonglong
% Description of change: Revision
%
switch nargin
    case 0
        disp('Here is no input argument.')
        a=10;
        b=10;
    case 1
        disp('Here is one input argument.')
        a=varargin{1};
        b=10;
    case 2
        disp('Here are two input arguments.')
        a=varargin{1};
        b=varargin{2};
    otherwise
        error('Unexpected inputs!')
end
out=a*b;
```

在命令窗口输入如下语句:

```
>> x=12; out=EXAMP03016(x)
```

运行结果如下:

```
Here is one input argument.
out = 120
```

在命令窗口输入如下语句:

```
>> x=12; y=11; out=EXAMP03016(x,y)
```

运行结果如下:

```
Here are two input arguments.
```

```
out = 132
```

【例 3.17】 varargout 示例。

与例 3.16 非常类似，不再深入分析。

EXAMP03017

```
function [varargout]=EXAMP03017(x)
% EXAMP03017 illustrates the usage of varargout.
% (阐述 varargout 用法)
% Function EXAMP03017 illustrates the usage of the function
% varargout. If there is no output argument entered by user, the result
% shows "No output argument given!'. If there is an output argument
% entered, the result is defined as the square of the input
% argment. If there are two output arguments entered, the result will
% consist of two parts, one is the square of the input argument defined
% as the first output argument, the other is the square root of the
% input argument defined as the second output argument. If the number
% of the output arguments is more than 2, an error message appears
% and the function stops.
% (函数 EXAMP03017 阐述函数 varargout 的用法。如果没有输出变量被用户键入，
% 结果显示"No output argument given!"。如果有 1 个输出变量被键入，
% 结果定义为输入变量的平方。如果有 2 个输出变量被键入，第一个输出变量定义为
% 输入变量的平方，另一个定义为输入变量的平方根。如果有多于 2 个输出变量被
% 键入，给出错误信息停止函数运行)
%
% Record of revisions:
% Date: 2015-07-21
% Programmer: Wang Yonglong
% Description of change: Revision
%
switch nargout
   case 0
      disp('No output argument given!')
   case 1
      disp('One output argument given!')
      varargout{1}=x.^2;
   case 2
      disp('Two output arguments given!')
      varargout{1}=x.^2;
      varargout{2}=x.^(1/2);
   otherwise
      error('Unexpected inputs!')
end
```

在命令窗口输入下面语句：

```
>> clear
>> [x,y]=EXAMP03017(16)
```

运行结果如下：

```
Two output arguments given!
```

```
x = 256
y = 4
```

3.6.4 全局变量和永久变量

全局变量和永久变量相同之处,在于 MATLAB 软件都会为这两类变量提供永久的储存空间。全局变量与局域变量有明显区别,全局变量在多个函数中都有效可以共用,而区域变量只在定义域函数有效。全局变量会影响其他函数里面的变量设置,局域变量却不会。永久变量只对函数内的指令设置有效,被存储在 workspace 空间可以随时调用,除非被清楚变量命令 clear 清除掉。对于永久变量设置常见错误:如果对当前工作存储空间(workspace)中已经有的变量,再对其设定永久性是无效的;如果函数的输入变量或输出变量中已经存在的变量,也不能再行设定它的永久行,否则也是无效的。

【**例 3.18**】 全局变量示例。

在主函数输入 3 个变量,速度 v、加速度 g 和时间 t 的基础上,分别定义了 2 个全局变量,时间 time 和路程 s_pass。并实现了在子函数间全局变量的共用功能。本题简单,只是定义了 time, s_pass 为全局变量而已,下面最终程序很容易看懂,为了节省篇幅,就不再进行分析。

EXAMP03018

```
function EXAMP03018(v,g,t)
% EXAMP03018 calculates the distance and the mean velocity.
% (EXAMP03018 计算路程与平均速度)
% Function EXAMP03018 calculates the distance and the average
% velocity for a moving ball with initial velocity v, acceleration
% g through the time t. And plot the line of distance and time and
% and that of average velocity and time. In this function, there
% defines two global variables time and s_pass which both are used
% in subfunctions.
% (函数 EXAMP03018 计算初始速度为 v、加速度为 g、时间为 t 运动小球的路程和平均
%  速度。并且绘制路程和时间曲线、平均速度和时间曲线。该函数中定义了 2 个全局变量 time
%  和 s_pass,这两个变量也被应用在子函数中)
%
% Call sequence:
% EXAMP03018(v,g,t)
%
% Record of revisions:
% Date: 2015-07-21
% Programmer: Wang Yonglong
% Description of change: Revision
%
% Define variable time as a global one.
% (定义变量 time 为全局变量)
global time
% Define the values of time by the function linspace.
% (通过函数 linspace 对变量 time 进行赋值)
time=linspace(0,t,1000);
% Define new variables a and b by the input arguments v and g.
```

```
%  (通过输入变量v和g定义新变量a和b)
a=v;
b=g;
%  Call the subfunction sub_first.
%  (调用子函数sub_first)
s=sub_first(a,b);
%  Call the subfunction sub_second.
%  (调用子函数sub_second)
d=sub_second;
%  In a same figure, plot two lines, one is of distance s and time,
%  the other is of mean velocity d and time.
%  (在同一个图形窗口,绘制两条曲线,一条为路程和时间曲线,另一条是平均速度
%   和时间曲线)
plot(time,s,'b-')
hold on
plot(time,d,'r-')
hold off
ledend('distance','mean velocity')
%
function s_pass=sub_first(v,g)
%  Subfunction sub_first calculates the distance.
%  (子函数sub_first计算路程)
%  Define two global variables: time and s_pass.
%  (定义两个全局变量: time和s_pass)
global time s_pass
%  Calculate the distance as s_pass.
%  (计算路程记为s_pass)
s_pass=v*time+1/2*g*time.^2;
%
function v_a=sub_second
%  Subfunction sub_second calculates the average velocity.
%  (子函数sub_second计算平均速度)
%  Define two global variables: time and s_pass.
%  (定义两个全局变量: time和s_pass)
global time s_pass
%  Calculate the mean velocity.
%  (计算平均速度)
v_a=s_pass./time;
```

在命令窗口输入下面执行语句:

```
EXAMP03018(10,9,7)
```

运行结果如图3.10所示。

好习惯:
　　在声明全局变量时,最好全用大写字母表示,以示与局域变量和永久变量的区别。一般将全局变量声明在帮助解释之后,可执行语句之前,这样可以很好传递数据达到共用的目的。

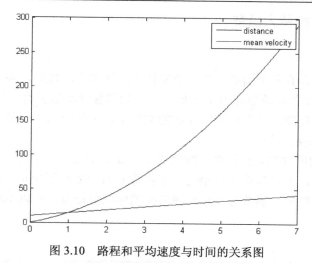

图 3.10　路程和平均速度与时间的关系图

永久变量设置得恰当可以节省程序运算时间。比如下面这个例子，现有函数 arrayToXLS：

```
function arrayToXLS(A, xlsfile, x1, x2)
% Define a persistent variable dblArray.
% (定义永久变量 dblArray)
persistent dblArray;
if isempty(dblArray)
    disp ('Writing spreadsheet file ...')
    % Write the matrix A to a excel file.
    % (将矩阵 A 写为一个 excel 文件)
    xlswrite(xlsfile, A);
end
disp('Reading array from spreadsheet ...')
% Read some data from the excel file.
% (从 excel 文件读取一些数据)
dblArray = xlsread(xlsfile, 'Sheet1', [x1 ':' x2]);
fprintf('\n');
```

第一次运行如下（用时 20.778383s）：

```
largeArray = rand(4000, 200);
tic, arrayToXLS(largeArray, 'myTest.xls','E254', 'J256'), toc
```

第二次运行如下（用时 1.590847s）：

```
tic, arrayToXLS(largeArray, 'myTest.xls','E257', 'J258'), toc
```

第三次运行如下（用时 1.049933s）：

```
tic, arrayToXLS(largeArray, 'myTest.xls','E259', 'J262'), toc
```

如果清除函数后再运行（用时 4.040753s）：

```
clear functions
tic, arrayToXLS(largeArray, 'myTest.xls','E263', 'J264'), toc
```

上面结果表明，特定情况下，永久变量设置可节省大量时间。这里用到的 tic 和 toc 命令为测试时间命令，tic 记下开始时间，toc 记下结束时间,两个时间之差即为程序运行所需时间。

3.6.5 子函数和私人函数

1. 子函数

单独文件中可以含有多个函数。当一个文件中出现多个函数时,最上面那个函数为正规函数(主函数),除此之外函数称为子函数。子函数和独立函数具有一样的特点和功能,只是它的使用范围比较局限,只局限于所存在的文件中。下面给出一个子函数示例。

【例3.19】 子函数示例。

定义函数 EXAMP03019,计算输入数组或矩阵 u 的平均值和中值。平均值和中值由子函数来完成。结构简单,主函数就 3 行语句;子函数 mean 就 1 行语句;子函数 median 包含 1 行排序,1 个 if 分支结构,偶数与奇数两种情况的中间值计算。所以本函数编写非常简单,直接给出程序如下。

EXAMP03019

```
function [avg, med]=EXAMP03019(u)
% EXAMP03019 find mean and median with internal function.
% (EXAMP03019 通过中间函数发现平均值和中间值)
% Function EXAMP03019 calculates the average and median of a data set
% by using subfunctions.
% (函数 EXAMP03019 通过子函数计算一组数据的平均值和中间值)
%
% Call sequence:
% (回调格式)
%     [avg,med]=EXAMP03019(u)
%
% Record of revision:
% Programmer: Chapman Stephen J.
% Description: original code
% Revised by Wang Yong-Long
%
% Define the length of u as n.
% (定义 u 的长度,记为 n)
n=length(u);
% Calculate the average of u by subfunction mean.
% (借助子函数 mean 计算 u 的平均值)
avg=mean(u,n);
% Calculate the median of u by subfunction median.
% (通过子函数 median 计算 u 的中间值)
med=median(u,n);
%
function a=mean(v,n)
% Subfunction mean calculates average.
% (子函数 mean 计算平均值)
a=sum(v)/n;
%
function m=median(v,n)
% Subfunction median calculates median.
% (子函数 median 计算 median)
% Sort the defined data v as w.
```

```
%    (排序已定义的数据v保存为w)
w=sort(v);
%    Distinguish even and odd number to calculate median.
%    (区别偶数和奇数数据，计算中间值)
if rem(n,2)==1
   m=w((n+1)/2);
else
   m=(w(n/2)+w(n/2+1))/2;
end
```

在命令窗口输入如下语句：

```
>> x=[2,3,12,34,12,12,23,21];
>> [a,m]=EXAMP03019(x)
```

运行结果如下：

```
a = 14.8750
m = 12
```

2. 私人函数

私人函数由于处在母函数的子目录下，所以才有该雅号。这些函数只有在母函数下才可见，才有效，在母函数外看不到它们，也不可用。它们的名字在其他分支仍然可用。

当在 M 文件中调用函数时，MATLAB 首先搜索该文件的子函数，如果没有继续搜索它的私有函数，如果仍然没有，才按照正常路径开始搜索其他资源库。所以私有函数在 MATLAB 编程中也占有重要的地位。

> **好习惯：**
> 用子函数和私有函数可以隐藏具有特殊功能的函数，它们只能被某些特定函数调用，可以避免意外调用，同时也可避免与 MATLAB 软件中自带函数发生冲突。

【例 3.20】 子函数和私有函数示例。

通过一个子函数和私有函数产生一个数组作为 x 轴坐标，通过另一个子函数和私有函数产生另一个数组作为 y 轴坐标，借助主函数给出正弦和余弦曲线。如果输入量不足给出错误提示。本题难度和例 3.19 相当，也不进行深入分析，下面给出两种函数结构，一种是子函数形式，另一种是私有函数形式，但目的相同，结果也一致，直接给出最终程序。但望读者能够根据最终程序分析出程序设计思路，并给出伪代码。

EXAMP03020（子函数）

```
function EXAMP03020(a,b)
%  EXAMP03020 plots a sine curve with x ranging from a to b.
%  (EXAMP03020绘制x取值范围为[a,b]的正弦曲线)
%
%  Call sequence:
%  (回调格式)
%      EXAMP03020(a,b)
%
%  Record of revision:
%  Programmer: Wang Yong-Long
```

```
%   Date: 2015-07-31
%   Description: Revision
%
%   Define the values of independent variable x.
%   (对自变量 x 进行赋值)
x=sub1(a,b);
%   Define the values of dependent variable y.
%   (对因变量 y 进行赋值)
y=sub2(x);
%   Plot the curve of x and y.
%   (绘制 x 和 y 的曲线)
plot(x,y,'bo','LineWidth',2)
    %
    function VECX=sub1(xmin,xmax)
    %  Subfunction sub1 produces the values of independent VECX.
    %  (子函数 sub1 产生自变量 VECX 的值)
        global VECX
        if nargin==2
            VECX=linspace(xmin,xmax,10*xmax);
        else
            msg='Too few input arguments !'
            error(msg)
        end
    %
    function VECY=sub2(VECX)
    %  Define the values of dependent variable VECY.
    %  (定义因变量 VECY 的值)
        global VECY
        VECY=sin(VECX);
```

在命令窗口输入如下语句:

```
>> EXAMP03020(0,15)
```

执行结果如图 3.11 所示。

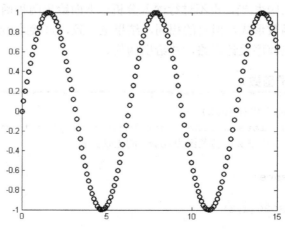

图 3.11 正弦曲线

EXAMP03021（私有函数）

```
function EXAMP03021(a,b)
%  EXAMP03021 illustrate the usage of provide function.
%  (EXAMP03021 子函数用于说明私有函数的用法)
%
%  Call sequence:
%  (回调格式)
%      EXAMP03021(a,b)
%
%  Record of revision:
%  Programmer: Wang Yong-Long
%  Date: 2015-07-31
%  Description: Revision
%
x=nest1(a,b);
y=nest2(x);
plot(x,y,'g^','LineWidth',2)
    function VECX=nest1(xmin,xmax)
    %  Private function nest1 produces the values of independent
    %  variable x.
    %  (私有函数 nest1 产生自变量 x 的值)
        global VECX
        if nargin==2
        VECX=linspace(xmin,xmax,15*xmax);
        else
        msg='Too few input arguments !'
        error(msg)
        end
    end
    function VECY=nest2(VECX)
    %  Private function nest2 produces the values of dependent
    %  variable y.
    %  (私有函数 nest2 产生因变量 y 的值)
        global VECY
        VECY=cos(VECX);
    end
end
```

在命令窗口输入如下语句：

```
>> EXAMP03021(0,15)
```

运行结果如图 3.12 所示。

注意：
脚本文件中一般可以省略 script filename 行，如果没有省略请注意文件名 filename 与保存文件名必须一致。本章所有脚本文件或函数文件的解释均含有英文和中文，有利于读者读懂 MATLAB 自带函数解释。

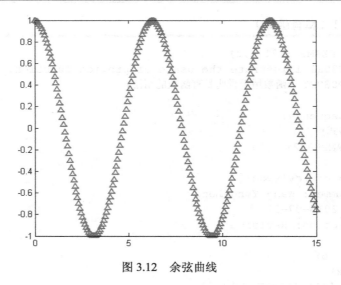

图 3.12 余弦曲线

小　　结

本章主要阐述了编写程序需要遵守的 Top-Down 设计模式，并借助伪代码进行分析说明，因为该设计模式的遵守对于编写大型程序是非常有意义的；随后对程序设计常用的三种结构：顺序结构、分支结构、循环结构分别进行表述并举例说明；最后对函数编写进行较详细阐述并举例说明。程序说明采用英文和中文解释，但愿对读者读懂 MATLAB 自带函数能有所帮助。

习　　题

3.1 请编写程序，计算用户输入任意 x 的相应函数值。给定函数表达式如下：

$$f(x)=\begin{cases} \sin(x) & x<-1 \\ \dfrac{\cos(1)-\sin(-1)}{2}x & -1\leqslant x \leqslant 1 \\ \cos(x) & x\geqslant 1 \end{cases}$$

3.2 请编写程序，能够根据用户输入的 x，y 两个变量值，给出下面函数表达式对应的解。给定函数表达式如下：

$$f(x,y)=\begin{cases} x-y & x\leqslant 0 \text{ 且 } y\leqslant 0 \\ \sin(x)\cos(x\times y) & x\leqslant 0 \text{ 且 } y\geqslant 0 \\ x+\exp(y) & x\geqslant 0 \text{ 且 } y\leqslant 0 \\ x\exp(y)-y\exp(xy) & x\geqslant 0 \text{ 且 } y\geqslant 0 \end{cases}$$

3.3 请编写程序，能够根据学生考试成绩进行分类投档。五所高校投档分数线如下：

$$\begin{cases} 清华大学 & 670分 \\ 北京航空航天大学 & 630分 \\ 北京工业大学 & 600分 \\ 北京联合大学 & 550分 \\ 海淀走读大学 & 450分 \end{cases}$$

3.4 请编写程序，为新生做向导。假设学校有三个院系分别为物理系、数学系、中文系，物理系位于西校区，中文系位于新校区，数学系位于北校区。当入学新生输入他所在院系就能得到位于哪个校区的明确答案。

3.5 请编写程序，求测量数据的平方均根。平方均根数学表达式如下：

$$平方均根 = \sqrt{\frac{1}{N}\sum_{i=1}^{N} x_i^2}$$

3.6 请编写程序，求解测量数据的调和平均值。调和平均值数学表达式如下：

$$调和平均值 = \frac{N}{\frac{1}{x_1} + \frac{1}{x_2} + \cdots + \frac{1}{x_n}}$$

3.7 请编写程序，求测量数据的平均值。平均值数学表达式如下：

$$s = \sum_{i=1}^{n} x_i$$

3.8 请编写程序，求解测量数据的几何平均值。几何平均值数学表达式如下：

$$几何平均值 = \sqrt[n]{\prod_{i=1}^{n} x_i}$$

3.9 第 n 个 Fibonacci 定义如下：

$$f(1) = 1$$
$$f(2) = 2$$
$$f(3) = f(2) + f(1) = 3$$
$$\vdots$$
$$f(n) = f(n-1) + f(n-2)$$

请编写程序，能够根据用户输入任何大于 2 的 n，给出第 n 个 Fibonacci。

3.10 AM 收音机前端射频接收器的简化调协电路如图 3.13 所示。该接收器是由 RLC 可调电路构成的，主要组件包含电阻 R、电容 C、电感 L，再将它们连接到一个天线上并接地。该调协电路能够方便选择 AM 频段的任意电台信号。这里要求 RLC 调协电路的谐振频率能够可调，当 RLC 调协电路的频率与某电台频率一致时，会发生共振，实现信号选取，接收电台信号。RLC 电路的谐振频率范围要能够覆盖各种 AM 频段的电台信号，对电感和电容可调提出可调范围要求。RLC 电路共振频率公式为

$$f_0 = \frac{1}{2\pi\sqrt{LC}}$$

式中 L——电感，单位为亨利（H）；

C——电容,单位为法拉第(F)。

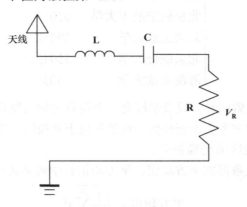

图 3.13 调协电路

请编写程序,能给出已知电感和电容的辐射频率,并对所编写的程序进行测试。取电感 $L=0.1\text{mH}$ 和电容 $C=0.2\text{nF}$,电阻 R 承载电压为辐射频率的函数,满足下面方程:

$$V_R = \frac{R}{\sqrt{R^2 + (\omega L - \frac{1}{\omega C})^2}} V_0$$

式中 $\omega = 2\pi f$,f 为频率,单位为赫兹(Hertz)。

假设 $L=0.1\text{mH}$,$C=0.25\text{nF}$,$R=50\Omega$ 和 $V_0=10\text{mV}$:

(1)画出电阻承载电压与辐射频率的关系图。找出电阻承载电压峰值所对应的频率,并给出峰值大小。这一个频率称为谐振频率。

(2)如果辐射频率相对谐振频率偏大 10%,那么电阻承载电压为多少?射频接收器是怎么选台的?

(3)辐射频率为多少时,电阻承载电压降低为峰值的 1/2?

3.11 请编写计算闰年的程序。众所周知,闰年需要满足以下条件:

(1)能被 400 整除的年份为闰年;

(2)能被 100 整除但不能被 400 整除的年份不是闰年;

(3)能被 4 整除但不能被 100 整除的年份为闰年;

(4)其余年份都不是闰年。

第 4 章 绘 图

本章将介绍关于绘制图形及属性设置的常用命令，共分 7 节。4.1 节介绍绘制二维图形常用的函数及属性设置；4.2 节介绍绘制三维图形常用的函数及属性设置；4.3 节介绍高维图形可视化；4.4 节介绍动画制作示例；4.5 节为应用实例；4.6 节介绍通过鼠标设置图形属性；4.7 节简单介绍图形句柄并举例说明。

4.1 二维绘图

4.1.1 函数 plot

函数 plot 是 MATLAB 中绘制二维图形最常用的命令函数，有多个属性参量，能够绘制各具特色的二维曲线。常用属性参量有 x 数组、y 数组（长度与 x 数组相等）、字符串（含有线的颜色、点符号、线型）、"LineWidth"线宽设置、"MarkerSize"点符号大小设置、"MarkerFaceColor"点符号填充颜色设置、"MarkerEdgeColor"点边框颜色设置，"MakerSize"点符号尺寸设置等。字符串中的线色、点符号、线型都有多种选择，参见表 4.1。

表 4.1 线色、点符号、线型

线色	描述	点符号	描述	线型	描述
b	蓝色	.	点	-	实线
g	绿色	o	圆	:	点虚线
r	红色	x	叉号	-.	点段虚线
c	蓝绿色，青色	+	加号	--	段虚线
m	红紫色，洋红	*	星号	(none)	没有线
y	黄色	s	正方形		
k	黑色	d	钻石形		
		v	倒三角		
		^	正三角		
		<	左三角		
		>	右三角		
		p	五角星		
		h	六角星		

plot 函数的主要命令格式介绍如下。

1. plot(x,'PropertyName',PropertyValue，…)

式中　x——绘制图表的数据；
　　　PropertyName——图表属性的字符选项；
　　　PropertyValue——对应图表属性的设置值。

当x不同时，绘制的图形不同。当x为实向量时，表示以该向量元素的下标为横坐标、元素的值为纵坐标画出一条连续曲线；当x为实矩阵时，则表示以该矩阵列的行下标为横坐标、每列元素值为纵坐标画出与列数同条数的曲线，曲线条数与列数相等；当x为复数矩阵时，所绘制的曲线以列为单位，矩阵元素的实部为横坐标、虚部为纵坐标，曲线条数与列数相等。

【例 4.1】 plot(x)示例。

EXAMP04001

```
% Script file: EXAMP04001.m
% Some examples illustrate the usage of
% plot(x,'PropertyName',PropertyValue).
% (针对plot(x,'PropertyName',PropertyValue)语句用法的几个示例)
% Clear variables in workspace.
% (清除工作空间中的变量)
clear
% Clear command window.
% (清除命令窗口)
clc
%
% Define an array as y1.
% (定义一个数组，记为y1)
y1=[0,2,3,6,9,5,2,1,3,6,9,11,6,3,1];
% Open a figure window.
% (打开一个图形窗口)
figure
% Plot the curve defined by y1.
% (绘制y1曲线)
plot(y1)
% Title the figure as Fig.1.
% (命名该图形为Fig.1)
title('Fig.1')
%
% Define two arrays a and t.
% (定义两个数组)
a=0:0.5:2;
t=0:0.1:3;
% In terms of the above two arrays, one can define a new two-dimensional
% array as y2.
% (根据上面两个数组，可以定义一个二维数组，记为y2)
y2=sin(3*t').*exp(-t')*a;
% Open a new figure window.
% (打开一个新图形窗口)
figure
% Plot curves depending on t and y2.
% (绘制由t和y2决定的曲线族)
plot(y2)
% Title the figure as Fig.2.
% (将该图形命名为Fig.2)
title('Fig.2')
%
% In terms of t, one can define a complex array as y3.
```

```
%  (根据变量t,可以定义一个复数数组,记为y3)
y3=t+exp(1/3*t).*sin(2*t+3)*I;
real_y3=real(y3);
imag_y3=imag(y3);
%  Open a new figure window.
%  (打开一个新图形窗口)
figure
%  Plot the line defined by y3.
%  (绘制y3定义的曲线)
plot(y3)
%  Title as Fig.3.
%  (命名为Fig.3)
title('Fig.3')
%  Open a new figure window.
%  (打开一个新图形窗口)
figure
%  Plot the curve defined by real_y3 and imag_y3.
%  (绘制real_y3和imag_y3定义的曲线)
plot(real_y3,imag_y3)
%  Title it as Fig.4.
%  (命名为Fig.4)
title('Fig.4')
```

运行结果如图4.1所示。

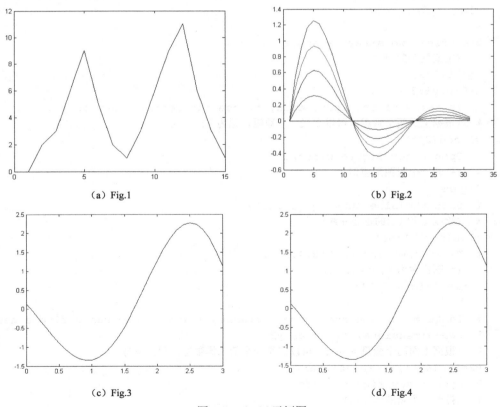

图4.1　plot(x)示例图

2. plot(x,y,'PropertyName',PropertyValue,…)

该函数的绘制参数与上面的函数格式相同，但在命令中增加了元素 y。x 和 y 都为需要绘制图形的数据数组。当 x 和 y 不同时，绘制的曲线也相应有所不同。当 x 和 y 是同维数的数组时，绘制曲线分别以 x 和 y 元素为横坐标和纵坐标；当 x 为一维数组，y 为矩阵，并且 y 的某一维数与 x 的维数相同时，此时绘制多条连续曲线，曲线的条数是矩阵 y 的另一维数，x 是曲线的横坐标；当 x 为矩阵，y 为一维数组时，此时曲线情形与上一情形相同，只是坐标轴方向变化；当 x 和 y 的维数相同时，此时以 x 和 y 的对应元素作为曲线的横坐标和纵坐标，曲线的条数和矩阵的列数相同。

【例 4.2】 plot(x,y)示例。

EXAMP04002

```
% Script file: EXAMP04002.m
% Some examples illustrate the usage of
% plot(x,y,'PropertyName',PropertyValue).
% (针对 plot(x,y,'PropertyName',PropertyValue)语句用法的几个示例)
% Clear variables in workspace.
% (清除工作空间中的变量)
clear
% Clear command window.
% (清除命令窗口)
clc
%
% Define two arrays a and t.
% (定义两个数组)
a=0:0.5:2;
t=0:0.01:3;
% In terms of the above array t, one can define a new as y1.
% (根据上面两个数组，可以定义一个数组，记为 y1)
y1=cos(2*t);
% Open a new figure window.
% (打开一个新图形窗口)
figure
% Plot the sine curve of t and y1.
% (绘制 t 和 y1 的正弦曲线)
plot(t,y1,'ro:')
% Title the plot as Fig.4.2-1.
% (将图命名为 Fig.4.2-1)
title('Fig.4.2-1')
%
% In terms of the above two variables a and t, one can define a new
% two-dimensional array as y2.
% (根据上面两个变量 a 和 t，可以定义一个新二维数组，记为 y2)
y2=sin(4*t').*exp(-t')*a;
% Open a new figure window.
% (打开一个新图形窗口)
figure
```

```
%   Plot curves depending on t and y2.
%   (绘制由 t 和 y2 决定的曲线族)
plot(t,y2)
%   Title the figure as Fig. 4.2-2.
%   (将该图形命名为 Fig.4.2-2)
title('Fig.4.2-2')
```

运行结果如图 4.2 所示。

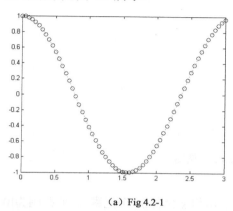
（a）Fig 4.2-1

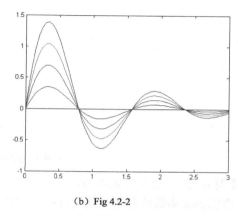
（b）Fig 4.2-2

图 4.2   plot(x,y)示例图

3. plot(x1,y1,x2,y2,'PropertyName',PropertyValue,…)

此命令与前面命令不同的地方在于，可以在窗口中同时绘制以 x1 和 y1、x2 和 y2 等分别为横坐标和纵坐标的曲线。实际上可以绘制任意多条曲线。

【例 4.3】   plot(x1,y1,x2,y2)示例。

EXAMP04003

```
%   Script file: EXAMP04003.m
%   Present two curves in one figure window.
%   (在一个图形窗口绘制两条曲线的示例)
%   Clear variables in workspace.
%   (清除工作空间中的变量)
clear
%   Clear command window.
%   (清除命令窗口)
clc
%   Define two arrays x1 and x2 as independent variables.
%   (定义两个数组 x1 和 x2 作为自变量)
x1=linspace(0,2*pi,100);
x2=linspace(0,3,100);
%   In terms of x1 and x2, one can define two dependent variables
%   y1 and y2.
%   (根据 x1 和 x2 定义两组因变量 y1 和 y2)
y1=1/3*x1.*sin(2*x1);
y2=x2.^2+2*x2;
```

```
%  In a figure window, plot two curves.
%  (在一个图形窗口绘制两条曲线)
plot(x1,y1,'r:',x2,y2,'b--')
```

运行结果如图 4.3 所示。

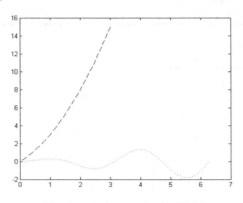

图 4.3 plot(x1,y1,x2,y2)示例图

【例 4.4】 绘制正弦曲线。

借助 plot 函数绘制正弦曲线，输入多个属性参数。数组 x 含有 30 个元素，取值范围为[0,2π]，数组 y 也含有相应的正弦值，曲线的颜色设置为红色，线型设置为虚线，点符号设置为圆，线宽为 2，点符号尺寸设置为 9，点符号填充颜色设置为绿色，符号边框颜色设置为蓝色。

EXAMP04004

```
%  Script file: EXAMP04004.m
%  Present an example to plot a sine curve.
%  (给出绘制正弦曲线的示例)
%  Clear variables in workspace.
%  (清除工作空间中的变量)
clear
%  Clear command window.
%  (清除命令窗口)
clc
%  Define an independent variable x with 30 elements from 0 to 2pi.
%  (定义一个自变量 x 含有 30 个元素，取值区间为从 0 到 2pi)
x=linspace(0,2*pi,30);
%  In terms of x, one can define a dependent variable y.
%  (根据 x 可以定义因变量 y)
y=sin(x);
%  Plot a requested plot.
plot(x,y,'ro-.','LineWidth',2,'MarkerSize',9,'MarkerFaceColor','g',...
'MarkerEdgeColor','b')
```

运行结果如图 4.4 所示。

【例 4.5】 在同一个窗口绘制多条曲线。

在同一个窗口绘制 3 条曲线，分别是正弦曲线、余弦曲线、正弦衰减曲线。另外，将正弦衰减曲线及其边界轮廓线在另一个窗口绘出。

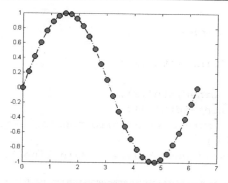

图 4.4 属性参数丰富的正弦曲线

EXAMP04005

```
% Script file: EXAMP04005.m
% Present an example to plot many curves in one figure window.
% (给出一个图形窗口绘制多条曲线的示例)
% Clear variables in workspace.
% (清除工作空间中的变量)
clear
% Clear command window.
% (清除命令窗口)
clc
% Define a variable x with 900 elements from 0 to 6pi.
% (定义一个变量 x 含有 900 个元素,取值范围为从 0 到 6pi)
x=linspace(0,6*pi,900);
% In terms of x, one define three vectors y1, y2 and y3 with 900 elements
% corresponding x, respectively.
% (根据 x,定义 y1、y2 和 y3 三个矢量,长度与 x 的长度相等)
y1=sin(x);
y2=cos(x);
y3=2.*sin(x.^3).*exp(-x/2);
% Plot the three plots.
plot(x,y1,x,y2,x,y3)
% Label the three curves with 'sin(x)', 'cos(x)' and
% 'damped-sine curve'.
% (标注 3 个曲线分别是 sin(x)、cos(x)和 damped-sine curve)
legend('sin(x)','cos(x)','damped-sine curve')
%

% Define a new variable x1 with 1000 elements from 0 to 6pi.
% (定义一个新变量 x1 含有 1000 个元素,取值范围从 0 到 6pi)
x1=linspace(1,6*pi,1000);
% Define three vectors y4,y5, and y6 with 1000 elements corresponding
% vector x1, respectively.
% (相应 x 定义 3 个矢量,y4、y5 和 y6)
y4=2*exp(-x1/5);
y5=-2*exp(-x1/5);
y6=2.*sin(x1*18).*exp(-x1/5);
% Open a new figure window.
% (打开一个新图形窗口)
```

```
figure
% Plot the three curves.
% (绘制这3条曲线)
plot(x1,y4,'g--','LineWidth',2);
hold on
plot(x1,y5,'g--','LineWidth',2);
plot(x1,y6,'r-','LineWidth',1);
% Label the three curves with corresponding names.
% (对3条曲线进行标注)
legend('Up outline','Down outline','Main line');
hold off
```

运行结果如图 4.5 所示。

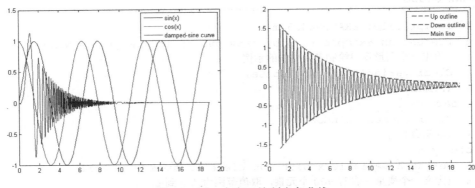

图 4.5　同一个窗口绘制多条曲线

> **注意：**
> 　　在同一个窗口绘制多条曲线，常用的有两种途径：一种途径是在 plot 命令输入参数中可增加自变量和因变量，完成绘图；另一种途径是借助 hold on 功能让绘图窗口一直处于当前状态，并且保持已绘制图形直至全部绘制结束。需要注意的是，最后加上 hold off 退出图形保持状态。

4.1.2　图形参数设置

1. 分格线控制

　　MATLAB 的默认设置为不画分格线，它的疏密取决于坐标刻度，如果想改变分格线的疏密，必须先定义坐标刻度。调用分格线格式控制描述参见表 4.2。

表 4.2　分格线控制

语句	描述
grid	是否画分格线的双向切换指令（使当前分格线状态翻转）
grid on	画出分格线
grid off	不画分格线
box	坐标形式在封闭式和开启式之间切换指令
box on	使当前坐标呈现封闭形式
box off	使当前坐标呈现开启形式

【例 4.6】 grid on 和 grid off，box on 和 box off 示例。

定义自变量 x，取值范围为[0,2pi]，定义 y_1=sin(x)，y_2=cos(x)。然后在窗口 1 和窗口 2 绘制正弦曲线，分格线分别开启和关闭；在窗口 3 和窗口 4 绘制余弦曲线，坐标形式为封闭式和开启式。

EXAMP04006

```
% Script file: EXAMP04006.m
% Present examples fot the usages of grid and box.
% (给出 grid 和 box 用法的示例)
% Clear variables in workspace.
% (清除工作空间中的变量)
clear
% Clear command window.
% (清除命令窗口)
clc
% Define a variable x with 100 elements from 0 to 2pi.
% (定义一个变量 x 含有 100 个元素，取值范围为从 0 到 2pi)
x=linspace(0,2*pi,100);
% In terms of x, one define three vectors y1 and y2 with 100 elements
% corresponding x, respectively.
% (根据 x, 定义 y1 和 y2 两个矢量，长度与 x 的长度相等)
y1=sin(x);
y2=cos(x);
% Set a new subplot window.
% (设定一个新子窗口)
subplot(2,2,1)
% Plot the sine curve of x and y1 with grid on.
% (绘制 x 和 y1 的正弦曲线，并设定 grid on)
plot(x,y1,'b','LineWidth',2)
grid on
% Set a new subplot window.
% (设定一个新子窗口)
subplot(2,2,2)
% Plot the sine curve of x and y1 with grid off.
% (绘制 x 和 y1 的余弦曲线，并设定 grid off)
plot(x,y1,'r:')
grid off
% Set a new subplot window again.
% (再设定一个新子窗口)
subplot(2,2,3)
% Plot the cosine curve of x and y2 with box on.
% (绘制 x 和 y2 的余弦曲线，并设定 box on)
plot(x,y2,'bo')
box on
% Set a new subplot window.
% (设定一个新子窗口)
subplot(2,2,4)
% Plot the cosine curve of x and y2 with box off.
% (绘制 x 和 y2 的余弦曲线，并且设定 box off)
plot(x,y2,'r-','LineWidth',2)
```

```
box off
```

运行结果如图 4.6 所示。

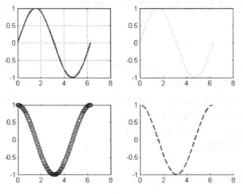

图 4.6 grid 和 box 的示例图

2．坐标轴

函数 axis 可以设置坐标系坐标刻度及其显示范围。坐标轴格式设定描述参见表 4.3。

表 4.3 坐标轴格式设定

语　句	描　述
axis([xmin, xmax, ymin, ymax])	设置坐标系 x 轴和 y 轴的显示范围
axis([xmin, xmax, ymin, ymax, zmin, zmax])	设置坐标系 x 轴、y 轴和 z 轴的显示范围
axis([xmin, xmax, ymin, ymax, zmin, zmax, cmin, cmax])	设置坐标系 x 轴、y 轴和 z 轴的显示范围，并且界定图形颜色刻度范围（不同颜色一般代表 c 的取值不同）
v = axis	在命令窗口中输入该命令，一般返回当前窗口坐标轴取值范围，对于二维图形由 4 个数值构成，对于三维图形由 6 个数值构成
axis auto	返回到坐标系 MATLAB 中的默认值
axis manual	固定当前坐标范围，即使输入 hold on 命令，在后面的绘制曲线也只是显示已经定义的 x 轴和 y 轴的坐标空间
axis tight	设置坐标显示范围与数据区间完全一致
axis fill	设置坐标空间与绘图窗口比例设置完全一致，也就是坐标系完全充满设定矩形空间，当然必须首先设置图形显示的矩形空间
axis equal	设置坐标各坐标轴等刻度显示，二维是 x 轴和 y 轴相等，三维是 x 轴、y 轴和 z 轴三轴相等。比如：绘制球时，如果不输入该命令行，给出的图形显示结果是个椭球体
axis image	该命令行等同于 axis equal，除非绘图框与数据完全符合
axis square	设置当前坐标轴外观尺度相等，不考虑坐标轴间取值范围的不同
axis normal	设置坐标系默认设置，比如已经运行的 axis square，axis equal 等将自动取消
axis off	关闭坐标系坐标刻度、标识符号及其背景设置
axis on	打开坐标系坐标刻度、标识符号及其背景设置

3．标题

函数 title 是为图形添加标题的常用命令。语法格式：title('文字')，可以将"文字"添加到当前坐标系的顶部，作为图形标题。对于输入的文字也可以设置很多属性，如字体、字号等，参见表 4.4。

表 4.4　标题文字的一些属性

语句	描述	语句	描述
\bf	黑体	\fontname{fontname}	定义字体
\it	斜体	\fontsize{fontsize}	定义字号
\rm	返回默认值		

在某些时候标题需要输入一些特殊符号，表 4.5 给出了一些常用特殊符号。

表 4.5　常用特殊符号

命令	符号	命令	符号	命令	符号
\alpha	α	\epsilon	ε	\infty	∞
\beta	β	\eta	η	\int	∫
\gamma	γ	\Gamma	Γ	\partial	∂
\delta	δ	\Delta	Δ	\leftarrow	←
\theta	θ	\Theta	Θ	\uparrow	↑
\lambda	λ	\Lambda	Λ	\rightarrow	→
\xi	ξ	\Xi	Ξ	\downarrow	↓
\pi	π	\Pi	Π	\div	÷
\omega	ω	\Omega	Ω	\times	×
\sigma	σ	\Sigma	Σ	\pm	±
\phi	φ	\Phi	Φ	\leq	≤
\psi	ψ	\Psi	Ψ	\geq	≥
\rho	ρ	\tau	τ	\neq	≠
\mu	μ	\zeta	ζ	\forall	∀
\nu	ν	\chi	χ	\exists	∃

4．二维图的其他参数设置函数

二维图常用参数设置函数描述参见表 4.6。

表 4.6　二维图常用参数设置函数

函数	描述
xlabel	xlabel('文本')　将文本添加到 x 轴 xlabel(AX,…)　可将文本添加到 x 轴坐标的指定位置
ylabel	ylabel('文本')　将文本添加到 y 轴 ylabel(AY,…)　可将文本添加到 y 轴坐标的指定位置
zlabel	zlabel('文本')　将文本添加到 z 轴 zlabel(AX,…)　可将文本添加到 z 轴坐标的指定位置
text	text(X,Y,'字符串')　在二维图形中，以点（X,Y）为起始点显示"字符串" text(X,Y,Z,'字符串')　在三维图形中，以点（X,Y,Z）为起始点显示"字符串"
gtext	gtext('字符串')　在图形窗口中，随着鼠标移动会显示一个大大的十字符号，当鼠标移动到某一位置时，单击鼠标，就会在此处显示"字符串"

(续表)

函数	描述
legend	legend(字符串 1,字符串 2,字符串 3,…,字符串 n) 一般根据绘制曲线的顺序，显示出相应"字符串 1"、"字符串 2"、"字符串 3"，等等，参数 n 可以取值 0、1、2、3、4，不同数值表示图例显示的位置不同，默认值为 1 legend(H,string1,string2,string3,…) 可以借助图形句柄，对相应图形窗口内曲线图形给出相应的图例说明
hold	hold on 设置当前图形窗口一直为当前图形窗口，再绘制图形曲线都显示在该坐标空间 hold off 取消设置当前图形窗口一直为当前窗口，绘制新图形曲线时，将自动打开一个新图形窗口，或者在当前窗口替换前一个图形
subplot	H = subplot(m,n,p)或 subplot(mnp) 将当前图形窗口拆分为 m×n 个子窗口，将第 p 个子窗口打开为当前窗口，可以在其中绘制图形（注意：子窗口顺序与矩阵元顺序不同，顺序为行从左到右，到达右边后再从左开始数）

【例 4.7】 函数 plot 参数设置示例。

绘制一条增益正弦曲线，并设定 x 轴、y 轴和 z 轴坐标标签，添加标题，添加图例说明，在坐标点（2, 2×sin(2)）处添加带有左箭头的"Okey!"文字，同时打开网格线、坐标系边框线。

EXAMP04007

```
% Script file: EXAMP04007.m
% Present an example to illustrate how to set properties.
% (给出设置属性的示例)
% Clear variables in workspace.
% (清除工作空间中的变量)
clear
% Clear command window.
% (清除命令窗口)
clc
% Define a variable x by the function linspace.
% (通过函数 linspace 定义变量 x)
x=linspace(0,30,500);
% In terms of x, an increasing sine function y is defiend.
% (基于 x, 一个增益正弦函数 y 被定义)
y=x.*sin(x);
% Plot the curve defined by x and y with 1.5 points of the line width.
% (绘制被 x 和 y 定义的曲线，线宽为1.5)
plot(x,y,'LineWidth',1.5);
% Title the curve as '增益正弦曲线'.
% (曲线命名为'增益正弦曲线')
title('增益正弦曲线');
% Label the x axis as 'alpha'.
% (标注 x 轴为'alpha')
xlabel('\alpha');
% Label the y axis as 'sin(\alpha)'.
% (标注 y 轴为'sin(\alpha)')
ylabel('Sin(\alpha)');
% Legend the curve as 'sin(alpha)'.
% (插入线形标注)
legend('sin(\alpha)',3);
% Insert 'Okey!' in the figure window.
% (将文字'Okey!'插入图形窗口中)
```

```
text(2,2*sin(2),'\leftarrow Okey!');
%  Present the grid in the figure window.
%  (图形窗口中显示网格线)
grid on
%  Show the box of axes.
%  (显示坐标轴框)
box on
```

运行结果如图 4.7 所示。

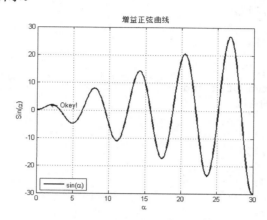

图 4.7 其他参数标注示例图

【例 4.8】 拆分窗口函数 subplot 应用示例。

将一个图形窗口拆分为 6 个子窗口，其中 4 个为比较小的窗口，分别为 4×3 窗口序列的第 1、3、4、6 个子窗口，分别绘制 tan(x)、ctan(x)、sin(x)-2 和 cos(x)-2 曲线。在未绘制的 2、5 处合为一个窗口，为 2×3 窗口序列的第 2 个窗口，在其中绘制 cos(x)-1 曲线。剩下的半空间为 2×1 窗口序列的第 2 个窗口，在其中绘制 sin(x)-1 曲线。

EXAMP04008

```
%  Script file: EXAMP04008.m
%  Present an example for the usage of subplot function.
%  (给出 subplot 函数用法的示例)
%  Clear variables in workspace.
%  (清除工作空间中的变量)
clear
%  Clear command window.
%  (清除命令窗口)
clc
%  Define a variable vector x with 60 elements by linspace
%  (通过 linspace 定义一个变量矢量 x 含有 60 个元素)
x=linspace(0,2*pi,60);
%  In terms of x, define variables vectors y1, y2, y3, y4, y5, and y6.
%  (根据 x，定义 y1, y2, y3, y4, y5 和 y6)
y1=sin(x);
y2=cos(x);
y3=sin(x)./(cos(x)+eps);
y4=sin(x);
```

```
y5=cos(x)./(sin(x)+eps);
y6=cos(x);
%  Set 6 subplot figures
%  (设置6个子窗口)
subplot(2,3,2);
%  Plot the curve of x and y2.
%  (绘制x和y2的曲线)
plot(x,y2);
title('cos(x)-1');
axis([0,6.3,-1,1]);
xlabel('x');
ylabel('cos(x)');
%
subplot(2,1,2);
%  Plot the curve of x and y1.
%  (绘制x和y1的曲线)
plot(x,y1);
title('sin(x)-1');
axis([0,6.3,-1,1]);
xlabel('x');
ylabel('sin(x)');
%
subplot(4,3,1);
%  Plot the curve of x and y3.
%  (绘制x和y3的曲线)
plot(x,y3);
title('tan (x)');
axis([0,6.3,-40,40]);
xlabel('x');
ylabel('tan (x)');
%
subplot(4,3,4);
%  Plot the curve of x and y4.
%  (绘制x和y4的曲线)
plot(x,y4);
title('sin(x)-2');
axis([0,6.3,-1,1]);
xlabel('x');
ylabel('sin(x)');
%
subplot(4,3,3);
%  Plot the curve of x and y5.
%  (绘制x和y5的曲线)
plot(x,y5);
title('cot (x)');
axis([0,6.3,-40,40]);
xlabel('x');
ylabel('cot (x)');
%
subplot(4,3,6);
%  Plot the curve of x and y6.
```

```
%   (绘制 x 和 y6 的曲线)
plot(x,y6);
title('cos(x)-2');
axis([0,6.3,-1,1]);
xlabel('x');
ylabel('cos(x)');
```

运行结果如图 4.8 所示。

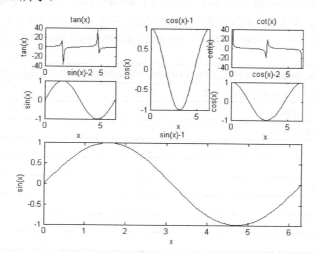

图 4.8　subplot 图形窗口拆分示例图

4.1.3　特殊二维图形绘制函数

特殊二维图形绘制函数描述参见表 4.7。

表 4.7　特殊二维图形绘制函数

函数	描　　述
bar	竖直条形图 ● bar(x,y)：x 是横坐标向量，y 为向量或者矩阵。y 为向量时，每个元素对应一个竖条；y 为 m 行 n 列矩阵时，将绘制出 m 组竖条，每组包含 n 条 ● bar(y)：横坐标使用默认值为 1：1：m ● bar(x,y,width)或 bar(y,width)：width 指定竖条的宽度，默认值为 0.8，注意如果宽度大于 1，条和条之间将重合 ● bar(x,y,'grouped')：产生组合的条形图 ● bar(x,y,'stacked')：产生堆叠的条形图 ● bar(x,y,'linespec')：指定条的颜色，其中 linespec 表示图形的线形、颜色、数据点等
barh	水平条形图
bar3	三维竖直条形图
bar3h	三维水平条形图
hist rose	hist 和 rose 函数可以建立直方图，其中 hist 在直角坐标系中建立直方图，rose 在极坐标中建立直方图 ● N = hist(y)：使用 10 个等距离分布的区间来对向量 y 的分布进行统计，并返回每个区间上含有 y 元素的个数 ● hist(y, n)：使用 n 个区间进行统计 ● hist(y, x)：使用中心位置在 x 元素处的小区间进行统计

(续表)

函数	描述
stairs	阶梯图形
stem	茎状图 • stem(x,y,'filled')或 stem(y,'filled')：产生带有实心数据标记的茎状图 • stem(x,y,'linespec')或 stem(y,'linespec')：使用 linespec 指定的线型、颜色绘制茎状图
fill	fill(x,y,c) 在 x 轴与 y 曲线包围区域用 c 颜色填充。c 填充颜色可以是单色'r'(红)、'g'(绿)、'b'(蓝)、'c'(青)、'm'(紫)、'y'(黄)、'w'(白)、'k'(黑)，或者根据三基色比例进行调配，可以输入[r, g, b]，分别为红绿蓝
feather	feather(U,V) 以 U 为 x 轴向速度取值，V 为 y 轴向速度取值，然后在等距的基点向由相应 U 和 V 元素确定的方向发射
polar	polar(THETA, RHO) 基于极向角度 THETA 和径向长度 RHO 提供数据完成极坐标图绘制
fplot	快速绘图命令，与 plot 函数功能相似
comet	彗星轨迹状的图形
pareto	带有标准的直方图
pie	饼图
pie3	三维饼图
plotmatrix	矩阵折线图
ribbon	带状图
scatter	点图
zoom	图形缩放

【例 4.9】 一些特殊二维图示例。

首先定义一个数组 x，然后定义 y 值，借助上面某些函数绘制特殊二维图形。

EXAMP04009

```
% Script file: EXAMP04009.m
% Present some examples for some two-dimensional figure special
% functions.
% (给出一些二维图特殊函数的示例)
% Clear variables in workspace.
% (清除工作空间中的变量)
clear
% Clear command window.
% (清除命令窗口)
clc
% Define a vector x with 19 elements.
% (定义一个含有19个元素的矢量x)
x=(-90:10:90)*pi/180;
% Define y1 and y2 as 5*x.*exp(-0.5*x) and 2*ones(size(x)),
% respectively.
% (分别定义y1=5*x.*exp(-0.5*x)和y2=2*ones(size(x)))
y1=5*x.*exp(-0.5*x);
y2=2*ones(size(x));
% u and v denote the velocity in the x-direction and that in
% the y-direction.
% (u 和 v 分别表示 x 轴和 y 轴方向速度的分量)
[u,v]=pol2cart(x,y2);
```

```
%   Open a subplot window.
%   (打开一个子绘图窗口)
subplot(3,3,1);
%   In the subplot window plot a bar figure.
%   (在该子绘图窗口绘制一个条形图)
bar(x,y1);
%   Title the figure as "bar(x,y1)".
%   (将该图命名为"bar(x,y1)")
title('bar(x,y1)');
%
%   Open a new subplot window.
%   (打开一个新子绘图窗口)
subplot(3,3,2);
%   In the subplot window plot a horizontal bar figure.
%   (在该子绘图窗口绘制一个水平条形图)
barh(x,y1);
%   Title the figure as "barh(x,y1)".
%   (将该图命名为"barh(x,y1)")
title('barh(x,y1)');
%   Open a new subplot window again.
%   (再次打开一个新子绘图窗口)
subplot(3,3,3);
%   In the new subplot window plot a hist figure.
%   (在这一个新子绘图窗口绘制hist图)
hist(y1);
%   Title the figure as "hist(y1)".
%   (将该图命名为"hist(y1)")
title('hist(y1)');
%   Open a new subplot window again.
%   (再次打开一个新子绘图窗口)
subplot(3,3,4);
%   In the subplot window plot a stair figure.
%   (在该子绘图窗口绘制stair图)
stairs(x,y1);
%   Title the figure as "stairs(x,y1)".
%   (将该图命名为"stairs(x,y1)")
title('stairs(x,y1)');
%   Open a new subplot window again.
%   (再次打开一个新子绘图窗口)
subplot(3,3,5);
%   In the subplot window plot a stem graph.
%   (在该子绘图窗口绘制一个stem图)
stem(x,y1);
%   Title the graph as "stem(x,y1)".
%   (将上面图形命名为"stem(x,y1)")
title('stem(x,y1)');
%   Open a new subplot window again.
%   (再次打开一个新子绘图窗口)
subplot(3,3,6);
%   In the subplot window plot a filled graph.
%   (在该子绘图窗口绘制filled图形)
fill(x,y1,'g');
```

```
%   Title the above figure as "fill(x,y1)".
%   (将上面图形命名为"fill(x,y1)")
title('fill(x,y1)');
%   Open a new subplot window again.
%   (再次打开一个新子绘图窗口)
subplot(3,3,7);
%   In the subplot window plot a feather graph.
%   (在该子绘图窗口绘制feather图)
feather(u,v)
%   Define the axes as equal.
%   (定义坐标轴的属性为equal)
axis equal
%   Title the above figure as "feather(u,v)".
%   (将上面图形命名为"feather(u,v)")
title('feather(u,v)');
%   Open a new subplot window again.
%   (再次打开一个新子绘图窗口)
subplot(3,3,8);
%   In the subplot window plot a polar graph.
%   (在该子绘图窗口绘制polar图)
polar(x,2*sin(x));
%   Title the above figure as "polar(theta,2sin(theta))".
%   (将上面图形命名为"polar(theta,2sin(theta))")
title('polar(\theta,2sin(\theta))');
%   Open a new subplot window again.
%   (再次打开一个新子绘图窗口)
subplot(3,3,9);
%   In the subplot window plot a graph by the function fplot.
%   (在该子绘图窗口借助函数fplot绘制一个图形)
fplot('x1*sin(x1)',[-pi,pi]);
%   Title the above figure as "fplot(x1*sin(x1))".
%   (将上面图形命名为"fplot(x1*sin(x1))")
title('fplot(x1*sin(x1))');
```

运行结果如图 4.9 所示。

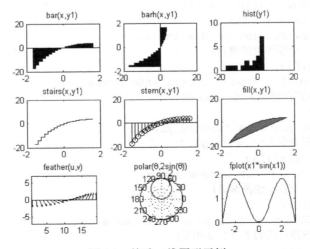

图 4.9 特殊二维图形示例

4.1.4 easy 二维绘图函数

easy 二维绘图函数描述参见表 4.8。

表 4.8 easy 二维绘图函数

函 数	描 述
ezplot	ezplot('f(x)')：绘制函数 "f(x)" 的图形，x 的默认取值范围为[-2π, 2π]
	ezplot('f(x,y)')：直接绘制函数 "f(x,y) = 0"，默认 x 和 y 取值范围都是[-2π, 2π]
	ezplot('f(x)', [a, b])：直接绘制函数 "f(x)"，自变量 x 取值范围为[a, b]
	ezplot('f(x,y)', [a, b])：直接绘制函数 "f(x, y) = 0"，默认 x 和 y 取值范围都是[a, b]
	ezplot('f(x,y)', [x1, x2, y1, y2])：直接绘制函数 "f(x, y) = 0"，默认 x 和 y 取值范围分别是[x1, x2]和[y1, y2]
	ezplot('fx(t)', 'fy(t)')：直接绘制参变量函数，默认 t 取值范围为[0, 2π]
	ezplot('fx(t)', 'fy(t)', [t1, t2])直接绘制参变量函数，参变量取值范围为[t1, t2]
ezpolar	ezpolar('f(theta)')：直接绘制极坐标曲线 "rho=f(theta)"，默认 theta 取值范围为[0, 2π]
	ezpolar('f(theta)', [a, b])：直接绘制极坐标曲线 "rho=f(theta)"，默认 theta 取值范围为[a, b]

【例 4.10】 请借助函数 ezplot 绘制函数曲线 $y=|\sin(x^2)|e^{2x}$，x 的取值范围为[0, 5]；借助 ezplot 函数绘制 $x=t\times\sin(t)$，$y=t\times\cos(t)$，t 的取值范围为默认值；借助 ezplot 函数绘制 $x=0.5\times t\times\sin(t)$，$y=0.5\times t\times\cos(t)$，t 的取值范围为[0, 4π]；借助 ezpolar 函数绘制 $rho=\sin(2\times t)\times\cos(3\times t)$，t 的取值范围为[0, π]。

EXAMP04010

```
% Script file: EXAMP04010.m
% Present some examples for the usage of some easy functions
% in two-dimensional space.
% (给出关于一些 easy 二维函数用法的示例)
% Clear variables in workspace.
% (清除工作空间中的变量)
clear
% Clear command window.
% (清除命令窗口)
clc
%
% Open a subplot window.
% (打开一个子绘图窗口)
subplot(2,2,1)
% Plot the curve defined by abs(sin(x.^2).*exp(2*x) with x ranging
% from 0 to 5 by the function ezplot.
% (借助函数 ezplot 绘制曲线 abs(sin(x.^2).*exp(2*x)，x 取值为从 0 到 5)
ezplot('abs(sin(x.^2)).*exp(2*x)',[0,5])
% Title the figure as 'ezplot(f,[a,b])'.
% (将该图命名为'ezplot(f,[a,b])')
title('ezplot(f,[a,b])')
%
% Open a new subplot window again.
% (再打开一个新子绘图窗口)
subplot(2,2,2)
```

```
%  Plot a curve parametrized by 't*sin(t)' and 't*cos(t)'.
%  (绘制't*sin(t)'和't*cos(t)'定义的曲线)
ezplot('t*sin(t)','t*cos(t)')
%  Title the figure as 'ezplot(f(t),g(t))'.
%  (命名该图为'ezplot(f(t),g(t))')
title('ezplot(f(t),g(t))')
%
%  Open a new subplot window again.
%  (再次打开一个新子绘图窗口)
subplot(2,2,3)
%  Ploat a curve parametrized by '0.5*t*sin(t)' and '0.5*t*cos(t)'
%  with t ranging from 0 to 4*pi.
%  (绘制'0.5*t*sin(t)'和'0.5*t*cos(t)'定义的曲线,t的取值范围
%  为[0,4*pi]).
ezplot('0.5*t*sin(t)','0.5*t*cos(t)',[0,4*pi])
%  Title the figure as 'ezplot(f(t),g(t),[a,b])'.
%  (命名该图形为'ezplot(f(t),g(t),[a,b])')
title('ezplot(f(t),g(t),[a,b])')
%
%  Open a new subplot window again.
%  (再次打开一个新子绘图窗口)
subplot(2,2,4)
%  Plot a curve defined by rho=sin(2*t)*cos(3*t) in polar coordinates,
%  t is ranging from 0 to pi.
%  (在极坐标系绘制由rho=sin(2*t)*cos(3*t)定义的曲线,t的取值范围为[0, pi])
ezpolar('sin(2*t)*cos(3*t)',[0,pi])
%  Title the figure as 'ezpolar(f(t),[a,b]).
%  (将该图形命名为'ezpolar(f(t),[a,b])')
title('ezpolar(f(t),[a,b])')
```

运行结果如图 4.10 所示。

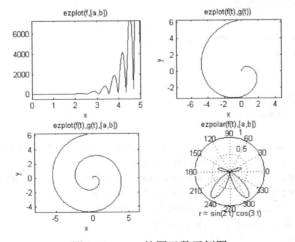

图 4.10 easy 绘图函数示例图

关于二维绘图函数是非常丰富的,不只局限于本节所述,限于篇幅这里不再赘述,如有兴趣可以借助 MATLAB 软件提供的帮助系统继续学习。比如,可根据需要自学 pie、area、quiver、loglog、semilogx、semilogy、contour、compass 等绘图函数。

4.2 三维绘图

4.2.1 函数 plot3

函数 plot3 是绘制三维空间曲线的常用函数。

语法格式：plot3(x, y, z, s)，其中 x, y 和 z 分别是第一维、第二维和第三维的数据，可以是矢量也可以是矩阵，但必须尺寸相等，s 是设置线型、颜色、数据点标记的字符串。

【例 4.11】 螺旋曲线。

EXAMP04011

```
%   Script file: EXAMP04011.m
%   Present two examples for the usage of plot3.
%   (显示 plot3 用法的两个示例)
%   Clear variables in workspace.
%   (清除工作空间中的变量)
clear
%   Clear command window.
%   (清除命令窗口)
clc
%   Define a vector t whose length is 100.
%   (定义长度为 100 的矢量，记为 t)
t=linspace(0,12*pi,100);
%   Plot a Helix curve by the function plot3, whose line color is blue,
%   whose line type is '-', and whose point marker is 'o' being filled
%   as green.
%   (借助 plot3 函数绘制一条螺旋曲线，线的颜色为蓝色，线型为'-'实线，点符号为'o'，
%   点符号的填充颜色为绿色)
plot3(t.*sin(t),t.*cos(t),t,'b-o','MarkerFaceColor','g',...
    'LineWidth',2);
%   Title the figure as '螺旋线'.
%   (将该图命名为'螺旋线')
title('螺旋线')
%   Label the x-axis as 'sin(t)', the y-axis as 'cos(t)', and
%   the z-axis as 't'.
%   (将 x 轴标注为'sin(t)', y 轴标注为'cos(t)', z 轴标注为't')
xlabel('sin(t)')
ylabel('cos(t)')
zlabel('t')
%   Display the grid background of axes.
%   (显示背景网格线)
grid on
%
%   Open a new figure window.
%   (打开一个新图形窗口)
figure
%   Redefine a new vector t1 whose length is 30.
%   (重新定义一个新矢量 t1，其长度为 30)
```

```
t1=linspace(0,6*pi,30);
% Plot 9 Helix curves in one figure window.
% (在一个图形窗口绘制 9 条螺旋曲线)
plot3(t1.*sin(t1),t1.*cos(t1),t1,'b-o','MarkerFaceColor','g');
% Hold on the present axes.
% (保留当前坐标系)
hold on
plot3(t1.*sin(t1)-60,t1.*cos(t1),t1,'r-o','MarkerFaceColor','b');
plot3(t1.*sin(t1),t1.*cos(t1)-60,t1,'b-o','MarkerFaceColor','g');
plot3(t1.*sin(t1),t1.*cos(t1)+60,t1,'b-o','MarkerFaceColor','g');
plot3(t1.*sin(t1)-60,t1.*cos(t1)+60,t1,'r-o','MarkerFaceColor','b');
plot3(t1.*sin(t1)-60,t1.*cos(t1)-60,t1,'r-o','MarkerFaceColor','b');
plot3(t1.*sin(t1)+60,t1.*cos(t1),t1,'b-o','MarkerFaceColor','c');
plot3(t1.*sin(t1)+60,t1.*cos(t1)-60,t1,'b-o','MarkerFaceColor','c');
plot3(t1.*sin(t1)+60,t1.*cos(t1)+60,t1,'b-o','MarkerFaceColor','c');
% Title the figure as 'Helixes'.
% (将该图命名为'Helixes')
title('Helixes')
% Label x-axis as 'Sin(t)'.
% (将 x 轴标注为'sint(t)')
xlabel('sin(t)')
% Label y-axis as 'Cos(t)'.
% (将 y 轴标注为'cos(t)')
ylabel('cos(t)')
% Lable z-axis as 't'.
% (将 z 轴标注为't')
zlabel('t')
% Display the background grid.
% (显示背景网格线)
grid on
% Hold off the present axes.
% (释放当前坐标系)
hold off
```

运行结果如图 4.11 所示。

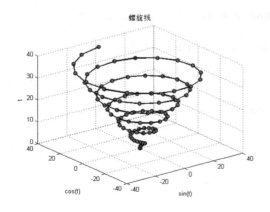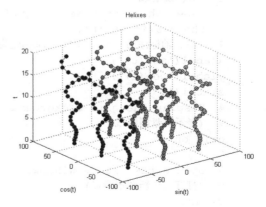

图 4.11 螺旋曲线

4.2.2 函数 patch

函数 patch 是一个建立补片图形对象的低级函数，补片对象是指坐标系中建立的多边形。如果多边形没有封闭，那么 patch 自动对它进行封闭。

语法格式：patch(x, y, z, c)，其中 x、y 和 z 分别是第一维、第二维和第三维的数据，可以是向量也可以是矩阵，但必须尺寸相等，c 用来指定颜色。

【例 4.12】 立方体示例图。

EXAMP04012

```
% Script file: EXAMP04012.m
% Present an example for the usage of fill and patch.
% (给出 fill 和 patch 用法的示例).
% Clear variables in workspace.
% (清除工作空间中的变量)
clear
% Clear command window.
% (清除命令窗口)
clc
% Define three arrays x、y and z.
% (定义三个数组)
x=[0 1 1 0 0 0;1 1 0 0 1 1;1 1 0 0 1 1;0 1 1 0 0 0];
y=[0 0 1 1 0 0;0 1 1 0 0 0;0 1 1 0 1 1;0 0 1 1 1 1];
z=[0 0 0 0 0 1;0 0 0 0 0 1;1 1 1 1 0 1;1 1 1 1 0 1];
% Plot the figure defined by the three arrays x、y and z.
% (绘制由 x、y 和 z 三个数组定义的图形)
plot3(x,y,z)
% Label x-axis as blod 'x'.
% (将 x 轴标注为黑体'x')
xlabel('x','fontweight','bold')
% Label y-axis as blod 'y'.
% (将 y 轴标注为黑体'y')
ylabel('y','fontweight','bold')
% Label z-axis as blod 'z'.
% (将 z 轴标注为黑体'z')
zlabel('z','fontweight','bold')
%
% Open a new figure window.
% (打开一个新图形窗口)
figure
% Patch all areas defined by the three arrays x、y and z.
% (填充由三个数组 x、y 和 z 定义的区域)
patch(x(:,1),y(:,1),z(:,1),'y');
patch(x(:,4),y(:,2),z(:,1),'b');
patch(x(:,1),y(:,6),z(:,6),'g');
```

运行结果如图 4.12 所示。

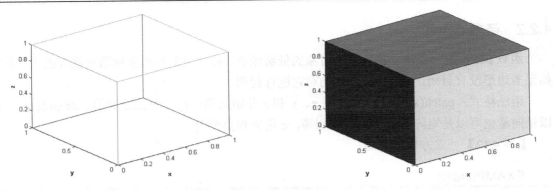

图 4.12 立方体示例图

4.2.3 三维网格图和曲面图函数

1. 函数 peaks

函数 peaks 是一个含有两个变量的样本函数。通常为函数 mesh、surf、pcolor、contour 等函数绘图提供数据。语法格式：Z = peaks，生成一个 49×49 矩阵。

Z = peaks(N)生成一个 $n×n$ 矩阵。

Z = peaks(V)生成一个 length(V)×length(V)矩阵。

Z = peaks(X,Y)生成一个与 X 和 Y 同行数同列数的矩阵。

如果不输入任何参数，那么默认的网格点由矩阵 X 和 Y 相对应的点构成，其中[X,Y] = meshgrid(V, V)。这里 V 是一个给定的矢量，默认含有从 -3 到 3 均匀分布的 49 个元素。为了对 peaks 函数有更加清楚的认识，可以在命令窗口输入 peaks 函数然后按回车键，得到如下结果：

$$z = 3*(1-x).^2.*exp(-(x.^2) - (y+1).^2)\cdots$$
$$- 10*(x/5 - x.^3 - y.^5).*exp(-x.^2-y.^2)\cdots$$
$$- 1/3*exp(-(x+1).^2 - y.^2)$$

（ $z = 3(1-x)^2 e^{(-x^2-(y+1)^2)} - 10(\frac{x}{5} - x^3 - x^5)e^{(-x^2-y^2)} - \frac{1}{3}e^{(-(x+1)^2-y^2)}$ ）

函数 peaks 相应图形如图 4.13 所示。

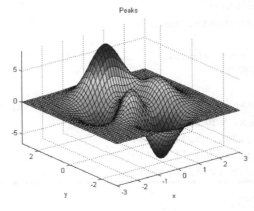

图 4.13 函数 peaks 相应图形

2. 函数 mesh

（1）函数 mesh 通常用来绘制三维网格面图。

语法格式：mesh(x, y, z, c) 基于 x 轴、y 轴和 z 轴三个坐标轴的矩阵值，以 c 色调为基准绘制出网格面图。如果调用语句改为 mesh(x, y, z)，那么 c 等于 z。同时，绘制网格图，还可以通过选择参数 views 设置视角，axis 设置坐标轴显示范围及其标度，还可以借助 colormap 对整体色调进行设置。

【例 4.13】 山峰网格面图。

首先通过 peaks 函数对 z 赋值，然后借助 mesh 函数完成镂空网格图绘制。

EXAMP04013

```
% Script file: EXAMP04013.m
% Present an example for the usage of mesh.
% (给出 mesh 用法的示例)
% Clear variables in workspace.
% (清除工作空间中的变量)
clear
% Clear command window.
% (清除命令窗口)
clc
% Define three variables x, y and z by peaks function.
% (借助函数 peaks 定义三个变量 x、y 和 z)
[x,y,z]=peaks(50);
% Plot the mesh surface of the previously defined peak.
% (绘制前面定义的山峰网格图)
mesh(x,y,z);
% Label x-axis as blod 'x'.
% (将 x 轴标注为黑体'x')
xlabel('x','fontweight','bold')
% Label y-axis as blod 'y'.
% (将 y 轴标注为黑体'y')
ylabel('y','fontweight','bold')
% Label z-axis as blod 'z'.
% (将 z 轴标注为黑体'z')
zlabel('z','fontweight','bold')
% Title the mesh as 'The mesh of peaks'.
% (将上面网格图命名为'The mesh of peaks')
title('Hidden off','fontsize',19)
%
% Open a new figure window.
% (打开一个新图形窗口)
figure
% Plot the mesh surface of the previously defined peak.
% (绘制前面定义的山峰网格图)
mesh(x,y,z);
% Hidden off to hidden surface elements.
% (Hidden off 用来隐去面元)
```

```
hidden on
% Label x-axis as blod 'x'.
% (将x轴标注为黑体'x')
xlabel('x','fontweight','bold','fontsize',16,'color','b')
% Label y-axis as blod 'y'.
% (将y轴标注为黑体'y')
ylabel('y','fontweight','bold','fontsize',16,'color','b')
% Label z-axis as blod 'z'.
% (将z轴标注为黑体'z')
zlabel('z','fontweight','bold','fontsize',16,'color','b')
% Title the mesh as 'The mesh of peaks'.
% (将上面网格图命名为'The mesh of peaks')
title('Hidden on','fontsize',19,'color','r')
```

运行结果如图4.14所示。

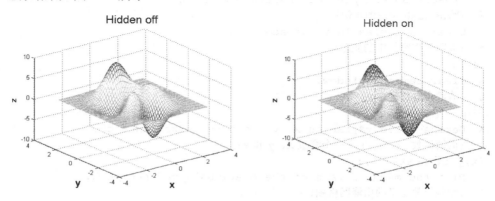

图4.14 山峰网格面图

（2）函数meshc通常用来绘制带有等高线的网格面图。

语法格式：meshc(…)使用方法与mesh(…)完全一致，在此不再赘述。

（3）函数meshz通常用来绘制带有帘幕的镂空网格面图。

语法格式：meshz(…)使用方法与mesh(…)也完全一致，只是多了帘幕，在此不再赘述。

【例4.14】 带等高线的网格图和带帘幕的网格图。

EXAMP04014

```
% Script file: EXAMP04014.m
% Present an example for the usage of meshc and meshf.
% (给出meshc和meshf用法的示例)
% Clear variables in workspace.
% (清除工作空间中的变量)
clear
% Clear command window.
% (清除命令窗口)
clc
% Define three variables x, y and z by peaks function.
% (借助函数peaks定义三个变量x、y和z)
[x,y,z]=peaks(50);
```

```
%  Plot the mesh surface with contours for the peaks.
%  (为上面山峰绘制带有等高线的网格图)
meshc(z);
%  Label x-axis as blod 'x'.
%  (将x轴标注为黑体'x')
xlabel('x','fontweight','bold')
%  Label y-axis as blod 'y'.
%  (将y轴标注为黑体'y')
ylabel('y','fontweight','bold')
%  Label z-axis as blod 'z'.
%  (将z轴标注为黑体'z')
zlabel('z','fontweight','bold')
%  Title the mesh as 'The meshc of peaks'.
%  (将上面网格图命名为'The meshc of peaks')
title('The meshc of peaks','fontsize',19)
%
%  Open a new figure window.
%  (打开一个新图形窗口)
figure
%  Plot the mesh surface with waterfall for the peaks.
%  (绘制带帘幕的网格曲面图)
meshz(z);
%  Title the mesh as 'The meshc of peaks'.
%  (将上面网格图命名为'The meshc of peaks')
title('The meshz of peaks','fontsize',19,'Color','b',…
'BackgroundColor','y')
```

运行结果如图 4.15 所示。

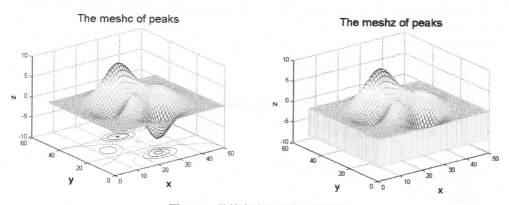

图 4.15 带等高线和帘幕的网格图

3. 函数 surf

（1）函数 surf 通常用来绘制颜色填充的网格面图。

语法格式如：surf(x, y, z, c) 表述与 mesh(x, y, z, c)表述句式基本一致。唯一区别是 mesh 只是网格面图，填充面元都是白色，而 surf 是被有色面元填充，在此不再赘述。其他参数设置也相同，只是多了 shading 参数。

【例 4.15】 surf 函数绘制曲面图。

EXAMP04015

```matlab
%   Script file: EXAMP04015.m
%   Present some examples for surf function.
%   (为函数 surf 提供几个示例)
%   Clear variables in workspace.
%   (清除工作空间中的变量)
clear
%   Clear command window.
%   (清除命令窗口)
clc
%   Define x, y and z by peaks command.
%   (通过 peaks 命令定义 x、y 和 z)
[x,y,z]=peaks(30);
%
%   Open a subplot figure window.
%   (打开一个子绘图窗口)
subplot(2,2,1)
%   Plot the surface of peaks by surf function.
%   (借助函数 surf 绘制 peaks 曲面)
surf(x,y,z);
%   Title the figure as '图 1 普通曲面图'.
%   (命名为'图 1 普通曲面图')
title('图 1 普通曲面图')
%
%   Open a subplot figure window.
%   (打开一个子绘图窗口)
subplot(2,2,2)
%   Plot the surface of peaks by surf function.
%   (借助函数 surf 绘制 peaks 曲面)
surf(x,y,z);
%   Set shading as flat.
%   (将 shading 参数值设为 flat)
shading flat
%   Title the figure as '图 2 平面阴影曲面图'.
%   (命名为'图 2 平面阴影曲面图')
title('图 2 平面阴影曲面图')
%
%   Open a subplot figure window.
%   (打开一个子绘图窗口)
subplot(2,2,3)
%   Plot the surface of peaks by surf function.
%   (借助函数 surf 绘制 peaks 曲面)
surf(x,y,z);
%   Set shading as interp.
%   (将 shading 参数值设为 interp)
shading interp
%   Title the figure as '图 3 插值阴影曲面图'.
%   (命名为'图 3 插值阴影曲面图')
title('图 3 插值阴影曲面图')
%
```

```
%   Open a subplot figure window.
%   (打开一个子绘图窗口)
subplot(2,2,4)
%   Define x1 and y1.
%   (定义 x1 和 y1)
x1=x(1,:);
y1=y(:,1);
%   Set some certain elements as not a number.
%   (设定一些特定元素为非数)
i=find(y1>0.8 & y1<1.2);
j=find(x1>-0.6 & x1<0.5);
z(i,j)=NaN;
%   Plot the surface of peaks by surf function.
%   (借助函数 surf 绘制 peaks 曲面)
surf(x,y,z);
%   Title the figure as '图 4 带洞的曲面图'.
%   (命名为'图 4 带洞的曲面图')
title('图 4 带洞的曲面图')
```

运行结果如图 4.16 所示。

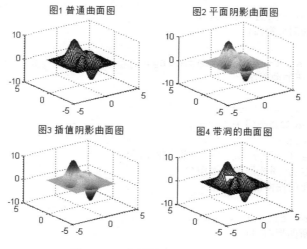

图 4.16　不同属性的 surf 曲面图

（2）函数 surfc 通常用来绘制带等高线的曲面图。

语法格式：surfc(…)表述与 surf(…)表述句式完全一致，不再赘述，只是多了等高线。

（3）函数 surfl 通常绘制带光照阴影的曲面图。

语法格式：surfl(…)表述与 surf(…)表述句式基本一致，在此不再赘述。

如果想给绘制图形加个光源，那么必须会用 MATLAB 中的函数 lighting，或者知道如何给函数 surfl 增加一个输入参数，用此来设定光源。如果这两种方法都不了解，请求助于 1.5 节阐述的 MATLAB 帮助系统。

（4）函数 surfnorm 通常绘制带法线的曲面。

语法格式：surfnorm(…)表述与 surf(…)表述句式基本一致，在此不再赘述。

【例 4.16】　带等高线和光照阴影的曲面图。

EXAMP04016

```
%   Script file: EXAMP04016.m
%   Present some examples for surfc and surfl functions.
%   (为函数 surfc 和 surfl 提供示例)
%   Clear variables in workspace.
%   (清除工作空间中的变量)
clear
%   Clear command window.
%   (清除命令窗口)
clc
%   Define x, y and z by peaks command.
%   (通过 peaks 命令定义 x、y 和 z)
[x,y,z]=peaks(50);
%   Plot the surface with contour.
%   (绘制带有等高线的曲面)
surfc(x,y,z);
%   Title the figure as '带等高线的曲面图'.
%   (将该图命名为'带等高线的曲面图')
title('带等高线的曲面图')
%
%   Open a new figure window.
%   (打开一个新图形窗口)
figure
%   Plot the surface with lighting.
%   (绘制带光照的曲面图)
surfl(x,y,z);
%   Set shading as interp.
%   (将 shading 参数值设为 interp)
shading interp
%   Title the figure as '带光照的曲面图'.
%   (将该图命名为'带光照的曲面图')
title('带光照的曲面图')
%
%   Open a new figure window.
%   (打开一个新图形窗口)
figure
%   Plot the surface with normals.
%   (绘制带法线的曲面图)
surfnorm(x,y,z);
%   Title the figure as '带法线的曲面图'.
%   (将该图命名为'带法线的曲面图')
title('带法线的曲面图')
```

运行结果分别如图 4.17 和图 4.18 所示。

4. 函数 waterfall

函数 waterfall 只是绘制沿 x 轴方向的等值线面。

语法格式：waterfall(…)表述与 mesh(…)表述句式基本一致，在此省略。

【例 4.17】 waterfall 示例图。

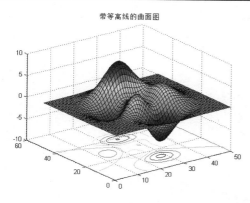 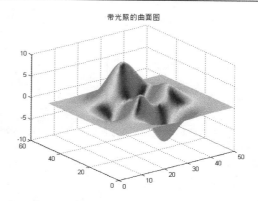

图 4.17　带等高线和带光照的曲面图

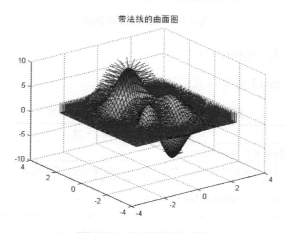

图 4.18　带法线的曲面图

EXAMP04017

```
%   Script file: EXAMP04017.m
%   Present an example for waterfall.
%   (为 waterfall 提供一个示例)
%   Clear variables in workspace.
%   (清除工作空间中的变量)
clear
%   Clear command window.
%   (清除命令窗口)
clc
%   Define x, y and z by peaks command.
%   (通过 peaks 命令定义 x、y 和 z)
[x,y,z]=peaks(50);
%   Plot the waterfall figure.
%   (绘制 waterfall 曲面图)
waterfall(x,y,z);
%   Title the figure as 'waterfall 曲面图'.
%   (将该图命名为'waterfall 曲面图')
title('waterfall 曲面图')
```

运行结果如图 4.19 所示。

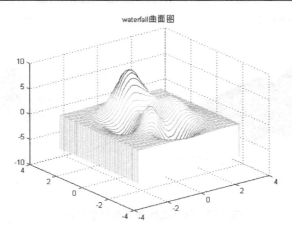

图 4.19 waterfall 示例图

4.2.4 函数 contour 和 contour3

函数 contour 和 contour3 命令格式描述参见表 4.9。

表 4.9 contour 和 contour3 命令格式

语 句	描 述
contour(z)	绘制 z 所表达的等值线图
contour(x, y, z)	借助 x 和 y 指定等值线的 x 和 y 坐标，绘制 z 所表达的等值线图
contour(z, n)或 contour(x, y, z, n)	借助参数 n 对等值线的条数进行设置
contour(z, v)或 contour(x, y, z, v)	向量 v 定义了等值线的数目和数值，即等值线的数值为向量元素值
[c, h]=contour(…)	返回 c 和 h，c 为等值线矩阵，h 为等值线句柄
contour3(x, y, z)	绘制 z 所表达的三维等值线
contour3(x, y, z, n)	借助参数 n 对等值线条数进行设置
[c, h]=contour3(…)	返回 c 和 h，c 为等值线矩阵，h 为等值线句柄
contourf(z, n)	绘制 z 所表达的三维等值线填充图，借助参数 n 设置线条数

【例 4.18】 等值线 contour、contour3 和 contourf 示例图。

EXAMP04018

```
%   Script file: EXAMP04018.m
%   Present some examples for contour, contour3 and contourf.
%   (为 contour、contour3 和 contourf 提供示例)
%   Clear variables in workspace.
%   (清除工作空间中的变量)
clear
%   Clear command window.
%   (清除命令窗口)
clc
%   Define x, y and z by peaks command.
%   (通过 peaks 命令定义 x、y 和 z)
```

```matlab
[x,y,z]=peaks(50);
% Plot the contour figure.
% (绘制等值线图)
contour(z);
% Title the figure as '(1) contour 示例图'.
% (将该图命名为'(1) contour 示例图')
title('(1) contour 示例图')
%
% Open a new figure window.
% (打开一个新图形窗口)
figure
% Plot the contour figure with n parameter.
% (绘制带参数 n 的等值线图)
contour(x,y,z,20)
% Title the figure as '(2) 20 条线的 contour 示例图'.
% (将该图命名为'(2) 20 条线的 contour 示例图')
title('(2) 20 条线的 contour 示例图')
%
% Open a new figure window.
% (打开一个新图形窗口)
figure
% Plot the three-dimensional contour figure.
% (绘制三维等值线图)
contour3(x,y,z)
% Title the figure as '(3) 三维 contour 示例图'.
% (将该图命名为'(3) 三维 contour 示例图')
title('(3) 三维 contour 示例图')
%
% Open a new figure window.
% (打开一个新图形窗口)
figure
% Plot the three-dimensional contour figure with n parameter.
% (绘制带参数 n 的三维等值线图)
contour3(x,y,z,20)
% Title the figure as '(4) 20 条线的三维 contour 示例图'.
% (将该图命名为'(4) 20 条线的三维 contour 示例图')
title('(4) 20 条线的三维 contour 示例图')
%
% Open a new figure window.
% (打开一个新图形窗口)
figure
% Plot the contour figure with labels.
% (绘制带标注的等值线图)
[c,h]=contour(x,y,z)
clabel(c,h)
% Title the figure as '(5) 带标注的 contour 示例图'.
% (将该图命名为'(5) 带标注的 contour 示例图')
title('(5) 带标注的 contour 示例图')
%
% Open a new figure window.
% (打开一个新图形窗口)
```

```
figure
%  Plot the filled contour figure.
%  (绘制带参数 n 的等值线图)
contourf(x,y,z)
%  Title the figure as '(6) 填充的 contour 示例图'.
%  (将该图命名为'(6) 填充的 contour 示例图')
title('(6) 填充的 contour 示例图')
```

运行结果如图 4.20 所示。

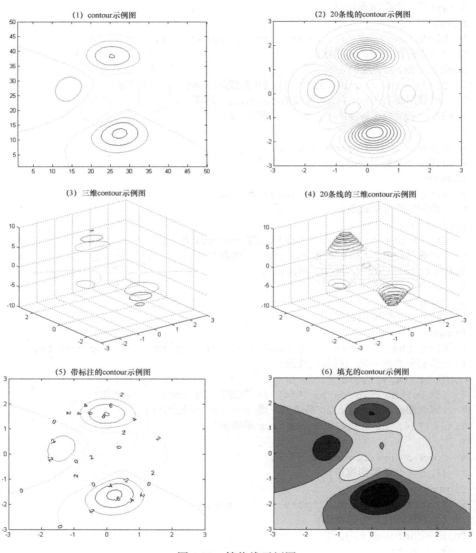

图 4.20　等值线示例图

4.2.5　函数 quiver

函数 quiver 矢量图（箭头簇）。

语法格式：quiver(x, y, u, v) 绘制矢量图，由许多小箭头构成，起始点是（x, y）对应格点值，箭头长度由相对应（u, v）值确定。

【例 4.19】 quiver 矢量示例图。

EXAMP04019

```
% Script file: EXAMP04019.m
% Present some examples for quiver and contour.
% (为 quiver 和 contour 提供示例)
% Clear variables saved in workspace.
% (清除工作空间中的变量)
clear
% Clear command window.
% (清除命令窗口)
clc
% Define x, y and z by peaks command.
% (通过 peaks 命令定义 x、y 和 z)
[x,y,z]=peaks(30);
% Define two velocity matrices of x-axis and y-axis with respect to
% z.
% (对应于 z 定义两个速度矩阵，分别表示 x 轴和 y 轴)
[vx,vy]=gradient(z,2,2);
% Plot the contour figure.
% (绘制等值线图)
contour(x,y,z,10);
% Hold on the present contour figure.
% (保留当前等值线图)
hold on
% Plot quiver figure.
% (绘制矢量图)
quiver(x,y,vx,vy);
hold off;
% Set the property of axis as image.
% (设置坐标轴属性为 image)
axis image
%
% Open a new figure window.
% (打开一个新图形窗口)
figure
% Define two matrices x1 and y1.
% (定义两个矩阵 x1 和 y1)
[x1,y1]=meshgrid([-2:0.25:2]);
% Define z1 depending on x1 and y1.
% (定义 z1 为 x1 和 y1 的函数)
z1=x1.*exp(-x1.^2-y1.^2);
% Plot the three-dimensional contour figure with 30 lines.
% (绘制三维 contour 图，含有 30 条线)
contour3(x1,y1,z1,30)
% Plot the surface with special properties.
% (绘制具有特定属性的曲面)
surface(x1,y1,z1,'EdgeColor',[0.7,0.7,0.7],'FaceColor','none')
% Disappear the grid mesh.
```

```
%  (不显示背景网格线)
grid off
%  Define the colormap of the figure as hot.
%  (定义图形的色调为热色调)
colormap hot
```

运行结果如图 4.21 所示。

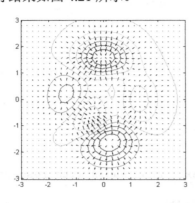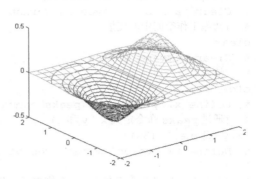

图 4.21　矢量图和等值线曲面图

4.2.6　easy 三维绘图函数

easy 三维绘图函数描述参见表 4.10。

表 4.10　easy 三维绘图函数

函　数	描　述
ezplot3	ezplot3(x, y, z)：绘制由 x(t)、y(t)和 z(t)定义的空间曲线，默认区间为[0, 2π] ezplot3(x, y, z, [a, b])：绘制由 x(t)、y(t)和 z(t)定义的空间曲线，定义区间为[a, b] ezplot3(x, y, z, [a, b], 'animate')：绘制由 x(t)、y(t)和 z(t)定义的循迹空间曲线，定义区间为[a, b]
ezmesh (ezmeshc)	ezmesh(f)：绘制 f(x,y)定义的曲面，默认 x 和 y 的取值范围都为[-2π, 2π] ezmesh(f, [x1, x2, y1, y2])：绘制 f(x,y)定义的曲面，x 的取值范围为[x1, x2]，y 的取值范围为[y1, y2] ezmesh(x, y, z)：绘制 x(s, t)、y(s, t)、z(s, t)定义的曲面，默认 s 和 t 的取值范围都为[-2π, 2π] ezmesh(x, y, z, [s1, s2, t1, t2])：绘制 x(s, t)、y(s, t)、z(s, t)定义的曲面，s 的取值范围为[s1, s2]，t 的取值范围为[t1, t2]
ezsurf (ezsurfc)	ezsurf(f)：绘制 f(x, y)定义的曲面，默认 x 和 y 的取值范围都为[-2π, 2π] ezsurf(f, [x1, x2, y1, y2])：绘制 f(x, y)定义的曲面，x 的取值范围为[x1, x2]，y 的取值范围为[y1, y2] ezsurf(x, y, z)：绘制 x(s, t)、y(s, t)、z(s, t)定义的曲面，默认 s 和 t 的取值范围都为[-2π, 2π] ezsurf(x, y, z, [s1, s2, t1, t2])：绘制 x(s, t)、y(s, t)、z(s, t)定义的曲面，s 的取值范围为[s1, s2]，t 的取值范围为[t1, t2] ezsurf(…,n)：n 用来设定网格的纵横数，默认值为 60
ezcontour (ezcontourf)	ezcontour(f)：绘制 f(x, y)定义的等值线，默认 x 和 y 的取值范围都为[-2π, 2π] ezsurf(f, [x1, x2, y1, y2])：绘制 f(x, y)定义的等值线，x 的取值范围为[x1, x2]，y 的取值范围为[y1, y2] ezsurf(…, n)：n 用来设定网格的纵横数，默认值为 60

【例 4.20】　easy 三维绘图函数示例图。

EXAMP04020

```matlab
%   Script file: EXAMP04020.m
%   Present some examples for easy three-dimensional functions.
%   (为三维easy绘图函数提供示例)
%   Clear variables in workspace.
%   (清除工作空间中的变量)
clear
%   Clear command window.
%   (清除命令窗口)
clc
%   Open a subplot figure window.
%   (打开一个子绘图窗口)
subplot(2,2,1)
%   Plot a spatial curve by ezplot3 command.
%   (通过ezplot3命令绘制一条空间曲线)
ezplot3(@cos,@(t)t.*sin(t),@sqrt);
%   Title the spatial curve as 'ezplot3'.
%   (将该图命名为'ezplot3')
title('ezplot3')
%
%   Open a new subplot figure window.
%   (打开一个新子绘图窗口)
subplot(2,2,2)
%   Plot a mesh surface by ezmesh command.
%   (通过ezmesh命令绘制一条空间曲线)
ezmesh(@(x,t)exp(-x).*cos(t),[-2,2,-4*pi,4*pi]);
%   Title the spatial curve as 'ezmesh'.
%   (将该图命名为'ezmesh')
title('ezmesh')
%
%   Open a new subplot figure window.
%   (打开一个新子绘图窗口)
subplot(2,2,3)
%   Plot a surface by ezsurf command.
%   (通过ezsurf命令绘制一条空间曲线)
ezsurf(@(x,y)x.*exp(-x.^2-y.^2),20);
%   Title the spatial curve as 'ezsurf'.
%   (将该图命名为'ezsurf')
title('ezsurf')
%
%   Open a new subplot figure window.
%   (打开一个新子绘图窗口)
subplot(2,2,4)
%   Plot a contour figure by ezcontour command.
%   (通过ezcontour命令绘制一条空间曲线)
ezcontour(@peaks,100);
%   Title the spatial curve as 'ezcontour'.
%   (将该图命名为'ezcontour')
title('ezcontour')
```

运行结果如图4.22所示。

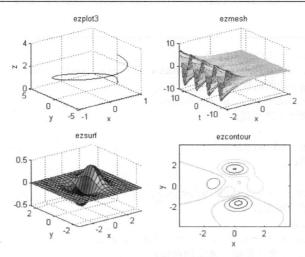

图 4.22 easy 三维绘图函数示例图

4.2.7 三维图形的参数设置

三维图形属性设置命令描述参加表 4.11。

表 4.11 三维图形属性设置命令

函　数	描　述
colormap	colormap(map)：设置当前图形的色图为 map，map 含有 jet、hsv、hot、cool、spring、summer、autumn、winter、gray、bone、copper、pink、lines 等 colormap('default')：设置当前图形的色图为默认的 jet map=colormap：返回当前色图属性
shading	• 阴影模式，通常用在由函数 surf、mesh、PColor、fill 和 fill3 产生的曲面 • shading flat：设置平面阴影模式 • shading interp：设置插值阴影模式 • shading faceted：设置面元阴影模式，为默认模式
material	• material shiny：设置绘制对象具有反光效果 • material dull：设置绘制对象不发亮或者粗糙 • material metal：设置对象材质为金属，金属一般是有光泽的 • material([ka kd ks])：设置绘制对象的环境光强、散射光强、镜面光强 • material([ka kd ks n])：设置绘制对象的环境光强、散射光强、镜面光强以及镜面指数 • material([ka kd ks n sc])：设置绘制对象的环境光强、散射光强、镜面光强、镜面指数以及镜面颜色反射 • material default 设置绘制对象的环境光强、散射光强、镜面光强、镜面指数以及镜面反射等参数为默认值
lighting	设置光照模式 • lighting flat：设置入射光为平行光，默认选项 • lighting gouraud：对定点颜色插补，对顶点勾画的面上颜色插补，用于表现曲面 • lighting phong：对顶点处的法线插值，计算各个像素的反光，生成较好的光照效果，花费时间稍长 • lighting none：关闭所有光源
light	创建光源 • light：对当前坐标系添加一光源，一切属性皆为默认值 • light(Para1, Value1, …)：对当前坐标系添加一光源，属性有 Value1 等设置

（续表）

函数	描述
camlight	创建或者设置光源位置 • camlight headlight：设置光源位于当前摄像机位置 • camlight right：在摄像机右上侧设置光源 • camlight left：在摄像机左上侧设置光源 • camlight(az, el)：将光源设置在(az, el)位置
spinmap	自旋色图 • spinmap(t)：t 指定循环旋转色图历时 3s • spinmap(inf)：指定循环时间，"Ctrl+C"组合键可以停止 • spinmap(t, inc)：t 指定循环时间，inc 用于指定快慢，默认值为 2，1 为慢，3 为快，如果取-2 代表反方向循环旋转色图
view	• view(az, el)：设置视角位置为（az, el） • view([x, y, z])：设置视角位置[x, y, z]

【例 4.21】 绘制具有特殊效果的山峰曲面。

EXAMP04021

```
% Script file: EXAMP04021.m
% Present an example for the usage of colormap and shading.
% (给出 colormap 和 shading 用法的示例)
% Clear variables in workspace.
% (清除工作空间中的变量)
clear
% Clear command window.
% (清除命令窗口)
clc
% Define three varialbes x, y and z.
[x, y, z]=peaks(100);
% Plot the surface of peaks.
% (绘制山峰图)
surf(z);
% Set shading as interp.
% (设置 shading 的属性为插值阴影)
shading interp
% Set the color map of the figure as hsv.
% (设置色图属性为 hsv)
colormap(hsv)
% Show a color bar.
% (显示色条)
Colorbar
```

运行结果如图 4.23 所示。

【例 4.22】 具有特定光效的椭球体。

首先借助 ellipsoid 函数定义 *x*, *y*, *z* 三个矩阵，然后根据这些数值矩阵，再借助 surf 函数绘制出椭球体，最后借助 material 和 lighting 两个函数对其绘制结果进行渲染，观察效果图。

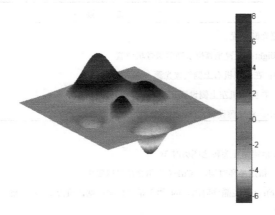

图 4.23 具有特殊视觉效果的山峰曲面图

EXAMP04022

```
%   Script file: EXAMP04022.m
%   Present an example for material and lighting.
%   (给出material和lighting用法的示例)
%   Clear variables in workspace.
%   (清除工作空间中的变量)
clear
%   Clear command window.
%   (清除命令窗口)
clc
%   Define three variables x, y, and z by ellipsoid function.
%   (通过ellipsoid函数定义三个变量x、y和z)
[x,y,z]=ellipsoid(0,0,0,5,3,1);
%   Plot the surface of a ellipsoid.
%   (绘制椭球体表面)
surf(x,y,z);
%   Set the material of the ellipsoid as metal.
%   (设置椭球体的材质为金属)
material metal
%   Set the lighting of the ellipsoid as phong mode.
%   (设置椭球体的光照性质为phong模式)
lighting phong
%   Set the axes as equal.
%   (设置坐标系为等轴)
axis equal
%   Set the color map of the ellipsoid as winter.
%   (设置椭球体的色图为winter)
colormap(bone)
%   Set the shading of figure as interp.
%   (设置图的shading属性为interp)
shading interp
```

运行结果如图 4.24 所示。

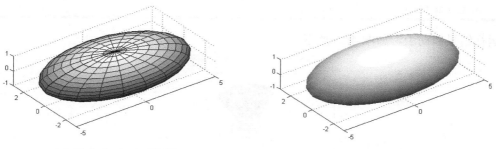

(a) 无 shading interp 示例图　　　　(b) 带有 shading interp 示例图

图 4.24　具有特定渲染效果的椭球体曲面图

【例 4.23】 绚丽的花瓶。

首先通过语句[x, y, z] = cylinder(2+sin(t))定义花瓶的空间点的三维坐标数值矩阵，然后借助 surf 函数完成绘制。属性由函数 material、shading、axis、view 和 spinmap 设定。

EXAMP04023

```
% Script file: EXAMP04023.M
% Present an example for the usage of spinmap.
% (给出 spinmap 用法的示例)
% Clear variables in workspace.
% (清除工作空间中的变量)
clear
% Clear command window.
% (清除命令窗口)
clc
% Define a variable as t.
% (定义一个变量,记为 t)
t=0:pi/50:2*pi;
% Define three matrices x, y, and z by cylinder command.
% (通过 cylinder 命令定义三个矩阵 x、y 和 z)
[x,y,z]=cylinder(2+sin(t),100);
% Plot the expectant figure.
% (绘制预期图形)
surf(x,y,z)
% Define its shading as interp.
% (定义它的 shading 属性为 interp)
shading interp
% Define its material as metal.
% (定义它的 material 属性为 metal)
material metal
% Define the axes as square.
% (定义坐标轴为等轴)
axis square
% Define the viewpoint as [-20,-20,11].
% (定义视点为[-20, -20, 11])
view([-20,-20,11])
% Define it being rotated slowly with 10 seconds.
% (定义 spinmap 历时 10s,渐变速度为慢)
```

```
spinmap(6,1);
```

运行结果似动画,如图 4.25 所示。

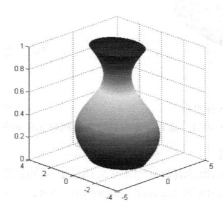

图 4.25 绚丽花瓶剪辑图

【例 4.24】 球的透视与非透视视图。

首先借助执行语句[x,y,z] = sphere(20)定义一个球的三维空间 x、y、z 的空间坐标数值矩阵,然后借助 mesh 和 surf 函数完成绘图,再借助 hidden 函数进行设定,并分析绘图结果。

EXAMP04024

```
% Script file: EXAMP04024.M
% Present some examples to illustrate how to set the properties
% of three-dimensional figures.
%  (给出一些设置三维图形属性的示例)
% Clear variables in workspace.
%  (清除工作空间中的变量)
clear
% Clear command window.
%  (清除命令窗口)
clc
% Define three matrices x, y, and z by cylinder function.
%  (通过 cylinder 函数定义三个矩阵 x、y 和 z)
[x,y,z]=sphere(30);
%
% Open a subplot window.
%  (打开一个子绘图窗口)
subplot(2,2,1);
% In the window, plot a spherial mesh.
%  (在该子窗口绘制网格球)
mesh(x,y,z);
% Define the axes as equal.
%  (定义坐标轴为等轴)
axis equal
% Title the figure as 'mesh of sphere'.
%  (命名该图为'mesh of sphere')
title('mesh of sphere')
%
% Open a new subplot window.
```

```
%   (打开一个新子绘图窗口)
subplot(2,2,2);
%   In the present window, plot a spherial mesh.
%   (在当前子窗口，绘制网格球)
mesh(x,y,z);
%   Set hidden as off.
%   (设置hidden属性为off)
hidden off
%   Define the axes as equal.
%   (定义坐标轴为等轴)
axis equal
%   Title the figure as 'mesh of sphere with hidden off'.
%   (将该图命名为'mesh of sphere with hidden off')
title('mesh of sphere with hidden off')
%
%   Open a new subplot again.
%   (再次打开一个新子绘图窗口)
subplot(2,2,3);
%   In the window, plot a spherical surface.
%   (在该窗口，绘制球面)
surf(x,y,z);
%   Set its shading as flat.
%   (设定shading属性为flat)
shading flat
%   Define the axes as equal.
%   (定义坐标轴为等轴)
axis equal
%   Title the figure as 'surface of sphere with shading flat'.
%   (将该图命名为'surface of sphere with shading flat')
title('surface of sphere with shading flat')
%
%   Open a new subplot again.
%   (再次打开一个新子绘图窗口)
subplot(2,2,4);
%   In the window, plot a spherical surface.
%   (在该窗口，绘制一个球面)
surf(x,y,z);
%   Set its shading as interp.
%   (设置shading属性为interp)
shading interp
%   Set its colormap as jet.
%   (设置colormap属性为jet)
colormap(autumn)
%   Define the axes as equal.
%   (定义坐标轴为等轴)
axis equal
%   Title the figure as 'surface of sphere with jet and interp'.
%   (将该图命名为'surface of sphere with jet and interp')
title('surface of sphere with jet and interp')
```

运行结果如图 4.26 所示。

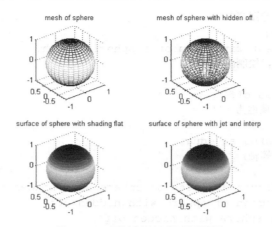

图 4.26 球的透视与非透视视图

如果读者欲对三维图形渲染参数有更广泛、更深入的了解，请借助 MATLAB 软件提供的帮助系统进行自学。

4.3 高维图形可视化

前面两节对二维可视化和三维可视化函数进行了简单讲述。二维图形只有一个自变量，三维图形只有两个自变量，当自变量增加时，如何可视化每个自变量对坐标处的标量数据的影响成为必须面对的问题。

函数 slice 为切片命令，可用于显示三维函数的切面图、等位线图等，从而在空间坐标中将数据显示出来。调用 slice 命令和相关变体命令，如 contourslice、streamslice 等的常用格式参见表 4.12。

表 4.12 slice 命令及变体命令的调用格式

语　　句	描　　述
slice(v, sx, sy, sz)	绘制函数 v=v(x, y, z)(阶数为 m*n*p)所确定的函数在 x、y 和 z 轴方向若干点上切片图，每个方向的切片多少由 sx、sy、sz 指定
slice(x, y, z, v, sx, sy, sz)	绘制函数 v=v(x, y, z)所确定的函数在 x、y 和 z 轴方向若干点上切片图，每个方向的切片多少由 sx、sy、sz 指定，x、y、z 为三维数组，用于指定 v 函数的坐标
slice(v,xi,yi,zi)	参数 xi，yi，zi 定义一个曲面，同时在绘制的曲面点上计算超立体 v 的数值，该命令能够显示由 xi，yi，zi 定义的曲面的切片图
slice(…, 'method')	'method'用于指定切片的内插值方法，常用的插值方法有 linear、cubic、nearest 等
contourslice(v, sx, sy, sz)	在切片内绘制等值线，调用格式与 slice 基本相同
streamslice(u,v)	在切片面内绘制流线

【例 4.25】　绚丽的光学效果图。

EXAMP04025

```
%  Script file: EXAMP04025.m
%  Present an example for the usage of slice function.
%  (给出关于 slice 函数用法的示例)
```

```
%   Clear variables in workspace.
%   (清除工作空间中的变量)
clear
%   Clear command window.
%   (清除命令窗口)
clc
%   Clear the current figure window.
%   (清除当前图形窗口)
clf
%   Define four matrices x, y, z, and v by the function flow.
%   (通过函数flow定义4个矩阵x、y、z和v)
[x,y,z,v]=flow;
%   Find the minimums of the 50×25×25 matrix x, y and z and save
%   as x1, y1 and z1, respectively, and save the maximums of x, y
%   and z as x2, y2 and z2, respectively.
%   (将矩阵x、y和z的最小值分别存为x1, y1和z1, 最大值分别存为x2, y2和z2)
x1=min(min(min(x)));
x2=max(max(max(x)));
y1=min(min(min(y)));
y2=max(max(max(y)));
z1=min(min(min(z)));
z2=max(max(max(z)));
%   Define sx with 5 elements.
%   (定义sx含有5个元素)
sx=linspace(x1+1.2,x2,5);
%   Define sy and sz as 0.
%   (定义sy和sz为0)
sy=0;
sz=0;
%   Plot slicers.
%   (绘制切片)
slice(x,y,z,v,sx,sy,sz);
%   Define viewpoint as [-33,36].
%   (定义视角为[-33,36])
view([-33,36]);
%   Set shading as interp.
%   (设置shading属性为interp)
shading interp;
%   Set colormap as hsv.
%   (设置colormap属性为hsv)
Colormap(hsv)
%   Set alpha as color.
%   (设置alpha属性为color)
alpha('color')
%   Set alphamap as rampdown.
%   (设置alphamap属性为rampdown)
alphamap('rampdown')
%   Set the increase value of alphamap as 1.
%   (设置alphamap增长值为1)
alphamap('increase',.1)
%   Display colorbar.
```

```
%  （显示色条）
colorbar
%  Turn off the axis.
%  （关闭坐标轴）
axis off
```

注释：

（1）函数 flow 通常用来产生流体流动的三维数据。

语法格式：

v = flow   产生 50×25×25 的数组 v。

v = flow(n)   产生 2n×n×n 的数组 v。

v = flow(x, y, z)   根据 x, y, z 坐标判断剖面速度然后存为数组 v。

[x, y, z, v] = flow(…)   相对于前面几个语法格式只是多了和 v 行数、列数、页数一致的坐标数组 x, y, z。

（2）函数 alpha 通常用来设定当前轴方向上绘制对象的透明性质，一般有选择项为"x"、"y"、"z"、"color"和"rand"。当选择面元为操作对象时，有"face-flat"、"face-interp"、"face-texture"、"face-opaque"和"face-clear"选项。如有不理解之处，可以借助 MATLAB 帮助系统阅读实际示例与操作演示。

（3）函数 alphamap 可用来设置图形透明度。

运行结果如图 4.27 所示。

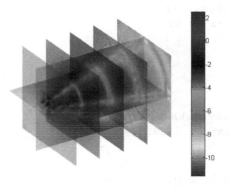

图 4.27   绚丽的光影

【例 4.26】   contourslice 和 streamslice 示例。

EXAMP04026

```
%  Script file: EXAMP04026.m
%  Present two examples for the usages of contourslice and streamslice.
%  (给出关于 contourslice 和 streamslice 函数用法的示例)
%  Clear variables in workspace.
%  (清除工作空间中的变量)
clear
%  Clear command window.
%  (清除命令窗口)
clc
%  Clear the current figure window.
```

```
%  (清除当前图形窗口)
 clf
% Define four matrices x, y, z, and v by the function flow.
%  (通过函数 flow 定义 4 个矩阵 x、y、z 和 v)
[x,y,z,v]=flow;
% Find the minimum and maximum of the 50×25×25 matrix x, and save
% as x1 and x2, respectively.
%  (将矩阵 x 最小值和最大值分别存为 x1 和 x2)
x1=min(min(min(x)));
x2=max(max(max(x)));
% Define sx with 4 elements.
%  (定义 sx 含有 4 个元素)
sx=linspace(x1+1.5,x2,4);
% Find the minimum and maximum of the 50×25×25 matrix v, and save
% as v1 and v2, respectively.
%  (将矩阵 x 最小值和最大值分别存为 x1 和 x2)
v1=min(min(min(v)));
v2=max(max(max(v)));
% Define sv with 20 elements.
%  (定义 sv 含有 20 个元素)
sv=linspace(v1+1,v2,20);
% Plot contour slicers.
%  (绘制切片)
contourslice(x,y,z,v,sx,0,0,sv);
% Set viewpoint as [-45,30]
%  (设置视角为[-45,30]).
view([-45,30]);
% Set shading as interp.
%  (设置 shading 属性为 interp)
shading interp;
% Set colormap as hsv.
%  (设置 colormap 属性为 hsv)
colormap(hsv)
% Display colorbar.
%  (显示色条)
colorbar
% Turn off the axis.
%  (关闭坐标轴)
axis off
%
% Open a new figure window.
%  (打开一个新图形窗口)
figure
% Define three variables x, y and z by the function peaks.
%  (通过函数 peaks 定义 3 个变量 x、y 和 z)
[x,y,z]=peaks(50);
% Plot the surface of peaks.
%  (绘制 peaks 的曲面)
surf(x,y,z)
% Set the shading of the surface as interp.
%  (设置 shading 属性为 interp)
shading interp
```

```matlab
% Set the hold of the figure as on.
% (设置 hold 属性为 on)
hold on
% Add contours to the surface of peaks.
% (对 peaks 曲面增加等值线)
[c,ch]=contour3(z,20);
% Set the edgecolor of the surface as blue.
% (设曲面的 edgecolor 属性为蓝色)
set(ch,'edgecolor','b')
% Define two variables u and v by the function gradient.
% (借助函数 gradient 定义两个变量 u 和 v)
[u,v]=gradient(z);
% Plot flowing lines, contours and slicers.
% (绘制流线、等值线和切片)
h=streamslice(-u,-v);
set(h,'color','k')
% For structure.
% (for 循环结构)
for i=1:length(h)
    zi=interp2(z,get(h(i),'xdata'),get(h(i),'ydata'));
    set(h(i),'zdata',zi);
end
% Display colorbar.
% (显示色条)
colorbar
% Set viewpoint as (30, 30).
% (设置视角为(30,30))
view(30,30);
% Set axis as tight.
% (设置 axis 属性为 tight)
axis tight
```

运行结果如图 4.28 所示。

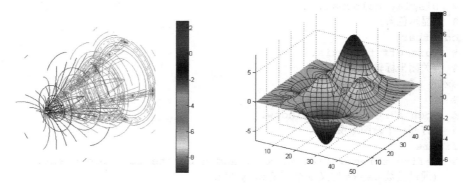

图 4.28　contourslice 和 streamslice 函数示例图

4.4　动画制作示例

本节主要借助函数 slice 和函数 movie 创建具有动画效果的示例。

4.4.1 电影程序编写

【例 4.27】 借助 slice 命令制作一部电影。

EXAMP04027

```
%  EXAMP04027.m
%  Show a movie.
%  (给出一个动画)
%  Clear variables in workspace.
%  (清除工作空间中的变量)
clear
%  Clear command window.
%  (清除命令窗口)
clc
%  Define three matrices x, y and z by meshgrid.
%  (通过meshgrid命令定义三个矩阵x、y和z)
[x,y,z]=meshgrid(-2:.2:2,-2:.25:2,-2:.16:2);
%  Define the volume depending on x, y and z.
%  (定义由x、y和z定义的体积)
v=x.*exp(-x.^2-y.^2-z.^2);
%  Define three variables xsp, ysp, zsp for a sphere.
%  (为球定义三个变量xsp、ysp和zsp)
[xsp,ysp,zsp]=sphere;
%  Plot some slicers to construct a certain situation.
%  (绘制一些切片构建一个特定环境)
slice(x,y,z,v,[-2,0,2],2,-2)
for i=-3:.03:3
    %  Define a handle hsp for the sphere.
    %  (定义球的句柄为hsp)
    hsp=surface(xsp+i,ysp,zsp);
    %  Rotate the defined sphere 90 degrees around x-axis.
    %  (将定义的球绕x轴旋转90°)
    rotate(hsp,[1 0 0],90)
    %  Take the XData of the sphere hsp as xd.
    %  (将球hsp的x数据存为xd)
    xd=get(hsp,'XData');
    %  Take the XData of the sphere hsp as yd.
    %  (将球hsp的y数据存为yd)
    yd=get(hsp,'YData');
    %  Take the XData of the sphere hsp as zd.
    %  (将球hsp的z数据存为zd)
    zd=get(hsp,'ZData');
    %  Delete the data of the sphere hsp.
    %  (删除球hsp的数据)
    delete(hsp)
    %  Set hold as on.
    %  (设置hold属性为on)
    hold on
    %  Plot slicers depending on the surface of moving sphere.
    %  (绘制依赖运动球曲面的切片)
    hslicer=slice(x,y,z,v,xd,yd,zd);
```

```
    % Set axis as tight.
    % (设置axis属性为tight)
     axis tight
    % Define the range of x as [-3,3].
    % (定义x的取值范围为[-3, 3])
     xlim([-3,3])
    % Define viewpoint as (-10,35).
    % (定义视角为[-10, 35])
     view(-10,35)
    % Draw the sphere.
    % (绘制球)
     drawnow
    % Delete the slicer.
    % (删除切片)
    delete(hslicer)
    % Set hold as off.
    % (设置hold属性为off)
    hold off
    % Set shading as interp.
    % (设置shading属性为interp)
    shading interp
end
```

运行结果如图4.29所示。

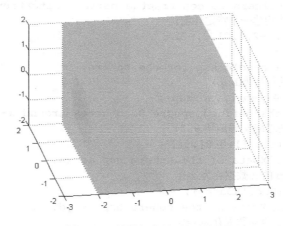

图4.29 电影程序最后一幅图

4.4.2 函数movie

函数 movie 可用来放映记录的电影画面。

语法格式：movie(M) 根据存在 M 中的电影画面数据进行播放，只播放一次。其中 M 的数据一般借助命令 getframe 来取得。

movie(M, N) 对存在 M 中的电影进行播放，播放 N 次。例如，电影一共由 4 帧组成，如果 N = [10,4,4,2,1]，则表示播放 10 次，按 4 帧→4 帧→2 帧→1 帧的模式进行播放。

movie(M, N, FPS) 对 M 进行播放，播放 N 次，播放速度为 FPS 帧/s。

【例 4.28】 旋转的山峰。

EXAMP04028

```
%  Script file: EXAMP04028.m
%  Present an example for the usage of movie.
%  (给出movie函数用法的示例)
%  Clear variables in workspace.
%  (清除工作空间中的变量)
clear
%  Clear the current figure window.
%  (清除当前图形窗口)
clf
%  Clear command window.
%  (清除命令窗口)
clc
%  Define three matrices x, y, and z by peaks.
%  (通过函数sphere定义三个矩阵x、y和z)
[x,y,z]=peaks(50);
%  Plot the surface of peaks with lighting.
%  (绘制带光照的山峰曲面)
surfl(x,y,z)
%  Set the range of x-, y- and z-axis.
%  (设置x、y和z轴的范围)
axis([-3 3 -3 3 -6 6])
%  Disappear the arrows of three axes.
%  (不显示坐标轴的箭头)
axis vis3d off
%  Set axes equal.
%  (设置坐标轴为等轴)
axis equal
%  Set shading as interp.
%  (设置shading属性为interp)
shading interp
%  Set colormap as copper.
%  (设置色图属性为copper)
colormap(copper)
%  Generate the frames of movie.
%  (产生电影的各帧)
for i=1:360
    %  Set the viewpoints of all the frames.
    %  (设置各帧的视角)
    view(-37.5+i,30)
    %  Save the current frame to m(i).
    %  (保存当前帧到m(i)中)
    m(i)=getframe;
end
%  Clear the current axes.
%  (清除当前坐标系)
cla
%  Show the movie.
%  (播放电影)
movie(m)
```

运行结果似动画，中间某一帧如图4.30所示。

图 4.30 旋转的山峰（中间某一帧）

【例 4.29】 鼓动的山峰。

EXAMP04029

```
%   Script file: EXAMP04029.m
%   Present an example for the usage of movie.
%   (给出 movie 函数用法的示例)
%   Clear variables in workspace.
%   (清除工作空间中的变量)
clear
%   Clear the current figure window.
%   (清除当前图形窗口)
clf
%   Clear command window.
%   (清除命令窗口)
clc
%   Define a variable z by the function peaks.
%   (借助 peaks 函数定义变量 z)
z = peaks;
%   Plot the surface of peaks by the function surf.
%   (借助 surf 函数绘制山峰的曲面)
surf(z);
%   Set axis as tight.
%   (设置 axis 属性为 tight)
axis tight
%   Set the nextplot of the current axis as replacechildren.
%   (设置当前坐标系的 nextplot 属性为 replacechildren)
set(gca,'nextplot','replacechildren');
%   Write the movie into F.
%   (将电影写入 F)
for j = 1:20
    surf(sin(2*pi*j/20)*z,z);
    F(j) = getframe;
end
%   Play the movie ten times.
%   (放映该电影 10 次)
movie(F,10)
```

4.5 应用实例

4.5.1 布朗运动

【例 4.30】 布朗运动。

EXAMP04030

```
%  Script file: EXAMP04030.m
%  This programe in MATLAB models Brown's motion.
%  (本程序模拟布朗运动)
%  Clear variables in workspace.
%  (清除工作空间中的变量)
clear
%  Clear the current figure window.
%  (清除当前图形窗口)
clf
%  Clear command window.
%  (清除命令窗口)
clc
%  Define the number of moving points n=100.
%  (定义运动的点的个数为n=100)
n=100;
%  Define the velocity of the points s=0.005.
%  (定义点的速度为s=0.005)
s=0.005;
%  Define the values of x-positions of the points.
%  (定义所有点的x坐标值)
x=rand(n,1)-0.5;
%  Define the values of y-positions of the points.
%  (定义所有点的y坐标值)
y=rand(n,1)-0.5;
%  Plot all the points.
%  (绘制所有点)
h=plot(x,y,'o');
%  Determine the range of axes.
%  (确定坐标轴范围)
axis([-1 1 -1 1]);
%  Set axes as square.
%  (设置axes属性为square)
axis square;
%  Do not show the background grid.
%  (不显示背景网格线)
grid off
%  Set some properties of the points.
%  (设置点的一些属性)
set(h,'EraseMode','Xor','MarkerSize',5,'MarkerFaceColor',…
'r','MarkerEdgeColor','r');
%  Let these points move arbitrarily.
%  (让这些点任意运动)
```

```
for i=linspace(1,10,5000)
    % Complete pending drawing events.
    % (重新绘点,同时擦除原先点)
    drawnow
    % Redefine the positions (x,y) of the points.
    % (重新定义所有点的坐标(x, y))
        x=x+s*randn(n,1);
    y=y+s*randn(n,1);
    % Take the newly defiend values to x and y.
    % (将新定义的值赋予 x 和 y)
        set(h,'XData',x,'YData',y);
end
```

布朗运动的运行结果如图 4.31 所示。

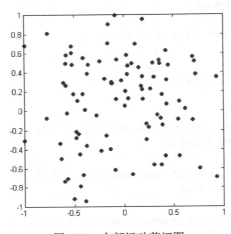

图 4.31 布朗运动剪切图

注释:
函数 drawnow 回调绘图数据完成图形窗口更新。
语法格式如下: drawnow
　　　　　　　drawnow expose
　　　　　　　drawnow update

4.5.2 相干波

【例 4.31】 相干波。

现有两列波函数表述如下:

$$\begin{cases} z_1 = \dfrac{A_0}{r_1}\cos[\omega(t-\dfrac{r_1}{v})] \\ z_2 = \dfrac{A_0}{r_2}\cos[\omega(t-\dfrac{r_2}{v})] \end{cases}$$

式中　$r_1 = \sqrt{(x-a)^2 + y^2}$;
　　　$r_2 = \sqrt{(x+a)^2 + y^2}$。

x 和 y 是 $z=0$ 平面上的点坐标。当 $A_0 = 0.5\text{m}$、$\omega = 6\text{rad/s}$、$v = 1.5\text{m/s}$ 和 $a = 5\text{m}$ 时，请借助 MATLAB 软件仿真绘制这两列波的干涉图像。

EXAMP04031

```
% Script file: EXAMP04031.m
% This programe in MATLAB models two coherent waves.
% (本程序模拟两列相干波)
% Clear variables in workspace.
% (清除工作空间中的变量)
clear
% Clear the current figure window.
% (清除当前图形窗口)
clf
% Clear command window.
% (清除命令窗口)
clc
% Initialize variables A0, w, v, and a.
% (初始化变量 A0、w、v 和 a)
A0=0.3;
w=4;
v=1;
a=4;
% Define two matrices of x and y by meshgrid.
% (通过 meshgrid 函数定义两个矩阵 x 和 y)
[x,y]=meshgrid([-4*pi:pi/20:4*pi],[-4*pi:pi/20:4*pi]);
% Define r1 and r2 depending on x, y and a.
% (定义 r1 和 r2 为 x、y 和 a 的函数)
r1=sqrt((x-a).^2+y.^2);
r2=sqrt((x+a).^2+y.^2);
% Value time t.
% (赋值时间 t)
t=0:0.1:10;
% For structure.
% (for 结构)
for ii=1:41
    % Define z depending on r1, t, v, A0, r2 and w.
    % (定义 z 为 r1、t、v、A0、r2 和 w 的函数)
    z=A0./r1.*cos(w.*(t(ii)-r1./v))+A0./r2.*cos(w.*(t(ii)-r2./v));
    % Plot the expectant surface.
    % (绘制预期曲面)
    surf(x,y,z);
    % Set colormap as summer.
    % (设置 colormap 属性为 summer)
    colormap(summer);
    % Set axis as tight.
    % (设置 axis 属性为 tight)
    axis tight;
    % Set shading as interp.
    % (设置 shading 属性为 interp)
    shading interp;
```

```
        %  Set viewpoint as [180, 90].
        %  (设置视角为[180, 90])
           view([180,90]);
        %  Set the range of axes as [-15, 15, -15, 15, -0.1, 0.1].
        %  (设置坐标轴范围为[-15, 15, -15, 15, -0.1, 0.1])
           axis([-15,15,-15,15,-0.1,0.1]);
        %  Turn off the axes.
        %  (关闭坐标系)
           axis off
        %  Write down the frame of movie.
        %  (记录电影帧的信息)
              m(ii)=getframe;
        end
        %  Play the movie.
        %  (放映电影)
        movie(m,5)
```

运行结果似动画,两列相干波最终停止画面如图 4.32 所示。

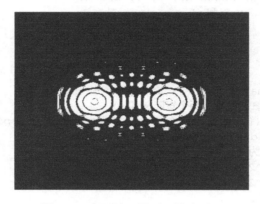

图 4.32 相干波(最后一帧动画)

4.5.3 带洞的峰面

【例 4.32】 带洞的峰面。

绘制一个带洞的峰面。

EXAMP04032

```
        %  Scipt file: EXAMP04032.m
        %  This program in MATLAB plots a surface with a hole.
        %  (本程序绘制带洞的峰面)
        %  Clear variables in workspace.
        %  (清除工作空间中的变量)
        clear
        %  Clear the current figure window.
        %  (清除当前图形窗口)
        clf
        %  Clear command window.
        %  (清除命令窗口)
```

```
clc
%  Define a matrix x by peaks.
%  (借助 peaks 函数定义一个矩阵 x)
x=peaks(50);
%  Define a hole in the peaks with NaN (not a number).
%  (借助 NaN(Not a number)在山峰上定义一个洞)
x(20:30,15:25)=NaN;
%  Plot the surface with a hole.
%  (绘制带洞的山峰曲面)
surf(x)
%  Set colormap as cool.
%  (设置 colormap 属性为 cool)
colormap(cool);
%  Set shading as interp.
%  (设置 shading 属性为 interp)
shading interp
```

运行结果如图 4.33 所示。

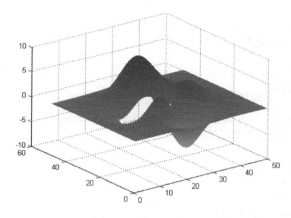

图 4.33　带洞的峰面

4.5.4　透视图

【例 4.33】　透视图。

请绘制二重透视的球。

EXAMP04033

```
%  Scipt file: EXAMP04033.m
%  This programe in MATLAB plots nested sphere.
%  (本程序绘制球的二重透视图)
%  Clear variables in workspace.
%  (清除工作空间中的变量)
clear
%  Clear the current figure window.
%  (清除当前图形窗口)
clf
%  Clear command window.
```

```
%   (清除命令窗口)
clc
%   Define matrices x, y and z by sphere.
%   (通过 sphere 函数定义矩阵 x、y 和 z)
[x,y,z]=sphere(35);
%   Define new two sets of x1, y1, z1 and x2, y2, z2.
%   (定义两套值,即 x1、y1、z1 和 x2、y2、z2)
x1=15*x;
y1=15*y;
z1=15*z;
x2=20*x;
y2=20*y;
z2=20*z;
%   Plot the smallest sphere.
%   (绘制最小的球)
surf(10*x,10*y,10*z);
%   Set shading as interp.
%   (设置 shading 属性为 interp)
shading interp
%   Set material as metal.
%   (设置 material 属性为 metal)
material metal
%   Create a light source.
%   (创建一个光源)
light
%   Keep ploting on the current figure window.
%   (将绘图保留在当前图形窗口)
hold on
%   Plot the second sphere.
%   (绘制第二个球)
mesh(x1,y1,z1);
%   Set colormap as hsv.
%   (设置 colormap 属性为 hsv)
colormap hsv
%   Plot the biggest shpere.
%   (绘制最大的球)
mesh(x2,y2,z2);
%   Free the present figure.
%   (释放当前窗口)
hold off
%   Set hidden as off.
%   (设置 hidden 属性为 off)
hidden off
%   Set axis as square.
%   (设置 axis 属性为 square)
axis square
%   Hide the axes.
%   (隐藏坐标系)
axis off
```

运行结果如图 4.34 所示。

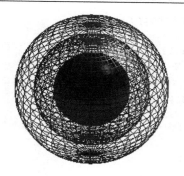

图 4.34 球的二重透视图

【例 4.34】 带有缺失网格的球的透视图。
请绘制二重透视并且缺失部分透视网格。

EXAMP04034

```
%   Scipt file: EXAMP04034.m
%   This programe in MATLAB plots nested sphere and meshes with holes.
%   (本程序绘制二重透视的球面，并且缺失部分透视网格)
%   Clear variables in workspace.
%   (清除工作空间中的变量)
clear
%   Clear the current figure window.
%   (清除当前图形窗口)
clf
%   Clear command window.
%   (清除命令窗口)
clc
%   Define matrices x, y and z by peaks.
%   (借助 peaks 函数定义矩阵 x、y 和 z)
[x,y,z]=sphere(50);
%   Redefine a new matrix z1 as z.
%   (重新定义一个新矩阵 z1 等于 z)
z1=z;
%   Define some elements as NaN.
%   (定义一些元素为 NaN)
z1(:,3:13)=NaN;
%   Plot the smallest sphere by surf.
%   (用 surf 函数绘制最小的球)
surf(1.5*x,1.5*y,1.5*z)
%   Label x-axis as 'x'.
%   (标注 x 轴为 x)
xlabel('x');
%   Label y-axis as 'y'.
%   (标注 y 轴为'y')
ylabel('y');
%   Label z-axis as 'z'.
%   (标注 z 轴为'z')
zlabel('z');
%   Set shading as interp.
```

```
%   (设置shading属性为interp)
shading interp
%   Set material as dull.
%   (设置material属性为dull)
material dull
%   Create a lighting source.
%   (创建一个光源)
light;
%   Set hold as on.
%   (设置hold属性为on)
hold on
%   Plot the second and third spherical surfaces by mesh.
%   (用mesh函数绘制第二个和第三个球面)
mesh(2*x,2*y,2*z1);
mesh(3*x,3*y,3*z1);
%   Set hidden as off.
%   (设置hidden属性为off)
hidden off
%   Set axis as equal.
%   (设置axis属性为equal)
axis equal
%   Set hold as off.
%   (设置hold属性为off)
hold off
%   Set axis as off.
%   (设置axis属性为off)
axis off
```

运行结果如图4.35所示。

图 4.35 带有缺失网格的球的透视图

4.5.5 能流图

【例 4.35】 圆形波导中的能流图。

圆形波导中能流密度平均值的数学表达式如下:

$$\overline{S}_{TE} = \frac{\beta\mu\omega H_0^2}{2k_c^2}\left[\frac{m^2}{k_c^2\rho^2}\left(J_m(k_c\rho)\right)^2\sin^2 m\varphi + \left(J_m'(k_c\rho)\right)^2\cos^2 m\varphi\right]e^{-i2\beta z}\boldsymbol{e}_z$$

$$\overline{S}_{TM} = \frac{\beta\varepsilon\omega}{2k_c^2}E_0^2\left[\frac{m^2}{k_c^2\rho^2}\left(J_m(k_c\rho)\right)^2\sin^2 m\varphi + \left(J_m'(k_c\rho)\right)^2\cos^2 m\varphi\right]e^{-i2\beta z}\boldsymbol{e}_z$$

请借助 MATLAB 软件完成图形绘制。

EXAMP04035

```
% Scipt file: EXAMP04035.m
% This programe in MATLAB plots figures of energy flow in circular
% waveguide.
% (本程序绘制圆形波导中的能流图)
%
% Records of revision
% Date: 2015-07-31
% Programmer: Wang Yonglong and Xia Changlong
% Description of changer: revised code
%
% Clear variables in workspace.
% (清除工作空间中的变量)
clear
% Clear the current figure window.
% (清除当前图形窗口)
clf
% Clear command window.
% (清除命令窗口)
clc
% Define two variables z and phi by linspace.
% (借助 linspace 函数定义两个变量记为 z 和 phi)
z=linspace(eps,1+eps,100);
phi=linspace(eps,2*pi+eps,100);
% Denote the size of z as a.
% (将 z 变量的尺寸表示为 a)
a=size(z);
% Denote the size of phi as b.
% (将 phi 变量的尺寸表示为 b)
b=size(phi);
% Create a pure zero matrix zz.
% (创建一个全零矩阵，记为 zz)
zz=ones(a(2),b(2));
% Create a new matrix phii which is the same as zz.
% (创建一个和 zz 完全一样的新矩阵 phii)
phii=zz;
% Revalue zz and phii by z and phi, respectively.
% (分别借助 z 和 phi 重新对 zz 和 phii 赋值)
for i=1:b(2)
    zz(:,i)=z;
end
for i=1:a(2)
    phii(i,:)=phi;
end
% Define two variables x and y depending on zz and phii.
% (定义两个变量 x 和 y 为 zz 和 phii 的函数)
x=zz.*cos(phii);
y=zz.*sin(phii);
```

```matlab
%[X,Y]=cart2plot(zz,phii)
% Value a variable m as 2.
% (将一个变量m赋值为2)
m=2;
% Offer the special values for Bessel funcitons.
% (给出贝塞尔函数的特定值)
mumn=[3.054,6.705,9.965,13.170];
% Determine to chose which element in mumn.
% (决定选择mumn中哪个元素)
n=1;
kc=mumn(n);
% Define a function y1 depending on Bessel function.
% (定义一个函数y1为贝塞尔函数的函数)
y1=0.5*(besselj(m-1,kc*z)-besselj(m+1,kc*z));
% Define a function yy1 as square y1.
% (定义yy1函数为y1的平方)
yy1=y1.^2;
% Define two functions y2 and y3.
% (定义两个函数y2和y3)
y2=(cos(m*phi)).^2;
y3=besselj(m,kc*z);
% Define function yy3 depending on y3.
% (定义yy3为y3的函数)
yy3=y3.^2./(kc*z).^2;
% Define two functions y4 and y5.
% (定义两个函数y4和y5)
y4=(sin(m*phi)).^2;
y5=(yy1'*y2)+m^2*(yy3'*y4);
% Open a subplot figure window.
% (打开一个子绘图图形窗口)
subplot(2,4,2*n-1)
% Plot the surface of y5.
% (绘制y5的曲面)
surf(x,y,y5);
% Set shading as interp.
% (设置shading属性为interp)
shading interp;
% Title the figure as (a).
% (将该图命名为(a))
title('(a)')
% Set axis as square.
% (设置axis属性为square)
axis square;
%
% Open a new subplot figure window.
% (打开一个新子绘图图形窗口)
subplot(2,4,2*n)
% Plot contours.
% (绘制等值线)
contour(x,y,y5)
% Title the figure as (a').
```

```
%   (将该图命名为(a'))
title('(a\prime)')
%   Set axis as square.
%   (设置axis属性为square)
axis square
%   Determine to chose which element in mumn.
%   (确定选择mumn中的元素)
n=2;
kc=mumn(n);
%   Redefine y1, yy1, y2, y3, yy3, y4, y5 with respect to new kc.
%   (对应新kc重新定义y1、yy1、y2、y3、yy3、y4、y5)
y1=0.5*(besselj(m-1,kc*z)-besselj(m+1,kc*z));
yy1=y1.^2;
y2=(cos(m*phi)).^2;
y3=besselj(m,kc*z);
yy3=y3.^2./(kc*z).^2;
y4=(sin(m*phi)).^2;
y5=(yy1'*y2)+m^2*(yy3'*y4);
%   Open a new subplot figure window.
%   (打开一个新子绘图图形窗口)
subplot(2,4,2*n-1)
%   Plot the surface.
%   (绘制曲面)
surf(x,y,y5);
%   Set shading as interp.
%   (设置shading属性为interp)
shading interp;
%   Title the figure as (b).
%   (将该图命名为(b))
title('(b)')
%   Set axis as square.
%   (设置axis属性为square)
axis square;
%
%   Open a new subplot figure window.
%   (打开一个新子绘图图形窗口)
subplot(2,4,2*n)
%   Plot contours.
%   (绘制等值线)
contour(x,y,y5)
%   Title the figure as (b').
%   (将该图命名为(b'))
title('(b\prime)')
%   Set axis as square.
%   (设置axis属性为square)
axis square
%   Determine to chose which element in mumn.
%   (确定选择mumn中的元素)
n=3;
kc=mumn(n);
%   Redefine y1, yy1, y2, y3, yy3, y4, y5 with respect to new kc.
```

```matlab
%    (对应新 kc 重新定义 y1、yy1、y2、y3、yy3、y4、y5)
y1=0.5*(besselj(m-1,kc*z)-besselj(m+1,kc*z));
yy1=y1.^2;
y2=(cos(m*phi)).^2;
y3=besselj(m,kc*z);
yy3=y3.^2./(kc*z).^2;
y4=(sin(m*phi)).^2;
y5=(yy1'*y2)+m^2*(yy3'*y4);
%   Open a new subplot figure window.
%   (打开一个子绘图图形窗口)
subplot(2,4,2*n-1)
%   Plot the surface.
%   (绘制曲面)
surf(x,y,y5);
%   Set shading as interp.
%   (设置 shading 属性为 interp)
shading interp;
%   Title the figure as (c).
%   (将该图命名为(c)).
title('(c)')
%   Set axis as square.
%   (设置 axis 属性为 square)
axis square;
%
%   Open a new subplot figure window.
%   (打开一个新子绘图图形窗口)
subplot(2,4,2*n)
%   Plot contours.
%   (绘制等值线)
contour(x,y,y5)
%   Title the figure as (c').
%   (将该图命名为(c'))
title('(c\prime)')
%   Set axis as square.
%   (设置 axis 属性为 square)
axis square
%
%   Determine to chose which element in mumn.
%   (确定选择 mumn 中的元素)
n=4;
kc=mumn(n);
%   Redefine y1, yy1, y2, y3, yy3, y4, y5 with respect to new kc.
%   (对应新 kc 重新定义 y1、yy1、y2、y3、yy3、y4、y5)
y1=0.5*(besselj(m-1,kc*z)-besselj(m+1,kc*z));
yy1=y1.^2;
y2=(cos(m*phi)).^2;
y3=besselj(m,kc*z);
yy3=y3.^2./(kc*z).^2;
y4=(sin(m*phi)).^2;
y5=(yy1'*y2)+m^2*(yy3'*y4);
%   Open a new subplot figure window.
```

```
%   (打开一个新子绘图图形窗口)
subplot(2,4,2*n-1)
%   Plot the surface.
%   (绘制曲面)
surf(x,y,y5);
%   Set shading as interp.
%   (设置 shading 属性为 interp)
shading interp;
%   Set colormap as winter.
%   (设置 colormap 属性为 winter)
colormap winter;
%   Title the figure as (d).
%   (将该图命名为(d))
title('(d)')
%   Set axis as square.
%   (设置 axis 属性为 square)
axis square;
%
%   Open a new subplot figure window.
%   (打开一个新子绘图图形窗口)
subplot(2,4,2*n)
%   Plot contours.
%   (绘制等值线)
contour(x,y,y5)
%   Title the figure as (d').
%   (将该图命名为(d'))
title('(d\prime)')
%   Set axis as square.
%   (设置 axis 属性为 square)
axis square
%
%
%   Value the elements of mumn for Bessel's function.
%   (对贝塞尔函数的 mumn 元素进行赋值)
mumn=[7.016,5.332,6.705,8.015,9.282,10.520];
%   Open a new figure window.
%   (打开一个新图形窗口)
figure
%   Set initial values for m and n.
%   (设定 m 和 n 的初始值)
m=0; n=1;
%   Take the nth element of mumn as kc.
%   (将 mumn 的第 n 个元素设置为 kc)
kc=mumn(n);
%   Define a function y1 of Bessel's function.
%   (定义一个函数 y1 为贝塞尔函数的函数)
y1=0.5*(besselj(m-1,kc*z)-besselj(m+1,kc*z));
%   Define yy1, y2, y3, yy3, y4, y5 functions.
%   (定义了 yy1、y2、y3、yy3、y4、y5 函数)
yy1=y1.^2;
y2=(cos(m*phi)).^2;
```

```
y3=besselj(m,kc*z);
yy3=y3.^2./(kc*z).^2;
y4=(sin(m*phi)).^2;
y5=(yy1'*y2)+m^2*(yy3'*y4);
% Open a new subplot figure window.
% (打开一个新子绘图图形窗口)
subplot(3,4,2*n-1)
% Plot the surface.
% (绘制曲面图形)
surf(x,y,y5);
% Set shading as interp.
% (设置shading属性为interp)
shading interp;
% Set axis as square.
% (设置axis属性为square)
axis square;
% Title the figure as (a).
% (将该图命名为(a))
title('(a)')
% Open a new subplot figure window.
% (打开一个新子绘图图形窗口)
subplot(3,4,2*n)
% Plot contours.
% (绘制等值线)
contour(x,y,y5)
% Title the figure as (a').
% (将该图命名为(a'))
title('(a\prime)')
% Set axis as square
% (设置axis属性为square).
axis square
%
% Revalue m and n.
% (重新对m和n进行赋值)
n=2; m=1;
% Take the nth element of mumn as kc.
% (将mumn的第n个元素设置为kc)
kc=mumn(n);
% Define y1, yy1, y2, y3, yy3, y4, y5 functions.
% (定义y1、yy1、y2、y3、yy3、y4、y5函数)
y1=0.5*(besselj(m-1,kc*z)-besselj(m+1,kc*z));
yy1=y1.^2;
y2=(cos(m*phi)).^2;
y3=besselj(m,kc*z);
yy3=y3.^2./(kc*z).^2;
y4=(sin(m*phi)).^2;
y5=(yy1'*y2)+m^2*(yy3'*y4);
% Open a new subplot figure window.
% (打开一个新子绘图图形窗口)
subplot(3,4,2*n-1)
% Plot the surface.
```

```
%   (绘制曲面)
surf(x,y,y5);
%   Set shading as interp.
%   (设置shading属性为interp)
shading interp;
%   Set axis as square.
%   (设置axis属性为square).
axis square;
%   Title the figure as (b).
%   (将该图命名为(b))
title('(b)')
%
%   Open a new subplot figure window.
%   (打开一个新子绘图图形窗口)
subplot(3,4,2*n)
%   Plot contours.
%   (绘制等值线)
contour(x,y,y5)
%   Title the figure as (b').
%   (将该图命名为(b'))
title('(b\prime)')
%   Set axis as square.
%   (设置axis属性为square)
axis square
%
%   Revalue m and n.
%   (重新对m和n进行赋值)
n=3; m=2;
%   Take the nth element of mumn as kc.
%   (将mumn的第n个元素设置为kc)
kc=mumn(n);
%   Redefine y1, yy1, y2, y3, yy3, y4, y5 functions.
%   (重新定义y1、yy1、y2、y3、yy3、y4、y5函数)
y1=0.5*(besselj(m-1,kc*z)-besselj(m+1,kc*z));
yy1=y1.^2;
y2=(cos(m*phi)).^2;
y3=besselj(m,kc*z);
yy3=y3.^2./(kc*z).^2;
y4=(sin(m*phi)).^2;
y5=(yy1'*y2)+m^2*(yy3'*y4);
%   Open a new subplot figure window.
%   (打开一个新子绘图图形窗口)
subplot(3,4,2*n-1)
%   Plot the surface.
%   (绘制曲面)
surf(x,y,y5);
%   Set shading as interp.
%   (设置shading属性为interp)
shading interp;
%   Title the figure as (c).
%   (将该图命名为(c))
```

```matlab
title('(c)')
% Set axis as square.
%   (设置axis属性为square)
axis square;
%
% Open a new subplot figure window.
%   (打开一个新子绘图图形窗口)
subplot(3,4,2*n)
% Plot contours.
%   (绘制等值线)
contour(x,y,y5)
% Title the figure as (c').
%   (将该图命名为(c'))
title('(c\prime)')
% Set axis as square.
%   (设置axis属性为square)
axis square
%
% Revalue m and n.
%   (重新对m和n进行赋值)
n=4; m=3;
% Take the nth element of mumn as kc.
%   (将mumn的第n个元素设置为kc)
kc=mumn(n);
% Redefine y1, yy1, y2, y3, yy3, y4, y5 functions.
%   (重新定义y1、yy1、y2、y3、yy3、y4、y5函数)
y1=0.5*(besselj(m-1,kc*z)-besselj(m+1,kc*z));
yy1=y1.^2;
y2=(cos(m*phi)).^2;
y3=besselj(m,kc*z);
yy3=y3.^2./(kc*z).^2;
y4=(sin(m*phi)).^2;
y5=(yy1'*y2)+m^2*(yy3'*y4);
% Open a new subplot figure window.
%   (打开一个新子绘图图形窗口)
subplot(3,4,2*n-1)
% Plot the surface.
%   (绘制曲面)
surf(x,y,y5);
% Set shading as interp.
%   (设置shading属性为interp)
shading interp;
% Title the figure as (d).
%   (将该图命名为(d))
title('(d)')
% Set axis as square.
%   (设置axis属性为square)
axis square;
%
% Open a new subplot figure window.
%   (打开一个新子绘图图形窗口)
```

```
subplot(3,4,2*n)
% Plot contours.
%  (绘制等值线)
contour(x,y,y5)
% Title the figure as (d').
%  (将该图命名为(d'))
title('(d\prime)')
% Set axis as square.
%  (设置 axis 属性为 square)
axis square
%
% Revalue m and n.
%  (重新对 m 和 n 进行赋值)
n=5; m=4;
% Take the nth element of mumn as kc.
%  (将 mumn 的第 n 个元素设置为 kc)
kc=mumn(n);
% Redefine y1, yy1, y2, y3, yy3, y4, y5 functions.
%  (重新定义 y1、yy1、y2、y3、yy3、y4、y5 函数)
y1=0.5*(besselj(m-1,kc*z)-besselj(m+1,kc*z));
yy1=y1.^2;
y2=(cos(m*phi)).^2;
y3=besselj(m,kc*z);
yy3=y3.^2./(kc*z).^2;
y4=(sin(m*phi)).^2;
y5=(yy1'*y2)+m^2*(yy3'*y4);
% Open a new subplot figure window.
%  (打开一个新子绘图图形窗口)
subplot(3,4,2*n-1)
% Plot the surface.
%  (绘制曲面)
surf(x,y,y5);
% Set shading as interp.
%  (设置 shading 属性为 interp)
shading interp;
% Title the figure as (e).
%  (将该图命名为(e))
title('(e)')
% Set axis as square.
%  (设置 axis 属性为 square)
axis square;
%
% Open a new subplot figure window.
%  (打开一个新子绘图图形窗口)
subplot(3,4,2*n)
% Plot contours.
%  (绘制等值线)
contour(x,y,y5)
% Title the figure as (e').
%  (将该图命名为(e'))
title('(e\prime)')
```

```matlab
%   Set axis as square.
%   (设置axis属性为square)
axis square
%
%   Revalue m and n.
%   (重新对m和n进行赋值)
n=6; m=5;
%   Take the nth element of mumn as kc.
%   (将mumn的第n个元素设置为kc)
kc=mumn(n);
%   Redefine y1, yy1, y2, y3, yy3, y4, y5 functions.
%   (重新定义y1、yy1、y2、y3、yy3、y4、y5函数)
y1=0.5*(besselj(m-1,kc*z)-besselj(m+1,kc*z));
yy1=y1.^2;
y2=(cos(m*phi)).^2;
y3=besselj(m,kc*z);
yy3=y3.^2./(kc*z).^2;
y4=(sin(m*phi)).^2;
y5=(yy1'*y2)+m^2*(yy3'*y4);
%   Open a new subplot figure window.
%   (打开一个新子绘图图形窗口)
subplot(3,4,2*n-1)
%   Plot the surface.
%   (绘制曲面)
surf(x,y,y5);
%   Set shading as interp.
%   (设置shading属性为interp)
shading interp;
%   Title the figure as (f).
%   (将该图命名为(f))
title('(f)')
%   Set colormap as winter.
%   (设置colormap属性为winter)
colormap winter;
%   Set axis as square.
%   (设置axis属性为square)
axis square;
%
%   Open a new subplot figure window.
%   (打开一个新子绘图图形窗口)
subplot(3,4,2*n)
%   Plot contours.
%   (绘制等值线)
contour(x,y,y5)
%   Title the figure as (f').
%   (将该图命名为(f'))
title('(f\prime)')
%   Set axis as square.
%   (设置axis属性为square)
axis square
```

运行结果如图4.36和图4.37所示。本示例只是给读者一些提示，MATLAB可以完成一些很难描述的数学表达式的图形化工作,值得深入研究,比如北京师范大学的彭芳麟老师已有较深入研究。

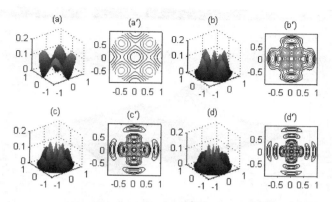

图 4.36 电磁波 $TE_{21,22,23,24}$ 四种模式的能流立体图与剖面图

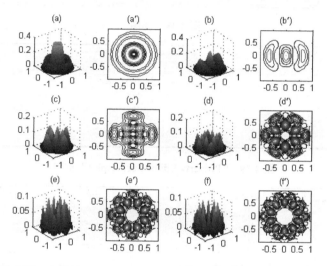

图 4.37 电磁波 $TE_{02,12,22,32,42,52}$ 六种模式的能流立体图与剖面图

4.6 鼠标对图形的操作

MATLAB 软件为用户提供了可借助鼠标设置图形的入口。如图 4.38 所示,可以选中快捷菜单中的 图标单击,然后将鼠标移动到图形窗口,双击任何部位,将会打开该部分的属性设置窗口,用户可以根据需要对其属性进行设置。

如果鼠标双击绘图窗口中的绘制图形的网格线,将会打开如图 4.39 所示的"Property Editor-Surfaceplot"窗口。

在图 4.39 中,左边模块包含"X Data Source:"、"Y Data Source:"、"Z Data Source:"和"Color:"选项,一般不进行设置。中间模块包含"Faces:"、"Edges:"和"Markers:"选项,可以进行选择。其中,"Faces:"下拉颜料桶可以选择填充每个小平面元的颜色或者选择具有特定阴影效果的填充色;"Edges:"可以选择线型、线宽、线的颜色;"Markers:"可选择符号点的形状、符号点的尺寸、符号填充颜色、符号边界颜色。不仅仅可以对所绘制图形的属性可进行更改设置，也可以对坐标系的属性进行重新设置。将鼠标移到坐标系的任意位置，双击其坐标线会打开如图 4.40 所示的窗口。

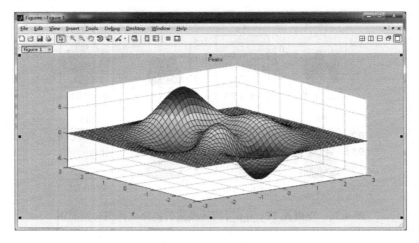

图 4.38　样图

图 4.39　"Property Editor-Surfaceplot"窗口

图 4.40　坐标系属性窗口

"More Properties…"按钮可以打开所有属性设置的对话窗口，可以进行更丰富的属性设置，如图 4.41 所示。有很多选项可以更新，比如"Title"可输入新标题；"Colors"可选择背景填充色、坐标轴线的颜色；"Grid"可对 x、y、z 轴各个方向网格线进行选择；还有对每个坐标轴也可进行任意设置，如"x Axis"，可设置 x 轴的标签、标度数值、坐标轴范围、是线性关系还是对数关系，还有倒序选项可让相应坐标轴由从小到大变为从大到小；"Ticks"按钮可以设置坐标轴的刻度标注。

将鼠标移动到没有图形，也没有线的背景区双击鼠标，会打开如图 4.42 所示的窗口。可以对绘制图形面元填充色进行设置，也可对坐标空间的背景色进行设置，还可以输入图形的名字，选择是否显示图的数量等。

第 4 章 绘 图

 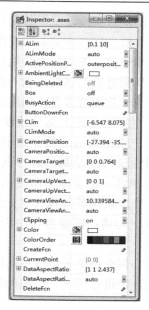

图 4.41 属性检测窗口

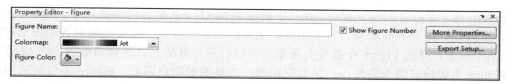

图 4.42 图形窗口的属性窗口

4.7 图形句柄

图形对象是借助 MATLAB 软件展示数据的绘图基元。绘图对象的每一个组成部分都有一个独具特色的句柄，可以借助图形句柄，通过 set 和 get 函数实现对每一基元属性设置。当然，这些属性也可在绘制图形过程中进行设定。这些图形对象的所有组员都有一定的等级层次，如图 4.43 所示。

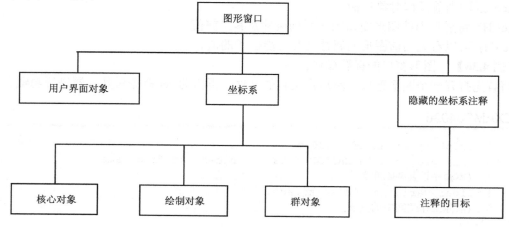

图 4.43 图形句柄的层次

图形句柄的层次实质是基于各图形对象的相关性完成的。举例说明，当在 MATLAB 中画一条线时，需要坐标系为这条线提供方向和参考坐标。接下来，坐标系需要一个图形窗口来显示坐标系和子图。图形所有常用部分基本都包含在坐标系类中。在通常情况下，我们不需要设置用户界面对象的属性，也不需要设置注释对象的属性。

绘制的对象一般由许多基本组成部分构成，但是它们有一些属性可以在成图后再进行设置。其中核心对象包含绝大部分绘制图形的描述元素，一般也涵盖坐标系的背景填充等。绘制对象是用户借助高级函数在图形窗口绘制的图像，当然也是整个图形窗口的重要组成部分，它的属性也可在绘制后进行重新设置。群对象通常可用来创建含多个对象的群，然后这些对象能够对一定操作做出相同的反应动作。当然也可通过将坐标系中任意子项加入一个群对象，然后实现整体操作（光除外），这其中也可包含其他群的对象。这段描述非常复杂，概括成一句话：图像窗口中所有构成组元，都可以借助图形句柄对其进行属性修改和更新等设置。

本节，我们仅介绍关于坐标系的图形句柄。用户界面交互对象的图形句柄将在第 6 章图形用户界面设计中进行阐述。隐藏的坐标系注释子部分将不进行介绍。

4.7.1 图形窗口

Figure 是一个打开的独立窗口，在其中显示着 MATLAB 软件绘制的图形。Figure 含有菜单栏、工具栏、用户界面对象、背景菜单、坐标轴和子坐标轴，以及其他的各类图形对象。MATLAB 对于打开的 Figure 个数没有限制，可以打开计算机允许的窗口数量。在 MATLAB 中，Figure 主要扮演两个角色：一是数据图形；二是图形用户界面。Figure 不仅展示用户绘制的图形，而且为用户操作绘制图形提供方便。换句话说，Figure 可以同时包含图形和图形用户界面两个部分。并且还可以借助"hold on"执行语句在同一个窗口绘制多个图形。

语法格式：figure 可以打开一个新图形窗口，并返回到它的图形句柄下。

figure(H) 将句柄为 H 的图形窗口设置为当前窗口，让该窗口置于所有窗口之上成为可视窗口。如果句柄为 H 的窗口不存在，将返回一个整数，同时创建一个句柄为 H 的新窗口，并打开。

gcf 返回当前图形窗口的句柄。

gca 返回当前坐标系句柄。

gco 返回当前操作对象句柄。

get(H) 将显示当前图形窗口的一系列属性及其当前值。

set(H) 可以看见当前图形的属性及其可能赋予的值。

【例 4.36】 图形窗口的标题操作。

请首先打开一个图形窗口，然后定义它的文件，再借助 set 命令对其标题进行隐藏。

EXAMP04036

```
%  Scipt file: EXAMP04036.m
%  This programe in MATLAB sets and hides the file name.
%    (本程序设置并隐藏文件名)
%  Clear variables in workspace.
%    (清除工作空间中的变量)
clear
%  Clear command window.
```

```
%  (清除命令窗口)
clc
%  Open a figure window as f1.
%  (打开一个图形窗口存为f1)
f1=figure;
sphere
axis equal
%  Define the name of the figure window as "Example 1 for HGO".
%  (定义图形窗口的名字为"Example 1 for HGO")
set(f1,'FileName','Example 1 for HGO')
%  Set the state of Number Title as "off".
%  (设置Number Title的状态为"off")
set(f1,'NumberTitle','off')
```

运行结果如图 4.44 所示。

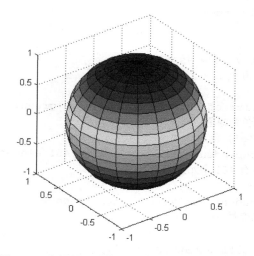

图 4.44　将标题"Example 1 for HGO"隐藏示例图

这仅仅是图形句柄的开始，并看不出什么优越性，下面通过对绘制图形元素进行属性设置看出优越性的实效。

4.7.2　核心对象

核心对象包括绘制基元，比如线、文本、多边形壳层（如 patch 函数绘制的）、特定对象表面（如 surf 函数绘制的）、坐标系的网格线、图像、光照，还有一些看不到但是会影响绘制对象视觉效果的，例如：

axes：坐标系对象定义显示图像的坐标系。

line：线是将所有定义数据连接起来的结果。

text：将文字显示在通过坐标指定的位置。

light：根据坐标系坐标定位方向光源（光将会影响片图和表面图）。

patch：填充多边形。一个片图可以含有多个面，每个面可以由不同的颜色填充，也可以由过渡颜色填充。

surface：基于(x, y)平面各坐标点对应的取值进行绘图，都是由四边图拼成曲面，每个面元是由相邻的四个函数值确定高度,由(x,y)数值确定平面位置的。这些面元都有一定的色彩。

image:二维矩阵图像,每个矩阵数值表示的是相应点的颜色。也可以认为是三维的数组,分别代表格点的 RGB 三基色。

【例 4.37】 对一条正弦曲线进行设置。

首先绘制一条正弦曲线,然后定义它的图形窗口属性,定义坐标、线型以及文本属性。

EXAMP04037

```
%   Scipt file: EXAMP04037.m
%   This programe in MATLAB sets some properties by handles.
%   (本程序通过句柄设置图形一些属性)
%   Clear variables in workspace.
%   (清除工作空间中的变量)
clear
%   Clear command window.
%   (清除命令窗口)
clc
%   Open a figure window as f1.
%   (打开图形窗口存为 f1)
f1=figure;
%   Define the color of interface window as green.
%   (定义界面窗口颜色为绿色)
set(f1,'Color','g')
%   Define an independent variable x with 461 elements.
%   (定义自变量 x 含有 461 个元素)
x=0:pi/230:2*pi;
%   Define a dependent variable y with respect to x.
%   (相应于 x 定义因变量 y)
y=sin(x);
%   Open an axes with a handle axes1.
%   (打开一个坐标系,句柄为 axes1)
axes1=axes;
%   Set the initial positions in Window and the width and the height
%   of the figure window.
%   (设置坐标系在图形窗口的起始位置、宽度和高度)
set(axes1,'position',[0.2,0.15,0.6,0.7])
%   Plot a sine curve denoted by line1.
%   (绘制一条正弦曲线,标记为 line1)
line1=plot(x,y);
%   Show a text at a certain position.
%   (在一个确定坐标位置显示一个文本)
t1=text(149*pi/230,sin(149*pi/230),'My God!');
%   Set hold as on.
%   (设置 hold 属性为 on)
hold on
%   Plot a small circle at the special position.
%   (在特定坐标位置绘制一个小圆)
plot(149*pi/230,sin(149*pi/230),'ro');
%   Set hold as off.
```

```
%（设置 hold 属性为 off）
hold off
% Title the figure as "Sine Curve" denoted by title1!
%（将该图命名为"Sine Curve",标注为 title1）
title1=title('Sine Curve');
% Define the color of the title as red.
%（定义该标题的颜色为红色）
set(title1,'Color','r');
% Set the fontweightof the title1 as bold, the fontsiez as 20.
%（设置 title1 的字体为黑体,字号为 20）
set(title1,'FontWeight','bold','FontSize',20)
% Define the xlabel and ylabel of axes1 as bold fontweight and 15 fontsize.
set(axes1,'FontWeight','bold','FontSize',15)
% Define the xtick of axes as [0, pi/2, pi, 3*pi/2, 2*pi].
%（定义 x 轴刻度为[0, pi/2, pi, 3*pi/2, 2*pi]）
set(axes1,'XTick',['0','pi/2','pi','3*pi/2','2*pi'])
% Define the color of x-axis as red.
%（定义 x 轴的颜色为红色）
set(axes1,'XColor',[1,0,0])
% Define the color of y-axis as green.
%（定义 y 轴的颜色为绿色）
set(axes1,'YColor',[0,0,1])
% Define the background color of axes as [0.5,1,1].
%（定义坐标系的背景颜色为[0.5, 1, 1]）
set(axes1,'Color',[0.5,1,1])
% Define the fontsize of the text t1 as 12.
%（定义文本 t1 的字号为 12）
set(t1,'FontSize',12)
% Define the linewidth of line1 as 2.
%（定义 line1 的线宽为 2）
set(line1,'LineWidth',2)
% Define the range of axes as [0,pi*2,-1,1].
%（定义坐标轴的范围为[0,pi*2,-1,1]）
axis([0,pi*2,-1,1])
```

运行结果如图 4.45 所示。

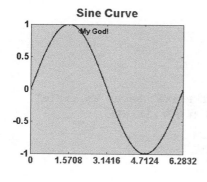
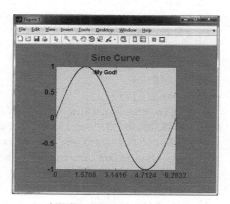

（a）保存的图形结果　　　　　　（b）抓图的图形结果

图 4.45　属性设置后的正弦曲线图

> **注意:**
> 定义坐标系属性的命令执行行必须放在核心图之后。如果搞错了顺序,将不会得到预期的效果图。例如:"set(axes1,'XColor',[1, 0, 0])"、"set(axes1,'YColor',[0, 0, 1])"、"set(axes1,'Color', [0.5, 1, 1])"、"set(axes1, 'XTick', [0, 3*pi/2])" 必须放在 "line1=plot(x, y);" 和 "plot(149*pi/230, sin(149*pi/230), 'ro');" 这两行之后。

【例 4.38】 设置面图属性。

本例在同一个窗口打开了 4 个子绘图窗口,每个子窗口存有不同图像。第一个窗口是峰面图,是借助函数 peaks 绘制,并且借助 "set" 命令对其坐标系 x、y、z 轴标签及图形标题进行设置的。第二个窗口显示一个特定图像,其属性被重新设置。第三个窗口图形由 "patch" 函数绘制。第四个窗口图像借助函数 "fill3" 绘制。

EXAMP04038

```
% Scipt file: EXAMP04038.m
% This programe in MATLAB sets some properties by handles.
% (本程序通过句柄设置图形一些属性)
% Clear variables in workspace.
% (清除工作空间中的变量)
clear
% Clear command window.
% (清除命令窗口)
clc
% Open a figure window saved as f1.
% (打开一个新图形窗口,保存为 f1)
f1=figure;
% Define the color of the interface of f1 as [0.9,1,1].
% (定义 f1 的界面颜色为[0.9, 1, 1])
set(f1,'Color',[0.9,1,1])
% Open a axes saved as axes1.
% (打开一个新坐标系保存为 axes1)
axes1=axes;
% Set the initial position, the width and the height for axes1.
% (设置 axes1 的起始位置、宽度和高度)
set(axes1,'Position',[0.1,0.5,0.4,0.4])
% Define a variable z by peaks.
% (借助 peaks 函数定义一个变量 z)
z=peaks(50);
% Plot the surface.
% (绘制曲面)
surf1=surf(z);
% Title the surface as "Surface of Peaks" saved as title1.
% (将该曲面命名为"Surface of Peaks",保存为 title1)
title1=title('Surface of Peaks');
% Set the color of title1 as blue.
% (设置 title1 的颜色为蓝色)
set(title1,'Color','b')
% Set fontweight as bold and fontsize as 15.
% (设置字体为黑体,字号为 15)
```

```
set(title1,'FontWeight','bold','FontSize',15)
%  Set shading as interp.
%  (设置 shading 属性为 interp)
shading interp
%  Set the background color of axes1 as [0.95,1,1].
%  (设置 axes1 背景颜色为[0.95, 1, 1])
set(axes1,'Color',[0.95,1,1])
%  Create a light at [5,0,2] with style being infinite.
%  (创建光源位置为[5, 0, 2]，模式为平行光)
light1=light('Position',[5,0,2],'Style','infinite');
%  Set material as shiny.
%  (设置 material 属性为 shiny)
material shiny
%  Appear xlabel as x.
x1=xlabel('x');
%  Label y-axis as y.
%  (标注 y 轴为 y)
y1=ylabel('y');
%  Label z-axis as z.
%  (标注 z 轴为 z)
z1=zlabel('z');
%  Set the fontsize of x-, y- and z-label as 12, color as red,
%  the fontweight as bold.
%  (设置 x, y 和 z 轴的字号为 12，字体为黑体)
set(x1,'FontSize',12,'Color','r','FontWeight','bold');
set(y1,'FontSize',12,'Color','r','FontWeight','bold');
set(z1,'FontSize',12,'Color','r','FontWeight','bold');
%  Set the color of x- and y-axis as red.
%  (设置 x 和 y 轴的颜色为红色)
set(axes1,'XColor','r','YColor','r','ZColor','r');
%  Set YTick and XTick as [0,20,40].
%  (设置 x 和 y 的刻度为[0, 20, 40])
set(axes1,'YTick',[0,20,40],'XTick',[0,20,40])
%
%  Open a new axes as axes2.
%  (打开一个新坐标系，记为 axes2)
axes2=axes;
%  Define the positon of axes2 position as [0.5,0.1,0.4,0.4].
%  (定义 axes2 坐标系的位置为[0.5,0.1,0.4,0.4])．
set(axes2,'Position',[0.5,0.1,0.4,0.4]);
%  Call an image from disc.
%  (从硬盘上调用一张图片)
A=imread('E:\MATLAB 修订版\书中实例程序\Waterlilies','JPG');
%  Show the loaded image.
%  (显示被上传的图片)
image2=image(A);
%  Title this picture as Water Lilies.
%  (将该图片命名为"Water Lilies")
title2=title('Water Lilies');
%  Set the color as green, the fontsize as 15, the fontweight as bold,
%  the background color as red for title2.
```

```
    % (设置title2的颜色为绿色,字号为15,字体为黑体,背景颜色为红色)
set(title2,'Color','g','FontSize',15,'FontWeight','bold','BackGround','r');
    % Set the color as blue for x- and y-axis.
    % (设置x和y轴的颜色为蓝色)
set(axes2,'XColor','b','YColor','b');
    % Set xtick as [0,200,400,600,800], ytick as [0,200,400,600].
    % (设置x的刻度为[0,200,400,600,800],y的刻度为[0,200,400,600])
set(axes2,'XTick',[0,200,400,600,800],'YTick',[0,200,400,600]);
    % Create a light source at position [5,0,0].
    % (在[5,0,0]处创建一个光源)
light2=light('Position',[5,0,0]);
    % Set material as shiny.
    % (设置material属性为shiny)
material shiny;
    %
    % Open a new axes again as axes3.
    % (再次打开一个新坐标系,存为axes3)
axes3=axes;
    % Set the position of axes3 as [0.6,0.65,0.3,0.3].
    % (设置axes3的位置为[0.6,0.65,0.3,0.3])
set(axes3,'Position',[0.6,0.65,0.3,0.3]);
    % Define some points with x-positions, y-positions and z-positions.
    % (借助x、y和z的坐标定义一些点)
x3=[0 0;0 1;1 1];
y3=[1 1;2 2;2 1];
z3=[1 1;1 1;1 1];
    % Define the color of the areas closed by the above points.
    % (定义被上面点包围区域的颜色)
tcolor(1,1,1:3)=[0 1 1];
tcolor(1,2,1:3)=[0 0 1];
    % Show the patch.
    % (显示这些片区)
patch(x3,y3,z3,tcolor)
    %
    % Open a new axes again as axes4.
    % (再次打开一个新坐标系,存为axes4)
axes4=axes;
    % Define the positions of axes4 as [0.1,0.1,0.3,0.3].
    % (定义axes4的位置为[0.1,0.1,0.3,0.3])
set(axes4,'Position',[0.1,0.1,0.3,0.3]);
    % Define points through x-positions, y-positions, and z-positions.
    % (通过x、y和z的坐标定义一些点)
x4=[0 1 1 2;1 1 2 2;0 0 1 1];
y4=[1 1 1 1;1 0 1 0;0 0 0 0];
z4=[1 1 1 1;1 0 1 0;0 0 0 0];
    % Define a vector used to descrive colors.
    % (定义一个用于描述颜色的矢量)
c4=[0.5000 0.8000 0.8000 0.5000;0.8000 0.5000 0.5000 0.1667;0.3330 0.3330 0.5000 0.5000];
    % Plot the filling picture with gradually changing color.
```

```
%  (绘制由渐变颜色填充的图形)
fill3(x4,y4,z4,c4)
%  Show color bar.
%  (显示色条)
colorbar
```

运行结果如图 4.46 所示。

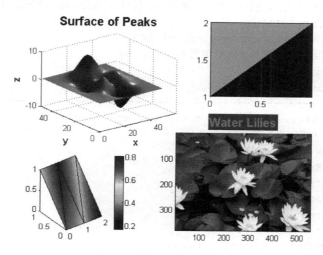

图 4.46 面图属性设置效果图

【例 4.39】 球谐函数图。

球谐函数如下：

$$Y_l^m(\theta,\varphi)=\sqrt{\frac{(l-|m|)!(2l+1)}{(l+|m|)!4\pi}}P_l^{|m|}(\cos\theta)\mathrm{e}^{\mathrm{i}m\varphi}$$

当 $l=6$、$m=2$ 时，绘制该球谐函数的图形。

EXAMP04039

```
%  Scipt file: EXAMP04039.m
%  This programe in MATLAB sets some properties by handles.
%  (本程序通过句柄设置图形一些属性)
%  Clear variables in workspace.
%  (清除工作空间中的变量)
clear
%  Clear command window.
%  (清除命令窗口)
clc
%  Define some necessary initial constants.
%  (定义一些必要的初始常数)
L=6;M=2;R=4;A=3;delta=pi/40;
theta0=0:delta:pi;
phi=0:2*delta:2*pi;
%  Define phi and theta by meshgrid.
%  (通过meshgrid函数定义phi和theta)
[phi,theta]=meshgrid(phi,theta0);
```

```
% Define Legendre function.
% (定义 Legendre 函数，存为 Ymn)
Ymn=legendre(L,cos(theta0));
Ymn=Ymn(M+1,:)';
LL=size(theta,1);
% B=repmat(A,M,N) creates a large matrix B consisting of an M-by-N
% tiling of copies of A. The size of B is [size(A,1)*M, size(A,2)*N].
% (B=repmat(A,M,N)创建一个更大矩阵B，尺寸为[size(A,1)*M,size(A,2)*N])
yy=repmat(Ymn,1,LL);
% Define the real and imagery of Y62.
% (定义球谐函数的实部和虚部分别为Reyy和Imyy)
Reyy=yy.*cos(M*phi);
Imyy=yy.*sin(M*phi);
%-----------------------------------------------------------------
% Define the real parts for x-, y-, and z-positions.
% (定义x, y和z坐标的实部)
ReM=max(max(abs(Reyy)));
Rerho=R+A*Reyy/ReM;
Rer=Rerho.*sin(theta);
Rex=Rer.*cos(phi);
Rey=Rer.*sin(phi);
Rez=Rerho.*cos(theta);
% Open a subplot figure window.
% (打开一个子绘图图形窗口)
subplot(1,2,1);
% Plot the surface of real spherical function.
% (绘制球谐函数实部的曲面)
surf(Rex,Rey,Rez);
% Open a light source.
% (打开光源)
light
% Set lighting as phong.
% (设置lighting属性为phong)
lighting phong
% Set axis as square.
% (设置axis属性为square)
axis square
% Set the range for axes.
% (设置坐标轴的范围)
axis([-5,5,-5,5,-5,5]);
% Set axis as off.
% (设置axis属性为off)
axis off
% Set viewpoint as (40,30)
% (设置viewpoint属性为(40,30)).
view(40,30);
% Title the figure as '球谐函数实部'.
% (将该图命名为'球谐函数实部')
title('球谐函数实部');
% Set shading as interp.
```

```
%   (设置shading属性为interp)
shading interp
%------------------------------------------------------------
%   Define the imagery parts of x-, y-, and z-positions.
%   (定义x, y和z轴的虚部)
ImM=max(max(abs(Imyy)));
Imrho=R+A*Imyy/(ImM+eps*(ImM==0));
Im=max(max(abs(Imyy)));
Imrho=R+A*Imyy/(ImM+eps*(ImM==0));
Imr=Imrho.*sin(theta);
Imx=Imr.*cos(phi);
Imy=Imr.*sin(phi);
Imz=Imrho.*cos(theta);
%   Open a new suplot figure window.
%   (打开一个新子绘图图形窗口)
subplot(1,2,2);
%   Plot the surface of imagery spherical function.
%   (绘制球谐函数虚部的曲面图)
surf(Imx,Imy,Imz);
%   Create a light source.
%   (创建一个光源)
light
%   Set lighting as phong.
%   (设置lighting属性为phong)
lighting phong;
%   Set axis as square.
%   (设置axis属性为square)
axis square
%   Set the range for axes.
%   (设置坐标轴的范围)
axis([-5,5,-5,5,-5,5]);
%   Set axis as off.
%   (设置axis属性为off)
axis off
%   Set viewpoint as (40,30).
%   (设置视角为(40,30))
view(40,30);
%   Title the figure as '球谐函数虚部'
%   (将该图命名为'球谐函数虚部')
title('球谐函数虚部');
%   Set shading as interp.
%   (设置shading属性为interp)
shading interp
%   Set material as metal.
%   (设置material属性为metal)
material metal
```

> **注意:**
> 复数函数也可借助 MATLAB 软件绘图。

运行结果如图 4.47 所示。

球谐函数实部　　　　　球谐函数虚部

图 4.47　球谐函数示例图

【例 4.40】 鼓动的 L 膜。

本示例基于解 L 形膜鼓动模拟而设计。L 形膜鼓动解的表达式，可表达为一维时间本征函数，二维空间本征函数的线性组合。二维空间函数是在膜函数初始化时求得的。这些本征函数的基本形式由 MathWorks 公司给出。L 形几何膜在数学上是特别有意义的，因为张力在接近于凹角时趋向于无穷大。在完成该运动的模拟过程中，引用了 Bessel 函数以及分数函数来达到解决边缘凹角区的奇异性。

EXAMP04040

```
% Scipt file: EXAMP04040.m
% This programe models the vibration of L-like membrane.
%  (本程序模拟 L 形膜的鼓动)
% Clear variables in workspace.
%  (清除工作空间中的变量)
clear
% Clear command window.
%  (清除命令窗口)
clc
% Eigenvalues.
%  (本征值)
lambda = [9.6397238445, 15.19725192, 2*pi^2, ...
    29.5214811, 31.9126360, 41.4745099, 44.948488, ...
    5*pi^2, 5*pi^2, 56.709610, 65.376535, 71.057755];
% Eigenfunctions.
%  (本征函数)
for k = 1:12
    L{k} = membrane(k);
end
% Get coefficients from eigenfunctions.
%  (从本征函数获取系数)
for k = 1:12
    c(k) = L{k}(25,23)/3;
end
% Set the properties of figure.
%  (设置图形属性)
```

```
fig = figure;
% Set color as [0.65,1,1] for the figure.
% (对上面图形窗口设置颜色为[0.65,1,1])
set(fig,'color',[0.65,1,1])
% Define a variable x.
% (定义一个变量x)
x = (-15:15)/15;
% Plot the surface with lighting.
% (绘制带光照的曲面)
h = surfl(x,x,L{1});
% Set lighting as phong.
% (设置lighting属性为phong)
lighting phong
% Get the values of viewpoint.
% (获取视角的值)
[a,e] = view;
% Reset the viewpoint.
% (重新设置视角)
view(a+270,e);
% Determine the range of axes.
% (确定坐标系坐标轴的显示范围)
axis([-1 1 -1 1 -1 1]);
% Set the pseudocolor scaling.
% (设置准色标)
caxis(26.9*[-1.5 1]);
% Set shading as interp.
% (设置shading属性为interp)
shading interp;
% Set colormap as hsv.
% (设置colormap属性为hsv)
colormap(hsv);
% Set axis as off.
% (设置axis属性为off)
axis off
% Buttons.
% (按钮)
uicontrol('pos',[20 20 60 20],'string','done','fontsize',12, ...
    'callback','close(gcbf)');
uicontrol('pos',[20 40 60 20],'string','slower','fontsize',12, ...
'callback','set(gcbf,''userdata'',sqrt(0.5)*get(gcbf,''userdata''))');
uicontrol('pos',[20 60 60 20],'string','faster','fontsize',12, ...
'callback','set(gcbf,''userdata'',sqrt(2.0)*get(gcbf,''userdata''))');
% Run.
% (运行)
% Defien the initial time as t=0.
% (定义初始时间为t=0)
t = 0;
% Define the time interval.
% (定义时间间隔)
```

```
dt = 0.025;
% Set userdata as dt for fig.
% (将图 fig 的 userdata 设定为 dt)
set(fig,'userdata',dt)
% while loop structure.
% (while 循环结构)
  while ishandle(fig)
  % Coefficients.
  %   (参数)
    dt = get(fig,'userdata');
    t = t + dt;
    s = c.*sin(sqrt(lambda)*t);
    % Amplitude.
    %  (振幅)
    A = zeros(size(L{1}));
    for k = 1:12
      A = A + 2*s(k)*L{k};
    end
    % Velocity.
    %  (速度)
    s = lambda .*s;
    V = zeros(size(L{1}));
    for k = 1:12
      V = V + s(k)*L{k};
    end
    V(16:31,1:15) = NaN;
    % Surface plot of height, colored by velocity.
    %  (曲面描绘高度，颜色代表速度)
    set(h,'zdata',A,'cdata',V);
    % Plot.
    %  (绘制)
    drawnow
  end;
```

运行本程序显示动画，其中某一帧截图如图 4.48 所示。

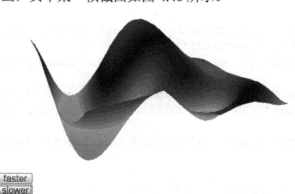

图 4.48　鼓动的 L 膜示例图

注释：

函数 membrane 用来产生 MATLAB 教学演示函数。

语法格式：L=membrane(k)，k<=12，用来生成 L 形膜出现 k 个峰的本征函数。

membrane(k) 只是绘出含有 k 个峰本征函数图，没有任何参数输出。

membrane 绘出含有一个峰的本征函数图，没有数据输出。

L=membrane(k,m,n,np) 不仅仅有数据输出，还设置了网格和精度参数。k 指定峰的个数；m 为三边点数默认值(15)，输出数据格式为(2×m+1)×(2×m+1)；n 为求和项的个数，默认值是 n=min(m,9)；np 是边求和项个数，默认值是 np=min(n,2)。

MATLAB 软件提供了大量的高级绘图命令，都可以绘制出漂亮的图形。并且为调节这些图形的属性也提供了多种方式。二维图形，提供了大量的参数设置函数，也可以借助图形句柄和 set 函数进行设置。对于三维图形，一样也提供了大量的参数设置函数。当然前面所阐述的实际操作不及 MATLAB 所提供九牛之一毛，如有兴趣，可借助帮助系统慢慢练习。

4.7.3 注释对象

用户创建注释对象的典型方法是借助绘图编辑工具栏或者插入菜单。然而，如果想创建注释对象，建议使用注释函数来实现。注释对象是在隐含坐标系创建的，并且充满了整个图示窗口。这能够辨别注释对象的具体位置（视图窗口的左下角是（0,0），右上角是（1,1））。

【例 4.41】 圈起带有注释框的子绘图窗口。

本示例中显示了如何创建矩形注释对象，并且用它来标注同一个图形中的两个子窗口。该例子运用了坐标位置属性和紧挨插入属性设置来定义注释对象的位置与大小。

EXAMP04041

```
%   Scipt file: EXAMP04041.m
%   This programe enclose the subplot figure window with illustrations.
%   (本程序可圈起带注释的子绘图图形窗口)
%   Clear variables in workspace.
%   (清除工作空间中的变量)
clear
%   Clear command window.
%   (清除命令窗口)
clc
%   Open a figure window with a handle f1.
%    (打开一个图形窗口句柄设为 f1)
f1=figure;
%   Define an independent variable as x.
%    (定义一个自变量 x)
x=-2*pi:pi/12:2*pi;
%   Define a dependent variable y with respect to x.
%    (相应于 x 定义一个因变量 y)
y=x.^2;
%   Open a subplot figure window.
%    (打开一个子绘图图形窗口)
subplot(2,2,1:2);
```

```
%   Plot a curve of y.
%   (绘制 y 曲线)
plot(x,y);
%   Open a new subplot figure window with a handle h1.
%   (打开一个新子绘图图形窗口句柄为 h1)
h1=subplot(223);
%   Define a new dependent variable y with respect to x.
%   (相应于 x 定义一个新因变量 y)
y=x.^4;
%   Plot the function y=x.^4.
%   (绘制 y=x.^4 曲线)
plot(x,y);
%   Open a new subplot figure window with a handle h2.
%   (打开一个新子绘图图形窗口句柄为 h2)
h2=subplot(224);
%   Define a new dependent variable y=x.^5.
%   (定义一个新因变量 y=x.^5)
y=x.^5;
%   Plot the function y=x.^5.
%   (绘制 y=x.^5 曲线)
plot(x,y)
%   Save the position of axes h1 as p1.
%   (将坐标系 h1 的位置数据存为 p1)
p1=get(h1,'Position');
%   Save the tightinset of axes h1 as t1.
%   (将坐标系 h1 的 TightInset 属性存为 t1)
t1=get(h1,'TightInset');
%   Save the position of axes h2 as p2.
%   (将坐标系 h2 的位置存为 p2)
p2=get(h2,'Position');
%   Save the tightinset of axes h2 as t2.
%   (将坐标系 h2 的 TightInset 属性存为 t2)
t2=get(h2,'TightInset');
%   Define the subtraction of p1(1) and t1(1) as x1.
%   (定义 p1(1)-t1(1) 为 x1)
x1=p1(1)-t1(1);
%   Save the subtraction of p1(2) and t1(2) as y1.
%   (定义 p1(2)-t1(2) 为 y1)
y1=p1(2)-t1(2);
%   Save the subtraction of p2(1) and t2(1) as x2.
%   (定义 p2(1)-t2(1) 为 x2)
x2=p2(1)-t2(1);
%   Save the subtraction of p2(2) and t2(2) as y2.
%   (定义 p2(2)-t2(2) 为 y2)
y2=p2(2)-t2(2);
%   Calculate the width of subplot(1,2,2) and save as w.
%   (计算 subplot(122) 的宽度,保存为 w)
w=x2-x1+t1(1)+p2(3)+t2(3);
%   Calculate the height of subplot(1,2,2) and save as h.
%   (计算 subplot(122) 的高度,保存为 h)
h=y2-y1+p2(4)+t2(2)+t2(4);
```

```
%   Draw the defined area with red color.
%   (将定义区域涂上红色)
annotation('rectangle',[x1,y1,w,h],...
    'FaceAlpha',.5,'FaceColor','r','EdgeColor','r');
```

运行结果如图 4.49 所示。

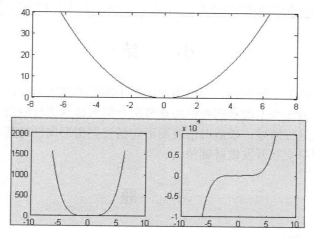

图 4.49　圈起带有注释框的子绘图窗口

4.7.4　总结

MATLAB 软件中绘制的所有元素都可称为图形对象。每个图形对象都有唯一的图形句柄，并且每个图像也有很多属性，每个属性都可能影响该图形的显示效果。

MATLAB 中图形对象有一定层次。当下一级对象被创建时，它就会继承很多上一级的属性。MATLAB 中最高级别的图是根，也就是终端计算机屏幕。屏幕上可以显示多个窗口，每个窗口都含有一定图形数据，并且每个窗口都有它自己的属性。

每个用户窗口可含有 4 类对象：绘制图形、背景、坐标系、文本。每个元素都有它自己的属性，这些属性可以借助 set 函数设置，也可以借助属性编译器完成，还可以通过属性监控窗口实现更新。MATLAB 绘制的每个图形对象都有唯一的句柄，这为重新设定图形属性带来了很大的方便。当前图形窗口、当前坐标系、当前操作对象都可分别被函数 gcf、gca 和 gco 调用。当然，任何对象的属性都可以借助 get 和 set 函数进行检查和修改。

MATLAB 软件为我们提供了数以百计的与绘图相关的函数，了解这些函数的基本属性与设置可通过网络进行了解学习。

> **好习惯：**
> （1）如果对所创建的图形属性进行检查或修改，建议给所绘制的图形定义一个句柄，便于后来 set 和 get 函数的设置与调用。
> （2）如果可能，尽量缩小想搜索的范围，可提高搜索速度。
> （3）如果想在特定位置放置一个窗口，将默认常规窗口放置到预定的位置比较容易。这与计算屏幕设置无关，可以直接借助函数 figure 参数设置完成。

MATLAB 常用图形属性设置函数描述参见表 4.13。

表 4.13 MATLAB 常用图形属性设置函数

函 数	描 述	函 数	描 述
figure	创建新图形窗口并设置为当前窗口	get	读取图形对象属性
axes	创建新坐标系并设置为当前坐标系	set	设置图形对象属性

小　　结

本章首先介绍了绘制二维图形、三维图形、高维图形的基本命令；基于 MATLAB 基本函数编写动画；并借用 5 个生动示例将前面章节所介绍的命令进行应用示范；然后对如何借助鼠标设置图形属性进行阐述；最后对借助图形句柄设置图形属性的基本用法进行讲解，为后面学习图形界面用户交互开发做好铺垫。

习　　题

4.1　绘制下面函数的曲线。

（1）$y = \dfrac{100}{1 - x + x^3}$

（2）$y = 2\sin x e^{-x^2}$

（3）$\dfrac{x^2}{16} + \dfrac{y^2}{36} = 1$

（4）$\begin{cases} x = t^2 \\ y = \sin(t) \end{cases}$

4.2　绘制下面极坐标函数的曲线。

（1）$\rho = \sin\theta + 3$

（2）$\rho = \sin\theta e^{\theta}$

（3）$\rho = \dfrac{5}{\cos\theta}\theta - 2$

（4）$\rho = 2\theta^3$

4.3　绘制下面三维图形。

（1）$\begin{cases} x = \cos t \\ y = \sin t \\ z = t \end{cases}$

（2）$\dfrac{x^2}{25} + \dfrac{y^2}{9} + \dfrac{z^2}{1} = 1$

4.4　用三个函数 plot、fplot 和 ezplot 分别对下面函数进行曲线绘制：

$$f(x) = \sin(x) e^{\frac{1}{5}x}$$

4.5 绘制半径为 5、金属材质、夏天色调、渐变填充、视角为[30,45]的球。请在纵横轴相等的坐标系下绘制。

4.6 绘制两个点电荷所形成的电势场。

4.7 绘制具有冬天色调、闪光材质、phong 光照模式的山峰。

4.8 绘制一个逐渐变大的立方体。

4.9 绘制 3 个点电荷形成的电场图,3 个点电荷电量分别为 Q、$-2Q$ 和 Q,位置分别在(0,-2)、(0,0)和(0,2)。

第 5 章　数值计算方法

本章将主要介绍使用 MATLAB 语言实现常见数值的解法，共 6 节。5.1 节讲述线性方程组数值解法；5.2 节介绍多项式插值与最小二乘曲线拟合；5.3 节阐述积分与微分数值计算方法；5.4 节表述矩阵的特征值与特征向量；5.5 节简述了常微分方程数值解法；5.6 节对非线性方程求根的数值计算方法进行简述。

5.1　线性方程组数值解法

在科学与工程计算中很多问题的数值求解，常常归结为求解线性方程组的问题。按照矩阵阶数大小和零元素多少，线性方程组可以分为两类：低阶稠密方程组和高阶稀疏方程组。关于线性方程组解法一般可以分为两类：直接方法和迭代方法。

5.1.1　直接方法

直接方法就是经过有限步骤的算术运算求得方程组精确解的方法（假设计算过程中没有舍入误差）。但实际计算中由于舍入误差的存在，这种方法也只能求得方程组的近似解。

1. 矩阵除法

对于一个 n 阶线性方程组 $Ax = b$，在 MATLAB 中，只需要一个反斜线符号运算符 "/" 或斜线符号运算符 "\" 就可解决。虽然只是一个符号，但它的内部包含许多算法，如 "/" 表示右除，A/B 等同于 $A \times B^{-1}$；"\" 表示左除，A/B 等同于 $A^{-1} \times B$。一般情况下，变量和方程的数目一样多，当方阵 A 为满秩时，原方程组有唯一解 $x = A^{-1}b$，其中 A^{-1} 是矩阵 A 的逆矩阵。反斜线符号运算符避免计算 A^{-1}，减少了计算量，同时反斜线符号运算符不仅可以求解一般的线性方程组，对于超定和欠定方程组同样有效。

【例 5.1】　求解线性方程组 $Ax = b$，其中：

$$A = \begin{bmatrix} 5 & -4 & 1 \\ -4 & 6 & -4 \\ 1 & -4 & 6 \end{bmatrix}, b = \begin{bmatrix} 2 \\ -1 \\ -1 \end{bmatrix}$$

解：在 MATLAB 命令窗口中输入执行语句如下：

```
>>A=[5 -4 1;-4 6 -4;1 -4 6];
>>b=[2 ;-1 ;-1];
>> x=A\b
x = 0.3333
   -0.1667
   -0.3333
>> sym(x)
%   (转化成分数)
```

```
     ans = 1/3
          -1/6
          -1/3
```

【例 5.2】 求解超定线性方程组 $Ax = b$，其中：

$$A = \begin{bmatrix} 5 & -4 \\ -4 & 6 \\ 1 & -4 \end{bmatrix}, b = \begin{bmatrix} 2 \\ -1 \\ -1 \end{bmatrix}$$

解：在 MATLAB 命令窗口中输入执行语句如下：

```
>>A=[5 -4;-4 6;1 -4];
>>b=[2 ;-1 ;-1];
>> x=A\b
x =  0.7319
     0.3696
```

注意：

综上所述，反斜杠运算符 "\" 是 MATLAB 内核中一个表示符号，它的内部包含许多个算法。下面是运算采取的步骤：

（1）MATLAB 检查，如果 A 是三角矩阵或其置换矩阵，就采用适合的回代求解；
（2）如果 A 是对称正定方阵，就采用 Choleshy 分解和回代求解；
（3）如果 A 是方阵，不满足（1）和（2），就采用 LU 分解和回代求解；
（4）如果 A 不是方阵，就用 QR 分解，并求最小二乘解。

2. 矩阵分解

在高等代数知识中，对于线性方程组 $Ax = b$，如果系数矩阵 A 是非奇异的，方程组有唯一解；同样，克莱姆法则证明了解的存在，并给出了求解的一个方法，但是克莱姆法则计算量很大。Gauss 消去法是对线性方程组的增广矩阵 $[A:b]$ 施以行初等变换，化 A 为三角矩阵，然后采用回代求解的方法。基于 Gauss 消去法，产生了几种分解的方法。

1）LU 分解

将系数矩阵 A 分解成一个上三角矩阵和一个下三角矩阵的乘积 $A = LU$，令 $Ux = y$，则原方程组可以写成 $Ly = b$，通过两次回代可以求出方程组的解。在 MATLAB 中，LU 分解是使用函数 lu 来实现的。MATLAB 的 lu 函数采用的是选列主元三角分解法，又称为 PLU 分解法，实现 LU=PA。该分解法要求的前提条件是 A 为非奇异矩阵。

【例 5.3】 使用 LU 分解，求解 $\begin{bmatrix} 1 & 2 & 3 \\ 1 & 3 & 5 \\ 1 & 3 & 6 \end{bmatrix} \begin{bmatrix} x_1 \\ x_2 \\ x_3 \end{bmatrix} = \begin{bmatrix} 2 \\ 4 \\ 5 \end{bmatrix}$。

解：在 MATLAB 命令窗口中输入执行语句如下：

```
>> A=[1 2 3;1 3 5;1 3 6]; %（对矩阵 A 赋值）
>> [L U]=lu（A）%（使用函数 lu 对矩阵 A 分解，即 PA=LU, A=LU）
L =  1    0    0
     1    1    0
     1    1    1
```

```
U =  1    2    3
     0    1    2
     0    0    1
```

此时，就可以通过回代来求解了。下面是一个解系数矩阵是三角矩阵的方程组，回代求解的 MATLAB 程序。

EXAMP05001

```
function [x]=backsub(L,b)
% BACKSUB solves a trigonometric equation with back substitution.
% （BACKSUB 通过回代方法求解系数矩阵为三角矩阵的方程组）
%
% Records of revision: 2nd
% Date: 2015-07-16
% Programmer: Z. Zhang
% Description of change: Original code
%
n=length(b);
%
if isequal(tril(L,-1),zeros(size(L)))
    x(n)=b(n)/L(n,n);
    for i=n-1:-1:1
        sum=0;
        for j=i+1:n
            sum=sum+L(i,j)*x(j);
        end
        x(i)=(b(i)-sum)/L(i,i);
    end
elseif isequal(triu(L,1),zeros(size(L)))
    x(1)=b(1)/L(1,1);
    for i=2:1:n
        sum=0;
        for j=1:1:i-1
            sum=sum+L(i,j)*x(j);
        end
        x(i)=(b(i)-sum)/L(i,i);
    end
else
    disp('The matrix is not a trigonometric equation')
end
```

首先将 MATLAB 当前目录转到 backsub.m 文件所在位置，然后在命令窗口输入 MATLAB 执行语句。运行结果如下：

```
>> clear
>> clc
>> A=[1 2 3;1 3 5;1 3 6];
>> B=[2 4 5];
>> [L,U]=lu(A);
>> y=backsub(L,B)
```

```
y =    2    2    1
>> x=backsub(U,y)
x = -1    0    1
```

> **注意：**
> 解方程组时，为了减少误差，一般都采用了选主元的方法，在上面程序中，没有选主元，对于一些特殊方程求解时可能误差较大。

2）Cholesky 分解

当系数矩阵 A 对称正定方阵时，可以使用平方根法来求解。存在唯一的主对角元素都是正数的下三角方阵 L，有 $A = LL^T$。MATLAB 为 Cholesky 分解有特定函数 chol。

【例 5.4】 使用 Cholesky 分解求解方程组 $\begin{bmatrix} 2 & -1 & -1 \\ -1 & 2 & 0 \\ -1 & 0 & 1 \end{bmatrix} \begin{bmatrix} x_1 \\ x_2 \\ x_3 \end{bmatrix} = \begin{bmatrix} 1 \\ 0 \\ 0 \end{bmatrix}$。

解： 当前路径为 backsub.m 所在位置，在命令窗口中输入执行语句如下：

```
>> A=[2 -1 -1 ;-1 2 0 ;-1 0 1];
>> B=[1;0;0];
>> L=chol(A)
L =  1.4142   -0.7071   -0.7071
        0    1.2247   -0.4082
        0        0    0.5774
>> y=backsub(L,B)   %（使用回代函数求 y）
y =0.7071        0        0
>> x=backsub(L',y)  %（使用回代函数求 x，式中 L'表示矩阵的转置）
x = 0.5000   0.2887   0.8165
>> sym(x)
ans = [ 1/2, 3^(1/2)/6, (2^(1/2)*3^(1/2))/3]
```

> **注意：**
> 矩阵分解中用到的函数 MATLAB 大都是自带函数，如果没有，需要读者自己编写。

5.1.2 迭代方法

线性方程组的迭代解法的基本思想是构造一个向量序列 $\{x^{(k)}\}$，使其收敛于某个极限向量 x，而 x 就是所要求的线性方程组 $Ax = b$ 的准确解。迭代解法的主要问题包括：构造迭代格式、迭代格式的收敛性和收敛速度、迭代格式的加速、迭代的误差估计。关于迭代方法的讨论，请读者参考数值分析的相关资料。本节借助 MATLAB 简要讨论几种迭代方法。

1. Jacobi 迭代

n 阶线性方程组 $Ax = b$，若系数矩阵 A 非奇异可分裂为 $A = D + L + U$，其中 $D = \text{diag}[a_{11}, a_{22}, \cdots, a_{nn}]$，$L$、$U$ 分别为矩阵 A 的严格下（上）三角部分。

迭代格式为 $x^{(k+1)} = Bx^{(k)} + f$。

式中 $B = D^{-1}(L+U) = I - D^{-1}A$；

$f = D^{-1}b$。

Jacobi 迭代函数程序如下。

EXAMP05002

```matlab
function x1=Jacobimethod(A,b,x0,Maxit,N)
% JACOBIMETHOD solves liner equations by Jacobi iterations.
% (JACOBIMETHOD 函数基于 Jacobi 迭代法求解线性方程组)
% Calling sequences: x1=Jacobimethod(A,bx0,Maxit,N)
%
% Records of revision:2nd
% Date: 2015-06-16
% Programmer: Z. Zhang
% Description of change: Original code
%
% A: a matrix consists of coefficients of linear equations.
% b: a vector consists of values for linear equations.
% x0: initial values for x.
% Maxit: the value of accuracy.
% N: the number of Jacobi iterations.
%
% (判断是否输入 N 值，在没有输入情况下给予补值为 500)
if nargin<5
    N=500;
else
    N=N;
end
n=length(b);
x1=zeros(n,1);
x2=zeros(n,1);
x1=x0;
k=0;
r=max(abs(b-A*x1));
xsequence=x0';
% (使用最近两次结果的无穷范数判断是否达到精度要求)
while r>Maxit
    for i=1:n;
        sum=0;
        for j=1:n
            if i~=j
                sum=sum+A(i,j)*x1(j);
            end
        end
        x2(i)=(b(i)-sum)/A(i,i);
    end
    r=max(abs(x2-x1));
    format long
    x1=x2;
    xsequence=[xsequence;x1'];
    k=k+1;
```

```
        if k>N
            warning('The number of iterations exceeds the limit!')
            return;
        end
    end
    xsequence %  (显示中间计算结果)
    x=x1;
```

> **注意：**
> 迭代方法中迭代的收敛性只和系数矩阵有关，和初值的选择没有关系，初值的选择任意。

【例 5.5】 用 Jacobi 迭代方法解线性方程组，要求 $\|x^{(k+1)} - x^{(k)}\| < 10^{-5}$ 时停止。

$$\begin{bmatrix} 4 & 2 & 3 \\ 1 & 3 & 1 \\ 1 & 1 & 4 \end{bmatrix} \begin{bmatrix} x_1 \\ x_2 \\ x_3 \end{bmatrix} = \begin{bmatrix} 3 \\ -1 \\ 4 \end{bmatrix}$$

解： 当前路径设为含有 Jacobimethod.m 文件位置，在命令窗口中输入执行语句如下：

```
>> A=[4 2 3;1 3 1;1 1 4];
>> b=[3;-1;4];
>> x0=[0;0;0];
>> Maxit=10^(-5);
>> x=Jacobimethod(A,b,x0,Maxit)
xsequence =
                   0                  0                  0
   0.750000000000000  -0.333333333333333   1.000000000000000
   0.166666666666667  -0.916666666666667   0.895833333333333
   0.536458333333333  -0.687500000000000   1.187500000000000
   0.203125000000000  -0.907986111111111   1.037760416666667
   0.425672743055556  -0.746961805555556   1.176215277777778
   0.241319444444445  -0.867296006944444   1.080322265625000
   0.373406304253472  -0.773880570023148   1.156494140625000
   0.269569679542824  -0.843300148292824   1.100118566442419
   0.346561149314598  -0.789896081995081   1.143432617187500
   0.287373578106916  -0.829997922167366   1.110833733170121
   0.331873661206092  -0.799402437092345   1.135656086015113
   0.297959154034838  -0.822509915740402   1.116882193971563
   0.323593312391528  -0.804947116002134   1.131137690426391
   0.304120290181274  -0.818243667605973   1.120338450902652
   0.318867995625998  -0.808152913694642   1.128530844356175
   0.307678323580190  -0.815799613327391   1.122321229517161
   0.316158884525825  -0.809999851032450   1.127030322436800
   0.309727183688625  -0.814396402320875   1.123460241626656
   0.314603019940445  -0.811062475105094   1.126167304658063
   0.310905759059000  -0.813590108199503   1.124114863791162
   0.313708906256380  -0.811673540950054   1.125671087285126
   0.311583455011183  -0.813126664513835   1.124491158673419
   0.313194963251854  -0.812024871228200   1.125385802375663
```

```
       0.311973083832353   -0.812860255209172   1.124707476994087
       0.312899519859021   -0.812226853608813   1.125221792844205
       0.312197082171253   -0.812707104234409   1.124831833437448
       0.312729677039118   -0.812342971869567   1.125127505515789
       0.312325856797942   -0.812619060851636   1.124903323707612
       0.312632037645109   -0.812409726835185   1.125073301013424
       0.312399887657525   -0.812568446219511   1.124944422297519
       0.312575906386616   -0.812448103318348   1.125042139640497
       0.312442446928801   -0.812539348675704   1.124968049232933
       0.312543637413153   -0.812470165387245   1.125024225436726
       0.312466913616078   -0.812522620949959   1.124981631993523
       0.312525086479837   -0.812482848536534   1.125013926833471
       0.312480979143164   -0.812513004437769   1.124989440514174
       0.312514421833254   -0.812490139885779   1.125008006323651
       0.312489065200151   -0.812507476052302   1.124993929513131
       0.312508290891302   -0.812494331571094   1.125004602713038
       0.312493713750769   -0.812504297868113   1.124996510169948
       0.312504766306596   -0.812496741306906   1.125002646029336
       0.312496386131451   -0.812502470778644   1.124997993750078
x =    0.312496386131451   -0.812502470778644   1.124997993750078
```

> **注意：**
> 程序中默认线性方程组使用 Jacobi 迭代是收敛的，关于判断迭代方法的收敛性，请读者参阅数值分析相关书籍。另外，程序中使用迭代最后两次向量的无穷范数作为迭代停止的条件，相关的讨论请读者查阅参考文献。

2. Gauss-Seidel 迭代

$$Ax = b$$
$$(L + U + D)x = b$$
$$(L + D)x = -Ux + b$$
$$x = -(L + D)^{-1}Ux + (L + D)^{-1}b$$

其中 $A = D + L + U$，$D = \text{diag}[a_{11}, a_{22}, \cdots, a_{nn}]$，$L$、$U$ 分别为矩阵 A 的严格下（上）三角部分。

迭代格式为：$x^{(k+1)} = -(L + D)^{-1}Ux^{(k)} + (L + D)^{-1}b$。

其中，L、D、U 定义如 Jacobi 迭代。

Gauss-Seidel 迭代函数程序如下。

EXAMP05003

```
function x=Gsmethod(A,b,x0,Maxit,N)
% GSMETHOD solves linear equations by Gauss-Seidel iterations.
% （GSMETHOD 借助 Gauss-Seidel 迭代方法解线性方程组）
% Calling sequence: x=Gsmethod(A,b,x0,Maxit,N)
%
% Records of revision:2nd
```

```
%   Date: 2015-06-16
%   Programmer: Z. Zhang
%   Description of change: Original code
%
%   A: a matrix consists of coefficients of linear equations.
%   b: a vector consists of values for linear equations.
%   x0: initial values for x.
%   Maxit: the value of accuracy.
%   N: the number of Jacobi iterations.
%
if nargin<5
N=500;
else
    N=N;
end
n=length(b);
x1=zeros(n,1);
xsequence=x0';
k=0;
r=max(abs(b-A*x1));
format long
while (r>Maxit)
    for i=1:n;
        sum=0;
        for j=1:n
            if j>i
                sum=sum+A(i,j)*x0(j);
            elseif j<i
                sum=sum+A(i,j)*x1(j);
            end
        end
        x1(i)=(b(i)-sum)/A(i,i);
    end
    r=max(abs(x1-x0));
    x0=x1;
    k=k+1;
    xsequence=[xsequence;x0'];
    if k>N
        warning('The number of iterations exceeds the limit! ')
        return;
    end
end
xsequence
x=x0;
```

当前路径转到含有 Gsmethod.m 文件处，然后在命令窗口输入执行语句如下：

```
>> A=[4 2 3;1 3 1;1 1 4];
```

```
>> b=[3;-1;4];
>> x0=[0 0 0]';
>> Maxit=10^(-5);
>> x=gsmethod(A,b,x0,Maxit)
xsequence =
                   0                   0                   0
   0.750000000000000  -0.583333333333333   0.958333333333333
   0.322916666666667  -0.760416666666667   1.109375000000000
   0.298177083333333  -0.802517361111111   1.126085069444444
   0.306694878472222  -0.810926649305556   1.126057942708334
   0.310919867621528  -0.812325936776620   1.125351517288773
   0.312149330421730  -0.812500282570168   1.125087738037109
   0.312434337757252  -0.812507358598121   1.125018255210217
   0.312489987891397  -0.812502747700538   1.125003189952285
   0.312498981386055  -0.812500723779447   1.125000435598348
x =  0.312498981386055  -0.812500723779447   1.125000435598348
```

通过例 5.5 使用 Jacobi 迭代和 Gauss-Seidel 迭代比较可以看出，在相同的精度要求下，使用 Jacobi 迭代需要 42 次，而 Gauss-Seidel 迭代 9 次就可以达到，收敛速度有所提升。

3. SOR 迭代

SOR（Successive Over-Relaxation Method）迭代是 Gauss-Seidel 迭代方法的一种加速方法，是解大型稀疏矩阵方程组的有效方法之一，计算公式简单，但需要较好的加速因子（最佳松弛因子）。

对于预求方程：

$$Ax = b$$

其中 $A = D + L + U$，$D = \text{diag}[a_{11}, a_{22}, \cdots, a_{nn}]$，$L$ 和 U 分别为矩阵 A 的严格下（上）三角部分，满足关系式：

$$Dx^{(k+1)} = (1-\omega)Dx^{(k)} + \omega(b - Lx^{(k+1)} - Ux^{(k)})$$

迭代格式为 $x^{(k+1)} = (D+\omega L)^{-1}[(1-\omega)D - \omega U]x^{(k)} + \omega(D+\omega L)^{-1}b$。

SOR 迭代函数程序如下。

EXAMP05004

```
function [xsequence,x]=Sormethod(A,b,x0,Omega,Maxit,N)
% SORMETHOD solves linear equations by SOR iterations.
% （SORMETHOD 通过 SOR 迭代求解线性方程组）
% Calling sequence:[x,k]=Sormethod(A,b,x0,Omega,Maxit,N)
%
% Records of revision:2nd
% Date: 2015-07-16
% Programmer: Z. Zhang
% Description of change: Original code
%
% A: a matrix consists of coefficients of linear equations.
% b: a vector consists of values for linear equations.
```

```
%   x0: initial values for x.
%   Omega: relaxing factor
%   Maxit: the value of accuracy.
%   N: the number of Jacobi iterations.
%
if nargin==3
    Omega=1;
    Maxit=10^(-5);
    N=500;
elseif nargin==4
    Maxit=10^(-5);
    N=500;
elseif nargin==5
    N=500;
end
format long
n=length(b);
x1=zeros(n,1);
xsequence=x0';
k=0;
r=max(abs(b-A*x1));
while (r>Maxit)
    for i=1:n
        sum=0;
        for j=1:i-1
            sum=sum+A(i,j)*x1(j);
        end
        for j=i+1:n
            sum=sum+A(i,j)*x0(j);
        end
        x1(i)=(1-Omega)*x0(i)+Omega*(b(i)-sum)/A(i,i);
    end
    r=max(abs(x1-x0));
    x0=x1;
    k=k+1;
    xsequence=[xsequence;x0'];
    if k>N
        warning('The number of iterations exceeds the limit!')
        return;
    end
end
x=x0;
```

当前路径转到含有 Sormethod.m 文件处，然后在命令窗口输入执行语句如下：

```
>> A=[4 2 3;1 3 1;1 1 4];
>> b=[3;-1;4];
>> x0=[0 0 0]';
```

```
>> Maxit=10^(-5);
>> Omega=1.057;
>> [xsequence,x]=Sormethod(A,b,x0,Omega,Maxit)
xsequence =
                    0                    0                    0
    0.792762084381334   -0.631659469695098    1.014444097635798
    0.277186173398851   -0.771415272740746    1.129778484022844
    0.289011646810939   -0.808250280374587    1.129811435832189
    0.307778879630020   -0.812774124125213    1.126045684112599
    0.312085078022061   -0.812706612421201    1.125104622125580
    0.312549913112956   -0.812542668508642    1.124992120437245
    0.312525951417496   -0.812503934621392    1.124994631234980
    0.312504855983512   -0.812499594990412    1.124998915867492
    0.312500368539015   -0.812499770960788    1.124999903900567
x =  0.312500368539015   -0.812499770960788    1.124999903900567
```

可以看出，使用 SOR 迭代和 Gauss-Seidel 迭代都是 9 次达到迭代精度要求，在以上程序中，迭代的速度并没有得到提升。

> **注意：**
> SOR 迭代方法中的最佳松弛因子 Omega 的选择很困难，数值分析已经证明超松弛迭代收敛的必要条件是松弛因子在 0 到 2 之间，并且如果系数矩阵是一个三对角矩阵，给出了一个最佳松弛因子的公式。

5.2 多项式插值与最小二乘曲线拟合

5.2.1 多项式插值

设 $f(x) \in C[a,b]$，已知 $f(x)$ 上 $n+1$ 个互异节点 $\{x_i, f(x_i)\}, i=0,1,\cdots,n$。求 $p(x) \in H = \text{span}(1, x, x^2, \cdots, x^n)$ 满足 $p(x_i)=f(x_i), i=0,1,2,\cdots,n$。

1. Lagrange 插值

已知 $n+1$ 个插值节点 $x_0, x_1, \cdots, x_n$ 及对应的函数值 $y_0, y_1, \cdots y_n$，使用 Lagrange 插值多项式公式，则对插值区间内任意 x 可以求得：

$$y(x) = \sum_{k=0}^{n} y_k \left(\prod_{\substack{j=0 \\ j \neq k}}^{n} \frac{x-x_j}{x_k-x_j} \right)$$

Lagrange 插值 MATLAB 函数程序如下。

EXAMP05005

```
function yy=Lagrange(x,y,x0)
% LAGRANGE is Lagrange form polynomial interpolation.
% （LAGRANGE 为 Lagrange 形式的多项式插值）
% Calling sequence: yy=Lagrange(x,y,x0)
```

```
%
% Records of revision:2nd
% Date: 2015-07-6
% Programmer: Z. Zhang
% Description of change: Original code
%
% x: a list of numbers.
% y: a list of numbers, its length is the same as x.
% x0: special values
% yy: a evaluated value with respect to x0.
%
xlen=length(x);
ylen=length(y);
if xlen~=ylen
    warning('The length of x and y is not equal!');
end
n=length(x0);
for i=1:n
z=x0(i);
s=0.0;
for k=1:xlen
    p=1.0;
    for j=1:ylen
        if j~=k
            p=p*(z-x(j))/(x(k)-x(j));
        end
    end
    s=p*y(k)+s;
end
yy(i)=s;
end
```

【例 5.6】 已知数据如下表，

x_i	0.56160	0.56280	0.56401	0.56521
y_i	0.82741	0.82659	0.82577	0.81495

试用 Lagrange 插值多项式求 $x = 0.5626, 0.5635, 0.5645$ 时的函数近似值。

解：在 MATLAB 命令窗口输入执行语句如下：

```
>> x=[0.56160;0.56280;0.56401;0.56521];
>> y=[0.82741;0.82659;0.82577;0.81495];
>> x0=[0.5626;0.5635;0.5645];
>> y0=lagrang(x,y,x0)
y0 =  0.8265    0.8268    0.8231
>> plot(x,y,'o',x0,y0,'k*')
>> legend('数据点','插值点')
```

运行结果如图 5.1 所示。

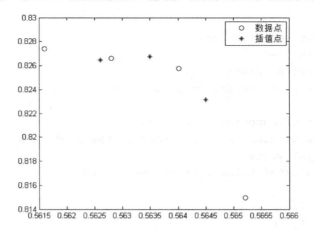

图 5.1 Lagrange 插值结果

2. 分段三次埃尔米特插值

对于代数插值来说，当插值多项式的次数很高时，逼近效果往往很不理想。高次的插值会带来龙格现象，许多有效的插值技术都基于分段的三次多项式。当数据点的函数值和一阶导数值给定时，可用埃尔米特插值法，如果没有给出导数值，将需要一些方法来限定斜率。为解决此类问题，MATLAB 提供了函数 pchip。函数 pchip（Piecewise Cubic Hermite Interpolating Polynomial）是分段三次埃尔米特插值多项式的首字母缩写。

3. 样条插值

已知 $f(x)$ 上 $n+1$ 个互异节点 $\{x_i, f(x_i)\}, i = 0, 1, \cdots, n$。要求构造一个三次样条函数 $S(x)$，满足下列条件：

(1) $S(x_i) = f(x_i), i = 0, 1, \cdots, n$；
(2) 在每个小区间 $[x_i, x_{i+1}]$ 上是 3 次多项式；
(3) $S(x) \in C^2[a, b]$。

常用的三次样条函数的边界条件有三种类型：

(1) $S'(x_0) = f'(x_0)$，$S'(x_n) = f'(x_n)$；
(2) $S''(x_0) = f''(x_0)$，$S''(x_n) = f''(x_n)$；
(3) $S_j(x_0) = S_j(x_n)$，$j = 0, 1, 2, \cdots$。

三次样条数据插值可以使用 spline 函数。

【例 5.7】 已知数据如下：

$$x = -3, -2, -1, 0, 1, 2, 3$$
$$y = -1, -1, -1, 0, 1, 1, 1$$

借助函数 pchip 和 spline 在 x0=−3：0.01：3 点上进行插值。

解： 在命令窗口输入执行语句如下：

```
>>x = -3:3;
>>y = [-1 -1 -1 0 1 1 1];
>>x0 = -3:0.01:3;
```

```
>> plot (x, y, 'o', x0, pchip (x,y,x0), '-', x0,...
    spline (x,y,x0),':','LineWidth',1)
>>legend ('data','pchip','spline',4)
```

运行结果如图 5.2 所示。

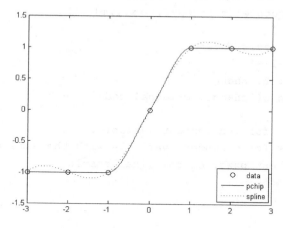

图 5.2 pchip 和 spline 插值结果

5.2.2 最小二乘曲线拟合

函数逼近是用一个简单的函数 $\phi(x)$ 来近似一个函数 $f(x)$，插值和拟合是一种函数逼近的方法。如何度量逼近的"好"和"坏"是和逼近误差的度量标准有关的，常用的度量标准有以下两种。

（1）一致逼近：

$$\|f(x)-\varphi(x)\|_\infty = \max_{x\in[a,b]}|f(x)-\varphi(x)|$$

（2）平方逼近：

$$\|f(x)-\varphi(x)\|_2 = \sqrt{\int_a^b |f(x)-\varphi(x)|^2 \,\mathrm{d}x}$$

曲线拟合的最小二乘法具体做法是对于给定的一组数据 $\{(x_i,y_i)\}_{i=0}^m$，要求在函数类：

$$\Phi = \mathrm{Span}\{\varphi_0(x),\varphi_1(x),\cdots,\varphi_n(x)\}(n\leqslant m)$$

中找到一个函数：

$$y = \varphi^*(x) = a_0^*\varphi_0(x) + a_1^*\varphi_1(x) + \cdots + a_n^*\varphi_n(x)$$

使误差的平方和最小：

$$\sum_{i=0}^m (\varphi^*(x_i)-y_i)^2 = \min_{\varphi\in\Phi}\sum_{i=0}^m (\varphi(x_i)-y_i)^2$$

最小二乘的曲线拟合解法由正规方程组求得。

多项式拟合 MATLAB 函数程序如下。

EXAMP05006

```
function c=Linearfit(x,y,m)
%   LINEARFIT accomplishes least square cure fitting.
%   （LINEARFIT 函数可以实现最小二乘法拟合）
%   Calling sequence: c=Linearfit(x,y,m)
%
%   Records of revision:2nd
%   Date: 2014-12-6
%   Programmer: Z. Zhang
%   Description of change: Original code
%
%   x: a vector for independent variable.
%   y: a vector for dependent variable with the same length of x.
%   m: the highest power of the final result.
%
nx=length(x);
ny=length(y);
if nx~=ny
error('The input data is wrong!')
end
A=zeros(m+1,m+1);
for i=0:m
    for j=0:m
        A(i+1,j+1)=sum(x.^(i+j));
    end
    b(i+1)=sum(x.^i.*y);
end
a=A\b';
c=fliplr(a');
yy=polyval(c,x);
plot(x,y,'k*',x,yy,'b+')
legend(2,2,'Data point','Least squres solution',4)
```

【例 5.8】 已知某种化学反应过程中温度与产品关系的数据如下表，

温度	100	110	120	130	140	150	160	170	180	190
产品	45	51	54	61	66	70	74	78	85	89

请用直线拟合得出函数关系。

解： 把这些离散点绘制在直角坐标系中，看出数据点大致在一条直线附近。根据题意，将当前路径转到含有 Linearfit.m 文件处，在命令窗口输入执行语句如下：

```
>> x=100:10:190;
>> y=[45 51 54 61 66 70 74 78 85 89];
>> c=Linearfit(x,y,1)%  （使用 1 次多项式来拟合数据点）
c =  0.4830    -2.7394
```

运行结果表示最后拟合的直线为 $0.4830x-2.7394$，如图 5.3 所示。

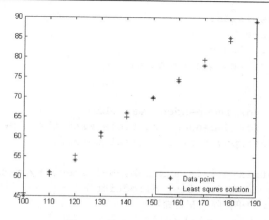

图 5.3 最小二乘拟合结果

多项式拟合数据点时，实际上是求解法方程来确定多项式的系数，当使用高次多项式时，方程往往是病态的，误差较大，需要借助正交基来处理。

如果 $p_0(x), p_1(x), \cdots, p_n(x)$ 是关于点集 $\{x_i\}_{i=0}^{m}$ 的正交多项式，借助下面关系求出：

$$\begin{cases} p_0(x) = 1 \\ p_1(x) = (x - a_1) p_0(x) \\ p_{i+1}(x) = (x - \alpha_{i+1}) p_i(x) - \beta_i p_{i-1}(x) \end{cases}$$

其中

$$\begin{cases} \alpha_i = \dfrac{(xp_i, p_i)}{(p_i, p_i)} = \dfrac{\sum\limits_{j=0}^{m} x_i p_i^2(x_j)}{\sum\limits_{j=0}^{m} p_i^2(x_j)} \\ \beta_i = \dfrac{(p_i, p_i)}{(p_{i-1}, p_{i-1})} = \dfrac{\sum\limits_{j=0}^{m} p_i^2(x_j)}{\sum\limits_{j=0}^{m} p_{i-1}^2(x_j)} \end{cases}$$

最后的拟合曲线为

$$p(x) = a_0^* p_0(x) + a_1^* p_1(x) + \cdots + a_n^* p_n(x)$$

其中系数：

$$a_i^* = \frac{(y, p_i)}{(p_i, p_i)} \quad (i = 0, 1, \cdots, n)$$

基于正交多项式求最小二乘曲线拟合函数 MATLAB 函数程序如下。

EXAMP 05007

```
function p=Orthpolyfit(x,y,N)
% ORTHPOLYFIT is least square cure fitting with orthogonal
% polynomial.
% （ORTHPLOYFIT 为基于正交多项式的最小二乘法拟合函数）
% Calling sequence: p=Orthpolyfit(x,y,N)
%
```

```
% Records of revision:
% Date: 2014-12-6
% Programmer: Z. Zhang
% Description of change: Original code
%
%
% x: a vector for independent variable.
% y: a vector for dependent variable with the same length of x.
% m: the highest power of the final result.
%
% Define a matrix psi is a N+1-by-N+1 identity matrix
% by flipping in left/right direction.
% (定义矩阵 psi 为单位阵的左右翻转)
psi=fliplr(eye(N+1,N+1));
p=zeros(1,N+1);
psi(2,N+1)=-sum(x)/length(x);
for k=2:N
    % (以 psi 的第 k-1 行作为多项式系数，计算多项式在 x 处的值，并赋给 t)
    t=polyval(psi(k,:),x);
    % (a 等于 x 与 t 中元素对应相乘然后乘 t 的转置，再除以 t 乘 t 的转置)
    t1=polyval(psi(k-1,:),x);
    % (b 等于 t 乘 t 的转置除以 t1 乘 t1 的转置)
    a=(x.*t)*t'/(t*t');
    b=(t*t')/(t1*t1');
    psi(k+1,:)=conv([1 -a],psi(k,2:N+1))-b*psi(k-1,:);
end
for k=0:N
    % (以 psi 的第 k+1 行作为多项式系数)
    t=polyval(psi(k+1,:),x);
    % y 乘 t 的转置，除以 t 乘 t 的转置
    p(k+1)=y*t'/(t*t');
end
p=p*psi;
```

【例 5.9】 已知观察数据如下表，求最小二乘拟合函数，并求出偏差平方和。

x_i	0	0.2	0.6	1.0	1.3	1.6	1.7	1.8	1.9	2.2	2.3	2.5	2.6
y_i	0	−2.5	−4.0	−5.7	−3.5	−2.0	−1.0	2.0	3.5	4.0	7.0	7.5	9.9
x_i	2.9	3.1	3.4	3.8	4.1	4.4	4.7	4.8	4.9	5.0	5.1	5.3	
y_i	5.9	11.9	13.5	13.0	11.9	9.0	6.5	4.0	1.5	0.0	−2.5	−5.0	

解： 将当前路径设置为含有 Orthpolyfit.m 文件的位置，在命令窗口输入执行语句如下：

```
>> x=[0 0.2 0.6 1.0 1.3 1.6 1.7 1.8 1.9 2.2 2.3 2.5 2.6 2.9 3.1 3.4 3.8 4.1 4.4 4.7 4.8 4.9 5.0 5.1 5.3];
>> y=[0 -2.5 -4.0 -5.7 -3.5 -2.0 -1.0 2.0 3.5 4.0 7.0 7.5 9.9 5.9 11.9 13.5 13.0 11.9 9.0 6.5 4.0 1.5 0.0 -2.5 -5.0];
>> nn=length(x);
>> N=3;
>> c=Orthpolyfit(x,y,N)
    c = -1.3382    8.8349   -5.4598   -0.8890
```

```
>> t=0:0.1:5.3;
>> u=polyval(c,t);
>> plot(t,u,'b--',x,y,'k+','LineWidth',2)
>> legend('least square fitting','data point',3)
```

运行结果如图 5.4 所示。

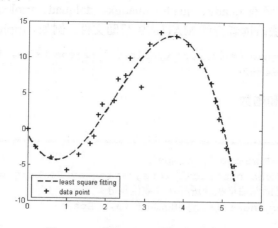

图 5.4　正交基最小二乘法拟合结果

5.3　积分与微分

5.3.1　数值积分

依据微积分原理，对于积分 $I = \int_a^b f(x)\mathrm{d}x$，可以使用牛顿-莱布尼兹公式求定积分：

$$I = \int_a^b f(x)\mathrm{d}x = F(b) - F(a)$$

但是必须求出 $f(x)$ 的原函数 $F(x)$，有时很困难甚至是不可能的，因此研究数值积分。

在区间 $[a,b]$ 上适当选取一些节点 x_i，使用 $f(x_i)$ 加权平均来近似平均高度，构造数值积分公式如下：

$$\int_a^b f(x)\mathrm{d}x \approx \sum_{i=0}^n A_i f(x_i)$$

根据节点的选择方法，数值分析中讲述的数值积分方法有 Newton-Cotes 公式、Romberg 算法、Gauss 公式等。

1. quad 函数

quad 函数是 MATLAB 自带函数，目的是使用自适应 Simpson 公式来求给定函数的定积分。函数调用格式如下：

q = quad（fun,a,b）

q = quad（fun,a,b,tol）

q = quad（fun,a,b,tol,trace）

【例 5.10】　求积分 $\int_0^{\frac{\pi}{6}} \sqrt{4-\sin^2(x)}\mathrm{d}x$。

解：借助隐函数和数值积分函数 quad，在命令窗口输入执行语句如下：

```
>> format long
>> Q=quad(@(x)(sqrt(4-sin(x).^2)),0,pi/6)
Q = 1.035763869648992
```

另外，MATLAB 中还有 quadv、quadl、quadgk、dblquad、triplequad、trapz 等求数值积分的函数，具体使用方法请读者查阅 MATLAB 帮助文件。例如，triplequad 函数求三重积分：

```
>> Q = triplequad(@(x,y,z)(y*sin(x)+z*cos(x)), 0, pi, 0, 1, -1, 1)
Q = 1.999999994362637
```

2. Romberg 方法和函数

EXAMP05008

```
function[s,n]=Romberg(a,b,Eps)
% ROMBERG solves numerical integration by Romberg quadrature.
% （ROMBERG 借助龙伯格积分进行计算数值积分）
% Calling sequence: [s,n]=Romber(a,b,Eps)
%
% Records of revision:2nd
% Date: 2015-06-10
% Programmer: Z. Zhang
% Description of change: Original code
% a: lower integral limit
% b: upper integral limit
% Eps: error bound
%
if nargin<3
    Eps=1e-6;
end
s=1;
s0=0;
k=2;
t=[];
t(1,1)=(b-a)*(f(a)+f(b))/2;
while abs(s-s0)>Eps
    h=(b-a)/2^(k-1);
    w=0;
    if h~=0
        for i=1:2^(k-1)-1
            w=w+f(a+i*h);
        end
        t(k,1)=h*(f(a)/2+w+f(b)/2);
        for j=2:k
            for i=1:(k-j+1)
                t(i,j)=(4^(j-1)*t(i+1,j-1)-t(i,j-1))/(4^(j-1)-1);
            end
        end
        s=t(1,k);
        s0=t(1,k-1);
```

```
            k=k+1;
            n=k;
        else
            s=s0;
            n=-k;
        end
end
%
% (根据被积函数的形式不同对函数 f 的表达式进行修改)
function f=f(x)
f=exp(-x.^2);
```

【例 5.11】 计算 $\int_0^1 e^{-x^2} dx$ 的值。

解：在 Romberg.m 文件中追加一个子函数，如下：

```
function f=f(x)
f=exp(-x.^2);
```

然后在命令窗口输入执行语句如下：

```
>> format long
>> Romberg(0,1)
ans = 0.746824133095094
```

5.3.2 数值微分

数值微分的方法对于求解常微分方程和偏微分方程等问题有很重要的应用，本小节讨论数值微分的方法。插值型求导公式是用插值多项式的导数来近似的：

$$f'(x) \approx P(x)$$

下面是常用的三点公式，设已给出三个节点 x_0、$x_1 = x_0 + h$、$x_2 = x_0 + 2h$ 的函数值，进行二次插值得到 $P(x)$，求导得：

$$\begin{cases} f'(x_0) \approx \dfrac{1}{2h}[-3f(x_0) + 4f(x_1) - f(x_2)] \\ f'(x_1) \approx \dfrac{1}{2h}[-f(x_0) + f(x_2)] \\ f'(x_2) \approx \dfrac{1}{2h}[f(x_0) - 4f(x_1) + 3f(x_2)] \end{cases}$$

基于 $n+1$ 个点的差分求解程序如下。

EXAMP05009

```
function [A,df]=Npointdiff(x,y,x0)
%  NPOINTDIFF is numerical differentiation with n+1 points.
%   (NPOINTDIFF 为带有 n+1 个点的数值微分).
%  Calling sequence: [A, df]=Npointdiff(x,y,x0)
%
%  Records of revision:2nd
```

```
%   Date: 2014-12-6
%   Programmer: Z. Zhang
%   Description of change: Original code
%
%   x: an independent variable.
%   y: a dependent variable.
%   x0: a specified point for x.
%
%   df: the differential for the fixed point
%
A=y;
N=length(x);
for j=2:N
    for k=N:-1:j
        A(k)=(A(k)-A(k-1))/(x(k)-x(k-j+1));
    end
end
X=x(1);
df=A(2);
prod=1;
n1=length(A)-1;
for k=2:n1
    prod=prod*(x0-x(k));
    df=df+prod*A(k+1);
end
```

【例 5.12】 已知函数 $y = f(x)$ 的下列数值：

x	2.5	2.6	2.7	2.8	2.9
y	12.185	13.4637	14.8797	16.4446	18.1741

计算 $x = 2.7$ 处的一阶导数值。

解：在 MATLAB 命令窗口输入执行语句如下：

```
>> x=[2.5 2.6 2.7 2.8 2.9];
>> y=[12.185 13.4637 14.8797 16.4446 18.1741];
>> [A,df]=Npointdiff(x,y,2.7)
A =   12.1850   12.7870    6.8650    1.9333    1.7083
df =  13.4735
```

5.4 矩阵的特征值与特征向量

特征值和特征向量的概念刻画了方阵的一些本质特征，这些概念不仅在理论上有重要的意义，在工程中经常会遇到特征值和特征向量的计算，诸如物理中的各种临界值问题，几何学、力学、控制理论等方面也有广泛的应用。

设 $A = [a_{ij}]$ 是 n 阶方阵，若有数 λ 和非零向量 x，使 $Ax = \lambda x$ 成立，则称 λ 是矩阵 A 的特

征值，x 是矩阵 A 的关于特征值 λ 的特征向量。

5.4.1 特征值函数

MATLAB 为求解特征值和特征向量的函数如下：
- eig( )：求特征值和特征向量。
- eigs( )：求大型稀疏矩阵的特征值和特征向量。
- qz( )：求广义的特征值分解。
- schur( )：Schur 分解。

【例 5.13】 求矩阵 $A = \begin{bmatrix} 2 & -1 & 0 \\ -1 & 2 & -1 \\ 0 & -1 & 2 \end{bmatrix}$ 的特征值和特征向量。

解：在 MATLAB 命令窗口输入执行语句如下：

```
>> A=[2 -1 0;-1 2 -1;0 -1 2];
>> [V, D] = eig(A)
V =  0.5000    -0.7071    -0.5000
     0.7071     0.0000     0.7071
     0.5000     0.7071    -0.5000
D =  0.5858          0          0
          0     2.0000          0
          0          0     3.4142
```

而 $A = \begin{bmatrix} 2 & -1 & 0 \\ -1 & 2 & -1 \\ 0 & -1 & 2 \end{bmatrix}$ 的准确特征值为 $\lambda_1 = 2 - \sqrt{2}$、$\lambda_2 = 2$、$\lambda_3 = 2 + \sqrt{2}$。

5.4.2 幂法和反幂法

幂法是求方阵的模的最大特征值及对应特征向量的一种向量迭代方法。

设 n 阶方阵 A 有 n 个线性无关的特征向量 $v_1, v_2, \cdots, v_n$，对应的特征值依次为 $\lambda_1, \lambda_2, \cdots, \lambda_n$ 且满足不等式 $|\lambda_1| > |\lambda_2| \geqslant \cdots |\lambda_n|$。当 k 充分大时，

$$A^k x^{(0)} \approx \lambda_1^k a_1 v_1。$$

也就是说，当 k 足够大时，$A^k x^{(0)}$ 可以近似作为属于方阵 A 对应于 λ_1 的特征向量。为了求得 λ_1 的近似值，记 $\max(x)$ 为 x 模最大的分量，则当 k 充分大时，有

$$\frac{\max(A^k x^{(0)})}{\max(A^{k-1} x^{(0)})} \approx \frac{\max(\lambda_1^k a_1 v_1)}{\max(\lambda_1^{k-1} a_1 v_1)} = \lambda_1$$

这就是幂法的基本思想，由此得到如下迭代公式：

$$x^{(k+1)} = A x^{(k)} \quad (k = 0, 1, 2, \cdots)$$

幂法数值求最大特征值与特征向量 MATLAB 函数程序如下：

EXAMP05010

```
function[Lambda,V]=Powermethod(A,x0,Eps,maxtimes)
%   POWERMETHOD solves eigenvalue and eigenvector with power method.
```

```
%    (POWERMETHOD 借助幂法计算最大特征值和特征向量)
% Calling sequence:
%        [Lambda,V]=Powermethod(A,x0,Eps,maxtimes)
%
% Records of revision:
% Date: 2014-11-12
% Programmer: Z. Zhang
% Description of changer: original code
%
% A: a matrix
% x0: initial vector
% Eps: the maximum error
% maxtimes: the maximum times
% Lambda: the maximum of the required eigenvalues
% V: the eigenvector with respect to Lambda
%
if nargin<2
    warning('Not enough imput arguments!')
elseif nargin==2
Eps=1e-6;
Maxtimes=500;
elseif nargin==3
    maxtimes=500;
end

Lambda=0;
cnt=0;
err=1;
state=1;
while (cnt<=maxtimes) & (state==1)
    Y=A*x0;
    [m j] = max(abs(Y));
    c1=m;
    dc=abs(Lambda-c1);
    Y=(1/c1)*Y;
    dv=norm(x0-Y);
    err=max(dc,dv);
    x0=Y;
    Lambda=c1;
    state=0;
    if err>Eps
        state=1;
    end
    cnt=cnt+1;
end
V=x0;
```

【例 5.14】 使用幂法求矩阵 $A = \begin{bmatrix} 0 & 11 & -5 \\ -2 & 17 & -7 \\ -4 & 26 & 10 \end{bmatrix}$ 的模最大的特征值和特征向量。

解：在 MATLAB 命令窗口输入执行语句如下：

```
>> A=[0 11 -5;-2 17 -7;-4 26 -10];
>> x0=[1;1;1];
>> [L V]=Powermethod (A,x0,10^(-6),500)
L = 4.0000
V = 0.4000
    0.6000
1.0000
```

在本例中，矩阵 A 的精确特征值为 $\lambda_1=4$、$\lambda_2=2$、$\lambda_3=1$，对应 $\lambda_1=4$ 的特征向量是 [0.4，0.6，1]，使用幂法经过 22 次迭代运算，达到精度要求。

反幂法是用来求方阵 A 的模最小的特征值及相应特征向量的迭代方法。若 A 可逆，则 A^{-1} 存在，用 A^{-1} 替代 A 进行上述的幂法就是所谓的反幂法。反幂法求最小特征值与特征向量 MATLAB 函数程序如下。

EXAMP05011

```
function [Lambda,V]=Invpowermethod (A,x0,alpha,Eps,maxtimes)
%  INVPOWERMETHOD solves eigenvalue and eigenvector by inverse
%  power method.
%  （POWERMETHOD 借助反幂法计算最小特征值和相应特征向量）
%  Calling sequence:
%        [Lambda,V]=Invpowermethod (A,x0,alpha,Eps,maxtimes)
%
%  Records of revision:
%  Date: 2014-11-12
%  Programmer: Z. Zhang
%  Description of changer: original code
%
%  A: a matrix
%  x0: initial vector
%  Eps: the maximum error
%  maxtimes: the maximum times
%  Lambda: the minimum of the required eigenvalues
%  V: the eigenvector with respect to Lambda
%
[n n]=size (A);
A=A-alpha*eye (n);
Lambda=0;
cnt=0;
err=1;
state=1;
while (cnt<=maxtimes) & (state==1)
    Y=A\x0;
    [m j] = max (abs (Y));
    c1=m;
    dc=abs (Lambda-c1);
    Y= (1/c1) *Y;
    dv=norm (x0-Y);
    err=max (dc,dv);
```

```
        x0=Y;
        Lambda=c1;
        state=0;
        if err>Eps
            state=1;
        end
        cnt=cnt+1;
end
Lambda=alpha+1/c1;
V=x0;
```

5.4.3 Jacobi 方法

Jacobi 方法是用来计算实对称矩阵的全部特征值和特征向量的一种迭代方法。由代数学的原理知道：若 n 阶方阵 A 是实对称矩阵，则存在正交矩阵 P：

$$P^{\mathrm{T}}AP = \mathrm{diag}(\lambda_1, \lambda_2, \cdots, \lambda_n) = \begin{bmatrix} \lambda_1 & & & \\ & \lambda_2 & & \\ & & \ddots & \\ & & & \lambda_n \end{bmatrix}$$

需要设法构造一系列简单的正交矩阵序列 $P_1, P_2, \cdots, P_n, \cdots$，使得：

$$\begin{cases} B_0 = A \\ B_k = P_k^{\mathrm{T}} B_{k-1} P_k \end{cases} (k=1,2,\cdots)$$

即

$$\lim_{k \to \infty} B_k = \lim_{k \to \infty} P_k^{\mathrm{T}} B_{k-1} P_k = \mathrm{diag}(\lambda_1, \lambda_2, \cdots, \lambda_n)$$

在实际应用中，当非对角线上的元接近于零时，迭代停止。此时：

$$B_k \approx B = \mathrm{diag}(\lambda_1, \lambda_2, \cdots, \lambda_n)$$

正交矩阵 P 的每个列向量是方阵 A 对应于特征值的特征向量的近似值。借助 Jacobi 方法计算特征值与特征向量的程序如下：

EXAMP 05012

```
function [V,D]=JacobimethodED(A,Eps)
%   JACOBIMETHODED solves eigenvalue and eigenvector by Jacobi
%   method.
%   （POWERMETHOD 借助 Jacobi 方法计算特征值和相应特征向量）
%   Calling sequence:
%       [V,D]=JacobimethodED(A,Eps)
%
%   Records of revision:
%   Date: 2014-11-12
%   Programmer: Z. Zhang
%   Description of changer: original code
%
%   A: a matrix
%   Eps: the maximum error
%   V: the matrix of eigenvectors
```

```
%    D: the diagonal n*n matrix of eigenvalues
%
[n,n]=size (A);
D=A;
V=eye (n);
state=1;
[n1 p]=max (abs (D-diag (diag (D))));
[n2 q]=max (n1);
p=p (q);
while state==1
    t=D (p,q) / (D (q,q) -D (p,p));
    c=1/sqrt (t^2+1);
    s=c*t;
    R=[c s;-s c];
    D ([p q],:) =R'*D ([p q],:);
    D (:,[p q]) =D (:,[p q]) *R;
    V (:,[p q]) =V (:,[p q]) *R;
    [n1 p]=max (abs (D-diag (diag (D))));
    [n2 q]=max (n1);
    p=p (q);
    if abs (D (p,q)) <Eps*sqrt (sum (diag (D) .^2) /n)
        state=0;
    end
end
D=diag (diag (D));
```

【例 5.15】 求矩阵 $A = \begin{bmatrix} 8 & -1 & 3 & -1 \\ -1 & 6 & 2 & 0 \\ 3 & 2 & 9 & 1 \\ -1 & 0 & 1 & 7 \end{bmatrix}$ 的特征值和特征向量。

```
>> A=[8 -1 3 -1;-1 6 2 0;3 2 9 1;-1 0 1 7];
>> [V D]=JacobimethodED (A,1e-5)
V = 0.5288    -0.5730     0.5823     0.2301
    0.5920     0.4723     0.1758    -0.6290
   -0.5360     0.2820     0.7925    -0.0712
    0.2875     0.6075     0.0447     0.7392
D = 3.2957     0          0          0
    0          8.4077     0          0
    0          0         11.7043     0
    0          0          0          6.5923
```

5.4.4 QR 方法

QR 方法是目前求一般方阵全部特征值的最有效并被广泛应用的方法之一。它是一种矩阵迭代法，从 $A_1 = A$ 出发，对 A_1 进行 QR 分解 $A_1 = Q_1 R_1$，然后将 Q_1, R_1 反序相乘得到 $A_2 = R_1 Q_1$，再以 A_2 替代 A_1，重复上述步骤得到方阵序列 $\{A_k\}$。

通过一系列正交相似变换得出方阵序列 $\{A_k\}$ 在一定条件下，可以证明当 $k \to +\infty$ 时，A_k 本质收敛于一个块对角矩阵，其对角子块是 1 阶或者 2 阶方阵，因此容易求出它们的特征值。

借助 QR 方法计算矩阵的特征值程序如下。

EXAMP05013

```
function D=Eigenqr(A,Eps)
% EIGENQR solves eigenvalues for a matrix by QR method.
% （EIGENQR借助QR方法求解矩阵的特征值）
% Calling sequence: D=Eigenqr(A,Eps)
%
% Records of revision:
% Date: 2014-11-12
% Programmer: Z. Zhang
% Description of changer: original code
%
% A: a matrix
% Eps: the maximum error
% D: the diagonal n*n matrix of eigenvalues
%
[n,n]=size(A);
m=n;
D=zeros(n,1);
B=A;
while m>1
    while (abs(B(m,m-1))>Eps)
        s=eig(B(m-1:m,m-1:m));
        [j,k]=min([abs(B(m,m)*[1 1]'-s)]);
        [Q,U]=qr(B-s(k)*eye(m));
        B=U*Q+s(k)*eye(m);
    end
    A(1:m,1:m)=B;
    m=m-1;
    B=A(1:m,1:m);
    D(m)=B;
end
D=diag(A);
```

【例 5.16】 使用 QR 方法求矩阵 $A=\begin{bmatrix} -3 & -5 & -1 \\ 13 & 13 & 1 \\ -5 & -5 & 1 \end{bmatrix}$ 的特征值。

解：将当前路径设为含有 Eigenqr.m 文件的位置，在命令窗口输入执行语句如下：

```
>> A=[-3 -5 -1 ;13 13 1; -5 -5 1];
>> format short
>> D=Eigenqr(A,1e-5)
D = 6.0000
 3.0000
 2.0000
```

5.5 常微分方程数值解法

在科学研究和工程计算技术问题中，经常需要求解常微分方程或者偏微分方程的定解问题。微分方程数值解法是指在一些离散点上给出定解问题的近似值。

5.5.1 欧拉（Euler）方法

对简单的一阶方程的初值问题：$\begin{cases} y' = f(x,y) \\ y(x_0) = y_0 \end{cases}$，由欧拉公式可得：

$$y_{n+1} = y_n + hf(x_n, y_n)$$

由公式编写的欧拉函数程序如下。

EXAMP05014

```
function [xx,yy]=Euler(f,a,b,y0,h)
% EULER solves the initial-value problem of ODE with Euler method.
% （EULER借助欧拉方法解有初始值的一般方程的问题）
% Calling sequence: [xx,yy]=Euler(f,a,b,y0,h)
%
% Records of revision:2nd
% Date: 2014-11-02
% Programmer: Z. Zhang
% Description of changer: original code
%
% f: the function entered as a string 'f'
% a: the left point
% b: the right point
% y0: the initial condition
% h: the step length
x=a:h:b;
n=length(x);
y=zeros(1,n);
y(1)=y0;
for i=1:n-1;
    y(i+1)=y(i)+h*feval(f,x(i),y(i));
end
xx=x;
yy=y;
plot(xx,yy,'rp')
xlabel('x'),ylabel('y')
title('Solution of ODE with Euler method')
```

【例 5.17】 使用欧拉方法求

$$\begin{cases} y'(x) = y(x) + (1+x)y^2(x) \\ y(x)|_{x=1} = -1 \end{cases} \quad 1 < x < 1.5$$

并与准确解 $y(x)=\dfrac{1}{x}$ 进行比较。

解：首先建立函数文件，取 $h=0.1$：

```
myfun.m
function f=myfun(x,y)
f=y+(1+x)*y^2
```

将当前路径设定在含有 Euler.m 和 myfun.m 文件的位置，在命令窗口输入执行语句如下：

```
[x y]=Euler('myfun',1,1.5,-1,0.1)
x =  1.0000    1.1000    1.2000    1.3000    1.4000    1.5000
y = -1.0000   -0.9000   -0.8199   -0.7540   -0.6986   -0.6514
```

运行结果如图 5.5 所示。

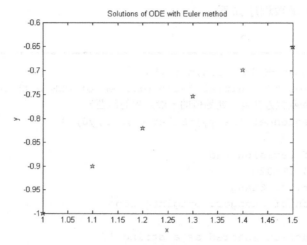

图 5.5　欧拉方法解一般方程

可以证明：使用欧拉方法的精度较低，只有一阶的代数精度。为提高精度，通常采用改进的欧拉方法：

$$\begin{cases} y_{n+1}=\dfrac{1}{2}(y_p+y_c) \\ y_p=y_n+hf(x_n,y_n) \\ y_c=y_n+hf(x_{n+1},y_n) \end{cases}$$

改进的欧拉方法解一般方程的函数程序如下。

EXAMP05015

```
function [xx,yy]=Modeuler(f,a,b,y0,h)
  % This program is to solve the initial-value problem of ODE with modified
Euler method.
  % MODEULER solves the initial-value problem of ODE with
  % modified Euler method.
  %  (EULER 借助改进的欧拉方法解有初始值的一般方程的问题)
  % Calling sequence: [xx,yy]=Modeuler(f,a,b,y0,h)
```

```
%
%   Records of revision:2nd
%   Date: 2014-11-02
%   Programmer: Z. Zhang
%   Description of changer: original code
%
%   f: the function entered as a string 'f'
%   a: the left point
%   b: the right point
%   y0: the initial condition
%   h: the step length
%
x=a:h:b;
n=length(x);
y=zeros(1,n);
y(1)=y0;
for i=1:n-1;
    yp=y(i)+h*feval(f,x(i),y(i));
    yc=y(i)+h*feval(f,x(i+1),yp);
    y(i+1)=(yp+yx)/2;
end
xx=x;
yy=y;
ey=-1./x
plot(xx,yy,'bp',xx,ey,'r*')
xlabel('x'),ylabel('y')
title('Solutions of ODE with modified Euler method')
```

对于例 5.17，使用改进的欧拉方法求解，在命令窗口输入执行语句如下：

```
>> [x y]=Modeuler('myfun',1,1.5,-1,0.1)
ey = -1.0000   -0.9091   -0.8333   -0.7692   -0.7143   -0.6667
x  =  1.0000    1.1000    1.2000    1.3000    1.4000    1.5000
y  = -1.0000   -0.9100   -0.8346   -0.7707   -0.7158   -0.6681
```

运行结果如图 5.6 所示。

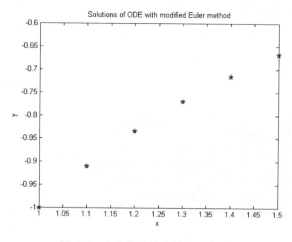

图 5.6　改进的欧拉方法解一般方程

因为改进的欧拉方法是二阶的，改进的欧拉方法比欧拉方法精度上有明显提高。

5.5.2 龙格库塔（Runge-Kutta）方法

为提高常微分方程数值解法的精度，间接使用 Taylor 展开的思想构造高阶的数值解法，四阶的龙格库塔方法（经典公式）如下：

$$\begin{cases} y_{n+1} = y_n + \dfrac{h}{6}(k_1 + 2k_2 + 2k_3 + k_4) \\ k_1 = f(x_n, y_n) \\ k_2 = f(x_n + \dfrac{h}{2}, y_n + \dfrac{h}{2}k_1) \\ k_3 = f(x_n + \dfrac{h}{2}, y_n + \dfrac{h}{2}k_2) \\ k_4 = f(x_n + h, y_n + hk_3) \end{cases}$$

经典公式的函数程序如下。

EXAMP05016

```
function [xx,yy]=Rungekutta(f,a,b,y0,h)
% RUNGEKUTTA solves the initial-value problem of ODE with
% Runge-Kutta method.
% （RUNGEKUTTA 借助龙格库塔方法求解一般方程）
% Calling sequence: [xx,yy]=Rungekutta(f,a,b,y0,h)
%
% Records of revision:2nd
% Date: 2014-11-12
% Programmer: Z. Zhang
% Description of changer: original code
%
% f: the function entered as a string 'f'
% a: the left point
% b: the right point
% y0: the initial condition
% h: the step length
%
x=a:h:b;
n=length(x);
y=zeros(1,n);
y(1)=y0;
for i=1:n-1;
%   feval: Execute the specified function.
%           （计算特定函数）
    k1=feval(f,x(i),y(i));
    k2=feval(f,x(i)+h/2,y(i)+h/2*k1);
    k3=feval(f,x(i)+h/2,y(i)+h/2*k2);
    k4=feval(f,x(i+1),y(i)+h*k3);
    y(i+1)=(k1+2*k2+2*k3+k4)*h/6+y(i);
end
xx=x;
```

```
yy=y;
ey=-1./x
plot (xx,yy,'kp',xx,ey,'k*')
xlabel (' x'),ylabel (' y')
title ('Solutionof ODE with Runge-Kutta method')
legend ('numal solution','exact solution',4)
```

对于例 5.17，使用龙格库塔方法求解，在命令窗口输入执行语句如下：

```
>> [x y]=Rungekutta ('myfun',1,1.5,-1,0.1)
ey =  -1.0000   -0.9091   -0.8333   -0.7692   -0.7143   -0.6667
x =    1.0000    1.1000    1.2000    1.3000    1.4000    1.5000
y =   -1.0000   -0.9091   -0.8333   -0.7692   -0.7143   -0.6667
```

运行结果如图 5.7 所示。

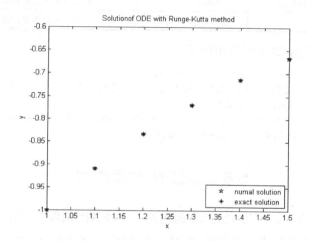

图 5.7　龙格库塔方法解一般方程

下面用 3 种方法对例题 5.17 进行求解，通过和精确解进行比较，结果如下：

i	x_i	精确值	欧拉方法误差	改进的欧拉方法误差	经典公式（龙格库塔方法）的误差
0	1	−1	0	0	0
1	1.1	−0.9091	0.0091	−0.0009	−0.0000024057
2	1.2	−0.8333	0.0134	−0.0013	−0.0000034166
3	1.3	−0.7692	0.0152	−0.0015	−0.0000037232
4	1.4	−0.7143	0.0157	−0.0015	−0.0000036769
5	1.5	−0.6667	0.0153	−0.0014	−0.0000034609

5.5.3　MATLAB 的相关函数

在 MATLAB 中有几个专门用于求解常微分方程的函数，如 ode45、ode113、ode15s、ode23s、ode23t、ode23tb，它们多是采用了 Runge-Kutta 方法。其中，ode23 系列为采用二阶、三阶 Runge-Kutta 方法，而 ode45 系列为采用四阶、五阶 Runge-Kutta 方法，一般来说，ode45 比 ode23 积分段少，运算速度更快一些。函数的详细用法，请读者查看 MATLAB 相关帮助文件。

几种常微分方程数值求解函数的比较参见表 5.1。

表 5.1　几种常微分方程数值求解函数比较

函　数	算　法	适用条件	精度	特　点
ode45	四阶、五阶龙格库塔法	非刚性方程	中	最常用的单步算法
ode113	可变阶算法	非刚性方程	低-高	多步算法，在相同的精度下，它可能比 ode45 或 ode23 更快一些，但不适用于不连续的系统
ode15s	可变阶算法	刚性方程	低-中	多步算法，方程刚性，用 ode45 很慢时
ode23	二阶、三阶龙格库塔法	非刚性方程	低	对于轻度的刚性方程，它比 ode45 更有效；对于相同的精度，它需要比 ode45 更小的步长
ode23s	基于改进的 ROSENBROCK 公式	刚性方程	低	单步算法，用 ode15s 解决不好的刚性方程
ode23t	自由内插实现的梯形规则	一般刚性	低	给出的解无数值衰减
ode23tb	TR-BDF2 算法	刚性方程	低	允许误差范围较大时比 ode15s 好

对例 5.17 使用上面函数求解语句如下：

```
>> [x y]=ode23 ('myfun',[1,1.5],-1)
x =  1.0000   1.0500   1.1000   1.1500   1.2000   1.2500   1.3000   1.3500
     1.4000   1.4500   1.5000
y =-1.0000  -0.9524  -0.9091  -0.8696  -0.8333  -0.8000  -0.7692
    -0.7407  -0.7143  -0.6896  -0.6667
```

5.6　非线性方程求根

求函数 $f(x)$ 的零点，即方程 $f(x)=0$ 的根是一个经典的问题。早在 16 世纪就找到了三次、四次代数方程的求根公式，但直到 19 世纪才证明 $n \geqslant 5$ 次的一般代数方程不能用代数公式求解。随着计算工具的发展，求满足精度要求的近似解方法取得了很大进展，本节主要介绍了几种常用的求解非线性方程的数值方法。

5.6.1　二分法

设 $f(x)=0$ 在区间 (a,b) 内有实根，若 $c=\dfrac{a+b}{2}$，$f(c)$ 可能有 $f(c)=0$，c 为方程的一个实根，输出根 c；$f(a)f(c)<0$，则有一个实根在区间 (a,c) 内，取 $a_1=a$、$b_1=c$；$f(c)f(b)<0$，则有一个实根在区间 (c,b) 内，取 $a_1=c$、$b_1=b$。

重复上述步骤，如果有限步求得根的值，则计算终止，否则每次有根区间长度减半，并得到一闭区间：$[a,b] \supset [a_1,b_1] \supset [a_2,b_2] \supset \cdots \supset [a_n,b_n] \supset \cdots$

由闭区间套定理，可得 $\lim\limits_{n \to \infty} a_n = \lim\limits_{n \to \infty} b_n = x^*$，实际计算时，可用精度 $\varepsilon > 0$ 来控制二分次数。二分法数值求解非线性方程的 MATLAB 函数程序如下。

EXAMP05017

```
%  EXAMP05017 solves nonlinear equation by bisection method.
%  （EXAMP05017 借助二分法求解非线性方程）
```

```
%
% Records of revision:2nd
% Date: 2015-07-12
% Programmer: Z. Zhang
% Description of changer: original code
%
clear
clc
f=input('请输入函数 f(x)=','s');
interv=input('请输入区间：');
err=input('请输入精度：');
a=interv(1);
b=interv(2);
yc=1;
n=0;
while ((b-a)>err) & (yc~=0)
c=(a+b)/2;
x=a;
% eval: Execute string with MATLAB expression.
%        (看作表达式执行)
ya=eval(f);
x=b;
yb=eval(f);
x=c;
yc=eval(f);
if ya*yc<0
b=c;
else
a=c;
end
x0=c;
end
n   %（显示计算的次数）
x0
```

【例 5.18】 用二分法求 $x^3+4x^2-10=0$ 在 $(1,2)$ 内的根，要求误差不超过 $\frac{1}{2}\times 10^{-5}$。

解：将当前路径设定为含有 EXAMP05017.m 文件的位置，在命令窗口输入执行语句如下：

```
>> EXAMP05017
请输入函数 f(x)=x^3+4*x^2-10
请输入区间：[1,2]
请输入精度：0.5*10^(-5)
n = 18
x0 = 1.3652
```

5.6.2 牛顿迭代法

解非线性方程的牛顿迭代法是一种将非线性函数线性化的方法。

令 $f(x) \approx f(x_n) + f'(x_n)(x-x_n)$，则线性方程 $f(x_n)+f'(x_n)(x-x_n)=0$ 的根就是非线性方程 $f(x)=0$ 的近似根。构造牛顿迭代格式为

$$x_{n+1} = x_n - \frac{f(x_n)}{f'(x_n)}$$

当迭代法收敛时，迭代序列的极限就是非线性方程的根，即 $\lim_{n \to \infty} x_n = x^*$。牛顿法数值求解非线性方程的 MATLAB 函数程序如下。

EXAMP05018

```
% EXAMP05018 solves nonlinear equation by Newton method.
% (EXAMP05018 借助牛顿法求解非线性方程)
%
% Records of revision:2nd
% Date: 2015-07-12
% Programmer: Z. Zhang and Wang Yong-Long
% Description of changer: original code
%
clear
clc
syms x
f=input('请输入函数 f(x)=','s');
f1=diff(f,x);
x0=input('请输入初始值：');
Eps=input('请输入精度：');
x=x0;
x1=x0-eval(f)/eval(f1);
n=0;
while abs(x1-x0)>Eps
    x0=x1;
    x=x0;
    fx0=eval(f);
    f1x0=eval(f1);
    x1=x1-fx0/f1x0;
    n=n+1;
end
n
x1
```

例 5.18 解：将当前路径设定为含有 EXAMP05018.m 文件的位置，在命令窗口输入执行语句如下：

```
>> EXAMP05018
请输入函数 f(x)=x^3+4*x^2-10
请输入初始值：1
请输入精度：0.5*10^(-5)
n = 4
x1 = 1.3652
```

上面结果显然说明牛顿迭代法远比二分法有效。

5.6.3 弦截法

在牛顿迭代法中，不但要求给出函数值，而且要求给出其导数值，当函数比较复杂时，求导的运算往往非常困难。离散化微分运算，使用差商近似微分，就得到了弦截法迭代公式：

$$x_{n+1} = x_n - \frac{f(x_n)}{f(x_n) - f(x_{n-1})}(x_n - x_{n-1})。$$

弦截法数值求解非线性方程的 MATLAB 函数程序如下。

EXAMP05019

```
% EXAMP05019 solves nonlinear equation by secant method.
% （EXAMP05019 借助弦截法求解非线性方程）
%
% Records of revision:2nd
% Date: 2015-07-12
% Programmer: Z. Zhang
% Description of changer: original code
%
clear
clc
syms x;
f=input ('Please input f(x)=');
n=input ('Please input the iteration times=');
x1=input ('Please input the initial value 1=');
x2=input ('Please input the initial value 2=');
for i=1:n
fx1=feval (inline(f),x1);
fx2=feval (inline(f),x2);
if abs (fx1-fx2) >1.0e-5
x=x2- (fx2/ (fx2-fx1)) * (x2-x1);
x1=x2;
x2=x;
x0= (x1+x2) /2
else
    break
end
end
```

例 5.18 解：将当前路径设定为含有 EXAMP05019.m 文件的位置，在命令窗口输入执行语句如下：

```
Please input f(x)=x^3+4*x^2-10
Please input the iteration times=9
Please input the initial value 1=0.5
Please input the initial value 2=1.5
x0 = 1.3944
x0 = 1.3246
x0 = 1.3629
x0 = 1.3653
x0 = 1.3652
```

小　结

本章主要介绍了线性方程组的矩阵分解直接方法和迭代方法；离散数据的插值和最小二乘曲线拟合；积分和微分的数值解法；矩阵特征值和特征向量的幂法、反幂法、Jacobi 方法和 QR 方法；以及用 Euler 方法和 Runge-Kutta 方法等解常微分方程；非线性方程求解的二分法、牛顿迭代法和弦截法。通过示例对算法进行分析和编程，算法表述清楚，能够使读者对 MATLAB 中许多自带函数的内在计算方法有了初步了解。

习　题

5.1　用矩阵三角分解方法解方程组：

$$\begin{cases} 2x_1 + x_2 - x_3 = 14 \\ 4x_1 - 5x_2 + 3x_3 = 18 \\ 6x_1 + 9x_2 - x_3 = 20 \end{cases}$$

5.2　用 Cholesky 分解方法解方程组：

$$\begin{cases} 3x_1 + 2x_2 + 3x_3 = 5 \\ 2x_1 + 2x_2 = 3 \\ 3x_1 + 12x_3 = 7 \end{cases}$$

5.3　设某实验数据如下表：

x_i	0	0.5	1	1.5	2	2.5
y_i	2.0	1.0	0.9	0.6	0.4	0.3

用最小二乘法求拟合曲线。

5.4　已知函数 $y = f(x)$ 的观测数据如下表：

x_i	-2	0	4	5
y_i	5	1	-3	1

试构造不超过三次的插值多项式，并计算 $f(-1)$ 的近似值。

5.5　用最小二乘方法对下表中的数据用经典公式 $y = ae^{bx}$ 进行拟合：

x_i	1	2	3	4	5	6	7	8	9	10
y_i	2650	1940	1495	1090	765	540	430	290	225	205

5.6　使用数值方法求函数 $f(x) = \dfrac{1}{(1+x^2)^2}$ 在 $x = 1.1$、1.2、1.3 处的导数值。$f(x)$ 的函数值如下表：

x	1.0	1.1	1.2	1.3	1.4
$f(x)$	0.2500	0.2268	0.2066	0.1890	0.1736

5.7 用 Romberg 方法计算积分 $\int_0^1 \frac{\sin(x)}{x} dx$，要求结果保留 5 位小数。

5.8 求矩阵 $\begin{bmatrix} 1.0 & 1.0 & 0.5 \\ 1.0 & 1.0 & 0.25 \\ 0.5 & 0.25 & 2.0 \end{bmatrix}$ 的全部特征值及其对应的特征向量。

5.9 使用 Euler 方法解初值问题 $\begin{cases} y' = -y - xy^2 \\ y(0) = 1 \end{cases}$ $(0 \leqslant x \leqslant 1)$，计算结果保留 6 位小数。

5.10 求解二阶微分方程的初值问题：
$$\begin{cases} y'' - y = x \\ y(0) = 0 \quad (0 \leqslant x \leqslant 1) \\ y'(0) = 1 \end{cases}$$

5.11 取步长 $h = 0.1$，用经典公式求解初值问题：
$$\begin{cases} y' = 2xy \\ y(0) = 1 \end{cases} (0 \leqslant x \leqslant 1)$$

第 6 章 图形用户界面设计（GUI）

图形用户界面（GUI）对于编程者和用户来讲是友好的图形界面。一个好的 GUI 可以让程序变得容易操作使用，比如有视觉效果较好的控件（如窗口、图标、按钮、文本、罗列框、滚动条、菜单等）。用户选中或激活这些对象，使得某个操作变化发生，从而实现用户和计算机或计算机程序之间的交互操作。GUI 应该运行在易理解、易预测的模式下，使得在用户完成一个操作时即可知道结果。比如，当鼠标移动到按钮上时，能够提示该按钮的基本功能。在 MATLAB 软件中，用户可以使用 guide 提供的多种设计模板，定制或创建自己的图形用户界面，在保存时会自动生成相应的 M 文件。当然，用户也可以借助图形句柄函数实现用户和计算机之间的交互操作，句柄图形有 uicontrol 对象、uimenu 对象、uicontextmenu 对象、uitoolbar 对象等。

本章内容有 5 节。6.1 节借助函数 guide 创建 GUI，6.2 节给出创建 GUI 示例，6.3 节给出 GUI 实例，6.4 节介绍常用 GUI 组件创建和设置，6.5 节介绍编译独立的应用程序。

6.1 借助函数 guide 创建 GUI

函数 guide 为创建 GUI 提供一系列工具，这些工具大大地简化了设计构建 GUI 程序。可以用函数 guide（Graphical User Interface Developed Environment）来创建 GUI，并且会自动生成一个 M 文件，里面包含了 GUI 控件的操作过程。借助 M 文件编译器，可在回调函数执行用户编写的 MATLAB 执行语句。

创建 GUI 步骤如下：首先在 MATLAB 命令窗口光标处输入 guide 命令，然后按回车键，打开新窗口，如图 6.1 所示。该窗口有两个选项卡，一个是"Create New GUI"（创建新的 GUI），另一个是"Open Existing GUI"（打开已有 GUI）。"创建新的 GUI"面板含有 4 个可选项，分别为"Blank GUI（Default）"（空白 GUI（默认））、"GUI with Uicontrols"（带控件的 GUI）、"GUI with Axes and Menu"（带坐标系和菜单的 GUI）和"Modal Question Dialog"（问答模板）。最下面有"Save new figure as:"可选项，如果选中，后面地址栏可以输入保存当前图形的地址及文件名。

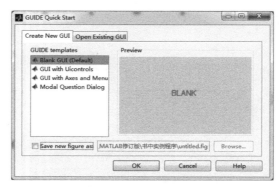

图 6.1 guide 快速启动

空白 GUI 窗口如图 6.2 所示，带控件的 GUI 窗口如图 6.3 所示，带坐标系和菜单的 GUI 窗口如图 6.4 所示，问答模板 GUI 窗口如图 6.5 所示。"打开已有 GUI"只是为已经创建的 GUI 打开提供途径。

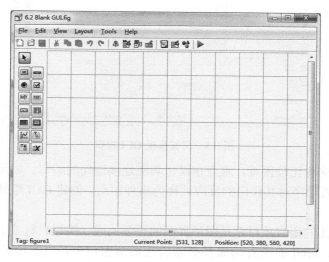

图 6.2　空白 GUI 窗口

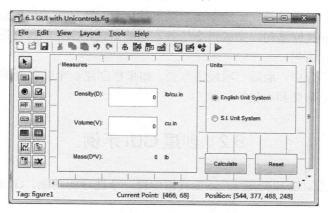

图 6.3　带控件的 GUI 窗口

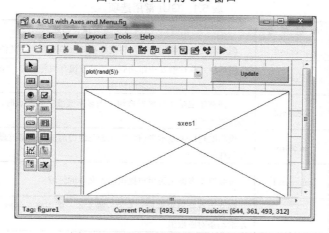

图 6.4　带坐标系和菜单的 GUI 窗口

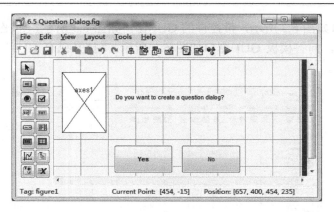

图 6.5　问答模板 GUI 窗口

GUI 对程序员来说比较困难，难点在操作响应连接语句的编写。如此操作方式都被称为众所周知的事件，而程序对其作出的反应被称为事件驱动。创建 GUI 组件的 3 个主要部分介绍如下。

（1）构成：GUI 的所有项目都是图形界面的重要构成（按钮、标签、可编辑框，等等）。

（2）图形窗口：GUI 的所有构成都必须放在图形窗口中，该图形窗口就显示在计算屏幕上。

（3）回调函数：对无论是鼠标的单击、菜单的选取，还是可选框的选中等操作都能作出反应，通过回调相应执行语句，完成指令任务。也就是说，GUI 的所有图形组件实际上就是正确及时回调可执行语句。

上面介绍的 3 部分中，最后一项最为重要。如果想创建一个性能好的 GUI 界面，首要就是回调执行语句的正确性和流畅性。

6.2　创建 GUI 示例

在创建 GUI 之前，必须知道 guide 工具箱中每一个选项的功能。GUI 工具箱常用选项描述参见表 6.1。

表 6.1　GUI 常用选项

基元	图标	创建命令	描述
按钮	OK	uicontrol	图形组元执行按钮功能，将鼠标在其上单击会产生功能作用
触发按钮		uicontrol	图形组元执行触发按钮功能。触发按钮有两种状态："on" 和 "off"，鼠标在其上单击会产生这两种状态的转换
广播按钮		uicontrol	广播按钮也是执行触发按钮，当显示圆圈中有黑点时，代表是打开状态，否则为关闭状态
检测框		uicontrol	检测框是用来显示当前状态是否已经检测过，检测过显示 "√"，没有检测的不显示任何符号
可编辑文本框	EDIT	uicontrol	可编辑文本框是为用户提供输入文本的区域
静文本框	TXT	uicontrol	静文本框是用来显示文本的。静文本一般用来显示控件基本功能，或者用来显示滑动条对应数值

(续表)

基 元	图 标	创 建 命 令	描 述
滑动条		uicontrol	滑动条也称为滚动条，一般提示用户可以在某一个范围之内任意取值，同时便于用户选取
罗列框		uicontrol	显示一系列选项的罗列框，用户可以选择一个和多个选项
常用菜单		uicontrol	常用菜单一般显示一系列选项，用户可以任意选取其中一个或几个条目让其执行
坐标系		axes	坐标系为 GUI 提供了显示图形的场所。这和 Axis 属性设置一样，程序员可借助属性设置参数进行任意设置。也可以为方便用户将其设置成按钮和图标，便于用户使用
镶嵌面板		uicontrol	镶嵌面板上可以由多个控件组成，可以给用户一目了然的感觉，一般是功能相关，或者是操作对象相近的操作指令图标、按钮等放在一个镶嵌面板内
按钮群		uicontrol	按钮群和镶嵌面板功能很类似，可用来管理外在具有可选择功能的按钮，如广播按钮、触发按钮等
ActiveX 组元		uicontrl	ActiveX 组元能在 GUI 界面上显示 ActiveX 控件
选择按钮			选择按钮是一个开关按钮，当打开时，可以根据用户选中的目标进行选取
菜单项		uimenu	创建菜单项。菜单项可通过鼠标光标单击进行激发
背景菜单		uicontextmenu	创建背景菜单，它能够显示在 GUI 图形界面上，可以通过鼠标单击右键触发

【例 6.1】 单击计数器。

本例为一个简单 GUI 设计，含一个按钮和一个文本框。完成的任务为单击一次按钮记录一次，同时文本框更新数据，显示单击的总次数。

分析与求解：

【第一步】 通过函数 guide 打开一个新空白 GUI 窗口，如图 6.6 所示。

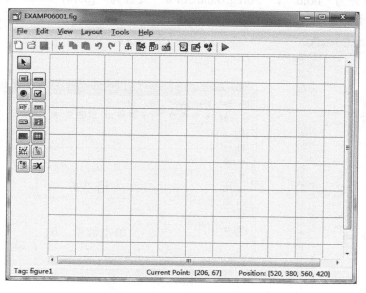

图 6.6 空白 GUI 窗口

【第二步】 首先单击 GUI 左边工具栏中的"pushbutton"按钮，然后在右下展示区域按住鼠标左键，拖动鼠标拉出来的形状就是按钮形状，当然还可以根据需要进行修改。文本框与按钮创建程序一样，在此不再赘述，其结果如图 6.7 所示。

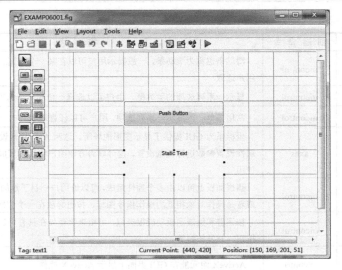

图 6.7 带按钮和文本框的初始 GUI 图

【第三步】 设置按钮属性,将鼠标光标移到按钮上单击右键,打开悬挂菜单并选中"Property Inspector"窗口,或者单击工具箱" "图标,打开"Property Inspector"窗口,如图 6.8 所示。

【第四步】 按钮的属性设置:"BackgroundColor"设置为"[0.89,0.94,0.9]","FontSize"设置为"19.0","FontWeight"设置为"demi","ForegroundColor"设置为"[0.85,0.16,0]","String"设置为"请单击!","Tag"设置为"MyButton1"。文本框属性设置:"BackgroundColor"设置为"[1,0.97,0.92]","FontSize"设置为"19.0","FontWeight"设置为"bold","ForegroundColor"设置为"[0.48,0.06,0.89]","String"设置为"单击总数:0","Tag"设置为"MyText1",其结果如图 6.9 所示。

图 6.8 按钮和文本框的属性设置窗口

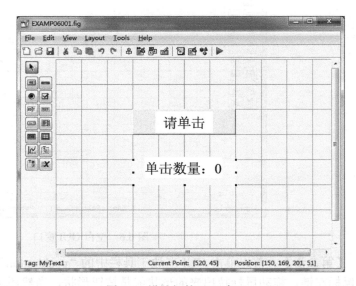

图 6.9 设计好的 GUI 窗口

【第五步】 将设计好的 GUI 面板保存为 EXAMP06001，自动打开该 GUI 的 M 文件。

【第六步】 现在，可以为按钮写入回调函数。该函数含有永久变量，可以用该变量来计数按钮被单击的次数。每当单击一次按钮，MATLAB 就将通过 MyButton1_Callback 调用一次函数。该函数应该能够随着单击按钮的次数不停更新文本内容，并将其赋值给 MyText1 的字符串属性。该部分输入的函数执行行以黑体的形式显示在 EXAMP06001 中。

EXAMP06001

```
function varargout = EXAMP06001(varargin)
% EXAMP06001 MATLAB code for EXAMP06001.fig
%   (EXAMP06001 为 EXAMP06001.fig 的 MATLAB 编码)
% EXAMP06001, by itself, creates a new EXAMP06001 or raises
% the existing singleton*.
%   (EXAMP06001 创建一个新文件 EXAMP06001 或者激活已存在的单独文件)
%
% H = EXAMP06001 returns the handle to a new EXAMP06001 or the handle
% to the existing singleton*.
%   (H=EXAMP06001 为定义 EXAMP06001 句柄)
%
% EXAMP06001('CALLBACK',hObject,eventData,handles,...) calls
% the local function named CALLBACK in EXAMP06001.M with the
% given input arguments.
%   (EXAMP06001('CALLBACK',hObject,eventData,handles,…) 回调
% CALLBACK 包含在 EXAMP06001.m 文件中的局域函数)
%
% EXAMP06001('Property','Value',...) creates a new EXAMP06001
% or raises the existing singleton*.  Starting from the left, property
% value pairs are applied to the GUI before EXAMP06001_OpeningFcn
% gets called.  An unrecognized property name or invalid value makes
% property application stop. All inputs are passed to
% EXAMP06001_OpeningFcn via varargin.
%   (EXAMP06001('Property','Value',…) 创建一个新文件 EXAMP06001 或者打开
% 现存的文件。从左开始，属性和选值成对被 EXAMP06001_OpeningFcn 子函数回调
% 前应用于 GUI。无效属性名和无效选值将停止运行。所有输入变量将通过 vararin
% 传递给子函数 EXAMP06001_OpeningFcn)
%
% *See GUI Options on GUIDE's Tools menu.  Choose "GUI allows only
% one instance to run (singleton)".
%   (查 guide 的工具菜单中的 GUI 选项，选择 GUI 只允许单示例运行)
%
% See also: GUIDE, GUIDATA, GUIHANDLES
% Edit the above text to modify the response to help EXAMP06001
% Last Modified by GUIDE v2.5 12-Aug-2015 17:34:25
%
% -----Begin initialization code - DO NOT EDIT
%         (不需要编辑的初始化编码开始)
% Set gui_Singleton=1.
%   (设置 gui_Singleton=1)
gui_Singleton = 1;
gui_State = struct('gui_Name', mfilename, ...
```

```
                     'gui_Singleton',  gui_Singleton, ...
                     'gui_OpeningFcn', @EXAMP06001_OpeningFcn, ...
                     'gui_OutputFcn',  @EXAMP06001_OutputFcn, ...
                     'gui_LayoutFcn',  [] , ...
                     'gui_Callback',   []);
%   When the number of input arguments is not zero and the first
%  is string, gui_CallBack can be defined by varargin{1}.
%  (当输入变量个数不为零，第一个为字符串时，gui_CallBack= varargin (1))
if nargin && ischar(varargin{1})
    gui_State.gui_Callback = str2func(varargin{1});
end
%  Executes the main function in two accesses.
%     (两种形式运行主函数)
if nargout
    [varargout{1:nargout}] = gui_mainfcn(gui_State, varargin{:});
else
    gui_mainfcn(gui_State, varargin{:});
end
% -----End initialization code - DO NOT EDIT
%         (不需要编辑的初始编码结束)
%
% --- Executes just before EXAMP06001 is made visible.
%        (只有在 EXAMP06001 可见情况下，**子函数 OpeningFcn** 才可运行)
function EXAMP06001_OpeningFcn(hObject,eventdata,handles,varargin)
% This function has no output args, see OutputFcn.
%  (该子函数没有输出变量，输出变量用 OutputFcn 子函数)
% hObject    handle to figure  （图形句柄）
% eventdata  reserved - to be defined in a future version of MATLAB
%              (用于存储未来定义的数据)
% handles    structure with handles and user data (see GUIDATA)
%              (带有句柄和用户数据的结构)
% varargin   command line arguments to EXAMP06001 (see VARARGIN)
%              (EXAMP06001 中的命令行中的变量)
% Choose default command line output for EXAMP06001
handles.output = hObject;
% Update handles structure
guidata(hObject, handles);
% UIWAIT makes EXAMP06001 wait for user response (see UIRESUME)
% uiwait(handles.figure1);
%
% --- Outputs from this function are returned to the command line.
%        (**子函数 OutputFcn** 给出输出结果)
function varargout=EXAMP06001_OutputFcn(hObject,eventdata,handles)
% varargout  cell array for returning output args (see VARARGOUT);
%              (给出输出变量)
% hObject    handle to figure  （图形句柄）
% eventdata  reserved - to be defined in a future version of MATLAB
%              (用于存储未来定义的数据)
% handles    structure with handles and user data (see GUIDATA)
%              (带有句柄和用户数据的结构)
% Get default command line output from handles structure
```

```
%     （从句柄结构获得默认的命令行输出结构）
varargout{1} = handles.output;
%
% --- Executes on button press in MyButton1.
%     （鼠标单击MyButton1按钮，执行子函数 MyButton1 Callback）
function MyButton1_Callback(hObject, eventdata, handles)
% hObject    handle to MyButton1 (see GCBO)（按钮图形句柄）
% eventdata  reserved - to be defined in a future version of MATLAB
%            （用于存储未来定义的数据）
% handles    structure with handles and user data (see GUIDATA)
%            （带有句柄和用户数据的结构）
%
% Declare and initialize variable to store the count.
%  （将数值保存为count并显示）
global count
if isempty(count)
    count=0;
end
% Update count （更新count变量）
count=count+1;
% Define new string （定义新字符串）
str=sprintf('单击总数：%d',count);
% Update the text （更新文字）
set(handles.MyText1,'String',str)
```

运行结果如图 6.10 所示。

【例 6.2】 计算质量的 GUI 窗口。

借助 guide 函数工具创建一个带有用户控件的 GUI 窗口，如图 6.11 所示。

图 6.10 单击计数器界面图

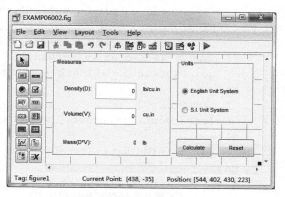

图 6.11 带有用户控件的 GUI 窗口

EXAMP06002

```
function varargout = EXAMP06002(varargin)
% EXAMP06002 MATLAB code for EXAMP06002.fig
% EXAMP06002, by itself, creates a new EXAMP06002 or raises
% the existing singleton*.
%
% H = EXAMP06002 returns the handle to a new EXAMP06002 or
```

```matlab
%       the handle to the existing singleton*.
%
%      EXAMP06002('CALLBACK',hObject,eventData,handles,...) calls
%      the local function named CALLBACK in EXAMP06002.m with the
%      given input arguments.
%
%      EXAMP06002('Property','Value',...) creates a new EXAMP06002
%      or raises the existing singleton*. Starting from the left, property
%      value pairs are applied to the GUI before EXAMP06002_OpeningFcn gets
%      called. An unrecognized property name or invalid value makes
%      property application stop. All inputs are passed to
%      EXAMP06002_OpeningFcn via varargin.
%
%      *See GUI Options on GUIDE's Tools menu. Choose "GUI allows only
%      one instance to run (singleton)".
%
% See also: GUIDE, GUIDATA, GUIHANDLES

% Edit the above text to modify the response to help EXAMP06002
%  （编辑上面文本也就是修改EXAMP06002的帮助解释）

% Last Modified by GUIDE v2.5 14-Aug-2015 10:17:19  （修改时间）
%
% -----Begin initialization code - DO NOT EDIT
%        （不需要编辑的初始化编码开始）
gui_Singleton = 1;
gui_State = struct('gui_Name',       mfilename, ...
                   'gui_Singleton',  gui_Singleton, ...
                   'gui_OpeningFcn', @EXAMP06002_OpeningFcn, ...
                   'gui_OutputFcn',  @EXAMP06002_OutputFcn, ...
                   'gui_LayoutFcn',  [] , ...
                   'gui_Callback',   []);
if nargin && ischar(varargin{1})
    gui_State.gui_Callback = str2func(varargin{1});
end

if nargout
    [varargout{1:nargout}] = gui_mainfcn(gui_State, varargin{:});
else
    gui_mainfcn(gui_State, varargin{:});
end
% -----End initialization code - DO NOT EDIT
%       （不需要编辑的初始编码结束）


% --- Executes just before EXAMP06002 is made visible.
%      （EXAMP06002.fig没有打开前可运行）
function EXAMP06002_OpeningFcn(hObject, eventdata, handles, varargin)
% This function has no output args, see OutputFcn.
% hObject    handle to figure
% eventdata  reserved - to be defined in a future version of MATLAB
% handles    structure with handles and user data (see GUIDATA)
% varargin   command line arguments to EXAMP06002 (see VARARGIN)
```

```
%
% Choose default command line output for EXAMP06002
handles.output = hObject;
% Update handles structure
guidata (hObject, handles);
initialize_gui (hObject, handles, false);
% UIWAIT makes EXAMP06002 wait for user response (see UIRESUME)
% uiwait (handles.figure1);
%
% --- Outputs from this function are returned to the command line.
%       (输出函数结果,返回命令行)
function varargout = EXAMP06002_OutputFcn (hObject, eventdata, handles)
% varargout  cell array for returning output args (see VARARGOUT);
% hObject    handle to figure
% eventdata  reserved - to be defined in a future version of MATLAB
% handles    structure with handles and user data (see GUIDATA)
%
% Get default command line output from handles structure
varargout{1} = handles.output;
%
% --- Executes during object creation, after setting all properties.
%       (所有属性设置完成后,创建对象期间运行)
function density_CreateFcn (hObject, eventdata, handles)
% hObject    handle to density (see GCBO)
% eventdata  reserved - to be defined in a future version of MATLAB
% handles    empty-handles not created until after all CreateFcns
%            called
% Hint: popupmenu controls usually have a white background on Windows.
%       See ISPC and COMPUTER.
% ispc: True for the PC (Windows) version of MATLAB.
%       (判定是否为 Windows 版 MATLAB)
% isequal: True if arrays are numerically equal.
%       (判定数值是否相等)
if ispc && isequal (get (hObject,'BackgroundColor'),…
 get (0, 'defaultUicontrolBackgroundColor'))
    set (hObject,'BackgroundColor','white');
end
%
function density_Callback (hObject, eventdata, handles)
% hObject    handle to density (see GCBO)
% eventdata  reserved - to be defined in a future version of MATLAB
% handles    structure with handles and user data (see GUIDATA)
% Hints: get (hObject,'String') returns contents of density as text
%        str2double (get (hObject,'String')) returns contents
%        of density as a double
%   (技巧: get (hObject,'String') 密度的返回值为文本
%          str2double (get (hObject,'String')) 密度的返回值为数值)
density = str2double (get (hObject, 'String'));
% isnan: True for Not-a-Number.
%       (判定是否为非数)
if isnan (density)
```

```matlab
    set (hObject, 'String', 0);
%   errordlg: Error dialog box.
%              （错误对话框）
    errordlg ('Input must be a number','Error');
end
% Save the new density value.
%    （保存更新的密度值）
handles.metricdata.density = density;
guidata (hObject,handles)
%
% --- Executes during object creation, after setting all properties.
function volume_CreateFcn (hObject, eventdata, handles)
% hObject    handle to volume (see GCBO)
% eventdata  reserved - to be defined in a future version of MATLAB
% handles    empty - handles not created until after all
%            CreateFcns called
% Hint: popupmenu controls usually have a white background on Windows.
%      （技巧：常用菜单空间通常背景设为白色）
%      See ISPC and COMPUTER.
% ispc: True for the PC (Windows) version of MATLAB.
%        （判定是否为 Windows 版 MATLAB）
% isequal: True if arrays are numerically equal.
%           （判定数值是否相等）
if ispc && isequal (get (hObject,'BackgroundColor'),…
 get (0,'defaultUicontrolBackgroundColor'))
    set (hObject,'BackgroundColor','white');
end
%
function volume_Callback (hObject, eventdata, handles)
% hObject    handle to volume (see GCBO)
% eventdata  reserved - to be defined in a future version of MATLAB
% handles    structure with handles and user data (see GUIDATA)
% Hints: get (hObject,'String') returns contents of volume as text
%        str2double (get (hObject,'String')) returns contents of volume
%        as a double
%   （技巧：get (hObject,'String') 体积的返回值为字符串,
%          str2double (get (hObject,'String')) 体积的返回值为数值）
volume = str2double (get (hObject,'String'));
% isnan: True for Not-a-Number.
%         （判定是否为非数）
if isnan (volume)
    set (hObject, 'String', 0);
    errordlg ('Input must be a number','Error');
end
% Save the new volume value
handles.metricdata.volume = volume;
guidata (hObject,handles)
%
% --- Executes on button press in calculate.
%       （单击计算按钮执行函数 calculate_Callback）
function calculate_Callback (hObject, eventdata, handles)
```

```
%     hObject    handle to calculate (see GCBO)
%     eventdata  reserved - to be defined in a future version of MATLAB
%     handles    structure with handles and user data (see GUIDATA)
%     Calculate the product of density and volume.
%     (计算密度和体积的乘积)
mass = handles.metricdata.density * handles.metricdata.volume;
%     Update and declare the mass
%     (更新并显示质量)
set (handles.mass, 'String', mass);

% --- Executes on button press in reset.
%     (单击重置按钮执行函数 reset_Callback)
function reset_Callback (hObject, eventdata, handles)
%     hObject    handle to reset (see GCBO)
%     eventdata  reserved - to be defined in a future version of MATLAB
%     handles    structure with handles and user data (see GUIDATA)
%     Reset the gui as its initial version.
%     (重置 gui 为初始形式)
initialize_gui (gcbf, handles, true);
%
% --- Executes when selected object changed in unitgroup.
%     (单位改变时,执行子函数 unitgroup_SelectionChangeFcn)
function unitgroup_SelectionChangeFcn (hObject, eventdata, handles)
%     hObject    handle to the selected object in unitgroup
%     eventdata  reserved - to be defined in a future version of MATLAB
%     handles    structure with handles and user data (see GUIDATA)
%
if (hObject == handles.english)
    set (handles.text4, 'String', 'lb/cu.in');
    set (handles.text5, 'String', 'cu.in');
    set (handles.text6, 'String', 'lb');
else
    set (handles.text4, 'String', 'kg/cu.m');
    set (handles.text5, 'String', 'cu.m');
    set (handles.text6, 'String', 'kg');
end
%
% --------------------------------------------------------------------
function initialize_gui (fig_handle, handles, isreset)
% If the metricdata field is present and the reset flag is false,
% it means that we are just re-initializing a GUI by calling it from
% the cmd line while it is up. So, bail out as we don't want to reset
% the data.
% isfield: True if field is in structure array.
%          (判定场量是否在结构数组中)
if isfield (handles, 'metricdata') && ~isreset
    return;
end
handles.metricdata.density = 0;
handles.metricdata.volume  = 0;
set (handles.density, 'String', handles.metricdata.density);
```

```
set (handles.volume, 'String', handles.metricdata.volume);
set (handles.mass, 'String', 0);
set (handles.unitgroup, 'SelectedObject', handles.english);
set (handles.text4, 'String', 'lb/cu.in');
set (handles.text5, 'String', 'cu.in');
set (handles.text6, 'String', 'lb');
% Update handles structure
% （更新结构句柄）
guidata (handles.figure1, handles);
```

在命令窗口运行 EXAMP06002 程序，然后输入 Density(D): 12 1b/cu.in 和 Volume(V): 32 cu.in，单击"Calculate"按钮，如图 6.12 所示。

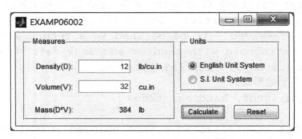

图 6.12 计算质量的 GUI 示例图

【例 6.3】 绘制带坐标系、菜单和按钮的典型 GUI 窗口。

设计一个典型 GUI 窗口，它含有一个下拉菜单，一个按钮和一个坐标系绘图窗口，能够完成如此功能：在下拉菜单中选一个菜单项，单击"Enter"按钮，坐标系窗口会绘制出相应图形。

分析与求解：

【第一步】 借助 guide 函数打开一个带有坐标系和菜单的 GUI 窗口，也就是选择"GUI with Axes and Menus"选项。

【第二步】 借助 Inspector Property 属性设置窗口定义所有组元的属性。例如，下拉菜单字号为 12，字的颜色为紫色，内容更改为 plot(rand(5))，plot(sin(1:0.01:25))，contour_clabel (peaks)，plot (membrane)，mesh (peaks)，surfl (peaks)；按钮的显示文字为"更新"，字号设置为 19，字的颜色为蓝色。

【第三步】 将设置完属性的 GUI 窗口保存为"EXAMP06003"。

【第四步】 在预先定义的菜单下输入 MATLAB 执行语句。

```
switch popup_sel_index
    case 1
        plot (rand (5));
    case 2
        plot (sin (1:0.01:25.99));
    case 3
        z=peaks (50);
        [C,h]=contour (z,5);
        clabel (C,h);
    case 4
        plot (membrane);
```

```
        case 5
            mesh(peaks);
        case 6
            surfl(peaks);
            shading interp
            lighting phong
            colormap(bone)
    end
```

同时，注意在菜单回调函数当前句柄的 string 设置必须与【第二步】中的修改完全一致。

EXAMP06003

```
function varargout = EXAMP06003(varargin)
% EXAMP06003 MATLAB code for EXAMP06003.fig
% EXAMP06003, by itself, creates a new EXAMP06003 or raises
% the existing singleton*.
%
% H = EXAMP06003 returns the handle to a new EXAMP06003 or the handle
% to the existing singleton*.
%
% EXAMP06003('CALLBACK',hObject,eventData,handles,...) calls the
% local function named CALLBACK in EXAMP06003.M with the given input
% arguments.
%
% EXAMP06003('Property','Value',...) creates a new EXAMP06003 or
% raises the existing singleton*. Starting from the left, property
% value pairs are applied to the GUI before EXAMP06003_OpeningFcn gets
% called. An unrecognized property name or invalid value makes
% property application stop. All inputs are passed to
% EXAMP06003_OpeningFcn via varargin.
%
% *See GUI Options on GUIDE's Tools menu. Choose "GUI allows only
% one instance to run (singleton)".
%
% See also: GUIDE, GUIDATA, GUIHANDLES
% Edit the above text to modify the response to help EXAMP06003
% Last Modified by GUIDE v2.5 14-Aug-2015 23:12:07
%
% Begin initialization code - DO NOT EDIT
gui_Singleton = 1;
gui_State = struct('gui_Name',mfilename,...
                   'gui_Singleton',gui_Singleton,...
                   'gui_OpeningFcn',@EXAMP06003_OpeningFcn,...
                   'gui_OutputFcn',@EXAMP06003_OutputFcn, ...
                   'gui_LayoutFcn',[] ,...
                   'gui_Callback',[]);
% nargin: Number of function input arguments.
%         （函数中输入变量的个数）
% ischar: True for character array (string).
%         （判定是否为字符串）
```

```matlab
% varargin: Variable length input argument list.
%          (输入变量单的变量长度)
if nargin && ischar(varargin{1})
    % str2func: Construct a function_handle from
    %           a function name string.
    %           (根据函数名字符串构建函数句柄)
    gui_State.gui_Callback = str2func(varargin{1});
end
% nargout: Number of function output arguments.
%          (函数输出变量的个数)
if nargout
    [varargout{1:nargout}] = gui_mainfcn(gui_State, varargin{:});
else
    gui_mainfcn(gui_State, varargin{:});
end
% End initialization code - DO NOT EDIT
%
% --- Executes just before EXAMP06003 is made visible.
%     (只有EXAMP06003可见,才运行子函数OpeningFcn)
function EXAMP06003_OpeningFcn(hObject,eventdata,handles,varargin)
% This function has no output args, see OutputFcn.
% hObject    handle to figure
% eventdata  reserved - to be defined in a future version of MATLAB
% handles    structure with handles and user data (see GUIDATA)
% varargin   command line arguments to EXAMP06003 (see VARARGIN)
%
% Choose default command line output for EXAMP06003
handles.output = hObject;

% Update handles structure
guidata(hObject, handles);
% This sets up the initial plot - only do when we are invisible
% so window can get raised using EXAMP06003.
% strcmp: Compare string.
%         (比较字符串)
% (如果hObject的Visible属性为off,绘制plot(rand(5))图形)
if strcmp(get(hObject,'Visible'),'off')
    plot(rand(5));
end
% UIWAIT makes EXAMP06003 wait for user response (see UIRESUME)
% uiwait(handles.figure1);
%
% --- Outputs from this function are returned to the command line.
%     (子函数OutputFcn给出该函数命令行的返回结果)
function varargout=EXAMP06003_OutputFcn(hObject,eventdata,handles)
% varargout  cell array for returning output args (see VARARGOUT);
% hObject    handle to figure
```

```
% eventdata  reserved - to be defined in a future version of MATLAB
% handles    structure with handles and user data (see GUIDATA)
% Get default command line output from handles structure
varargout{1} = handles.output;
%
% --- Executes on button press in pushbutton1.
%     (单击按钮pushbutton1的回调函数)
function pushbutton1_Callback(hObject, eventdata, handles)
% hObject    handle to pushbutton1 (see GCBO)
% eventdata  reserved - to be defined in a future version of MATLAB
% handles    structure with handles and user data (see GUIDATA)
% （激活坐标系）
axes(handles.axes1);
% （清除坐标系）
cla;
% （获取菜单值，根据菜单值执行相应语句）
popup_sel_index=get(handles.popupmenu1,'Value');
switch popup_sel_index
    case 1
        plot(rand(5));
    case 2
        plot(sin(1:0.01:25.99));
    case 3
        z=peaks(50);
        [C,h]=contour(z,5);
        clabel(C,h);
case 4
        plot(membrane);
    case 5
        mesh(peaks);
    case 6
        surfl(peaks);
        shading interp
        lighting phong
        colormap(bone)
end
% --------------------------------------------------------------------
function FileMenu_Callback(hObject, eventdata, handles)
% hObject    handle to FileMenu (see GCBO)
% eventdata  reserved - to be defined in a future version of MATLAB
% handles    structure with handles and user data (see GUIDATA)
% --------------------------------------------------------------------
function OpenMenuItem_Callback(hObject, eventdata, handles)
% hObject    handle to OpenMenuItem (see GCBO)
% eventdata  reserved - to be defined in a future version of MATLAB
% handles    structure with handles and user data (see GUIDATA)
% uigetfile: Standard open file dialog box.
```

```matlab
%                (通常打开文件对话框)
file = uigetfile ('*.fig');
if ~isequal (file, 0)
    open (file);
end
% --------------------------------------------------------------
function PrintMenuItem_Callback (hObject, eventdata, handles)
% hObject    handle to PrintMenuItem (see GCBO)
% eventdata  reserved - to be defined in a future version of MATLAB
% handles    structure with handles and user data (see GUIDATA)
% printdlg: Print dialog box.
%                (打印对话框)
printdlg (handles.figure1);
% --------------------------------------------------------------
function CloseMenuItem_Callback (hObject, eventdata, handles)
% hObject    handle to CloseMenuItem (see GCBO)
% eventdata  reserved - to be defined in a future version of MATLAB
% handles    structure with handles and user data (see GUIDATA)
% questdlg: Question dialog box.
%                (问题对话框,这里选项有'Yes','No', 'Yes').
selection = questdlg (['Close ' get (handles.figure1,'Name') '?'],...
                     ['Close ' get (handles.figure1,'Name') '...'],...
                     'Yes','No','Yes');
% (如果 selection 为'No'执行 if 结构中语句)
if strcmp (selection,'No')
    return;
end
% delete: Delete file or graphics object.
%           (这里是删除图形 figure1)
delete (handles.figure1)
%
% --- Executes on selection change in popupmenu1.
%       (变换菜单选项回调子函数 popupmenu1_Callback)
function popupmenu1_Callback (hObject,eventdata,handles)
% hObject    handle to popupmenu1 (see GCBO)
% eventdata  reserved - to be defined in a future version of MATLAB
% handles    structure with handles and user data (see GUIDATA)
% Hints: contents = get (hObject,'String') returns popupmenu1
%        contents as cell array contents{get (hObject,'Value')}
%        returns selected item from popupmenu1
%
% --- Executes during object creation, after setting all properties.
%       (创建菜单 popupmenu1 的回调子函数)
function popupmenu1_CreateFcn (hObject,eventdata,handles)
% hObject    handle to popupmenu1 (see GCBO)
% eventdata  reserved - to be defined in a future version of MATLAB
% handles    empty - handles not created until after all
%               CreateFcns called
```

```
%   Hint: popupmenu controls usually have a white background on Windows.
%         See ISPC and COMPUTER.
if ispc && isequal(get(hObject,'BackgroundColor'), ...
 get(0,'defaultUicontrolBackgroundColor'))
    set(hObject,'BackgroundColor','white');
end
set(hObject,'String',{'plot(rand(5))','plot(sin(1:0.01:25))',...
'contour_clabel(peaks)','plot(membrane)','mesh(peaks)',...
'surfl(peaks)'});
```

运行结果如图 6.13 所示。

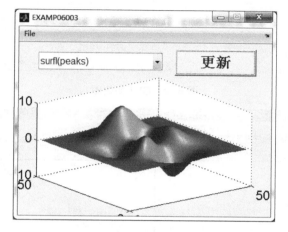

图 6.13　带坐标系、菜单和按钮的 GUI 窗口示例图

6.3　GUI 实　例

【**例 6.4**】　函数极限可视化。

绘制函数 $y = \dfrac{\sin x}{x}$ 在 $-100 \leqslant x \leqslant 100$、$-0.24 \leqslant y \leqslant 1.1$ 下的曲线，并观察和推断当 $x \to -\infty$、$x \to +\infty$、$x \to 0-$、$x \to 0+$、$x \to 0$ 时，函数的变化趋势，说明单、双侧极限的关系。

分析与求解：

通过绘制函数的动态曲线（彗星轨迹）和静态曲线极限图，推断当 $x \to -\infty$、$x \to +\infty$ 时函数的变化趋势。首先构思草图形成框架，在布局编辑器中布置控件，使用几何位置排列工具对控件的位置进行调整。主要构成部件有两个坐标系，分别用来显示动态彗星轨迹与函数极限图形；三个按钮分别为动画、绘图和退出。

【**第一步**】　在命令窗口输入 guide，选择"Blank GUI（Default）"选项，选中"Save new figure as"选项，地址栏为"E:\MATLAB 修订版\书中实例程序\EXAMP06004.fig"，然后单击"OK"按钮打开一个 GUI 设计窗口，如图 6.14 所示。同时保存 EXAMP06004.m 文件并打开。

【**第二步**】　借助左边快捷菜单，在 GUI 操作窗口绘制 2 个坐标系和 3 个按钮，如图 6.15 所示。

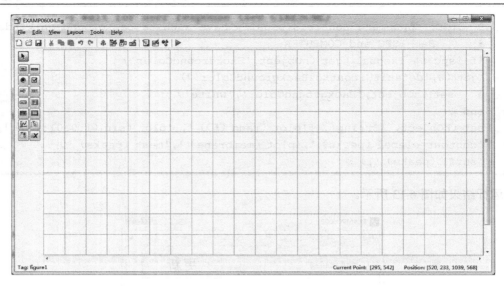

图 6.14　GUI 初始图

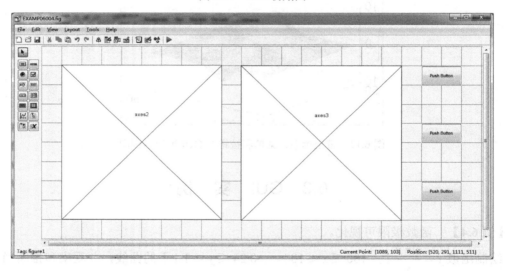

图 6.15　带有 2 个坐标系和 3 个按钮的 GUI 初始图

【第三步】　3 个按钮的属性设置如下：

（1）第一个按钮字符显示"动画"，字符颜色为蓝色，字体为"楷体_GB2321"、加粗，字号为 20（MATLAB 属性设置语句："FontName"为"楷体_GB2321"，"FontSize"为 20，"FontWeight"为"bold"，"ForegroundColor"为"蓝色"，"String"为"动画"，"Tag"为"donghua"）；

（2）第二个按钮字符显示"绘图"，字符颜色为蓝色，字体为"楷体_GB2321"、加粗，字号为 20（MATLAB 属性设置语句："FontName"为"楷体_GB2321"，"FontSize"为 20，"FontWeight"为"bold"，"ForegroundColor"为"蓝色"，"String"为"绘图"，"Tag"为"huitu"）；

（3）第三个按钮字符显示"退出"，字符颜色为红色，字体为"楷体_GB2321"、加粗，字号为 20（MATLAB 属性设置语句："FontName"为"楷体_GB2321"，"FontSize"为 20，

"FontWeight"为"bold","ForegroundColor"为"红色","String"为"退出","Tag"为"tuichu")。

3个按钮属性设置如图6.16所示。

图6.16　3个按钮属性设置窗口

3个按钮属性设置后的GUI界面如图6.17所示。

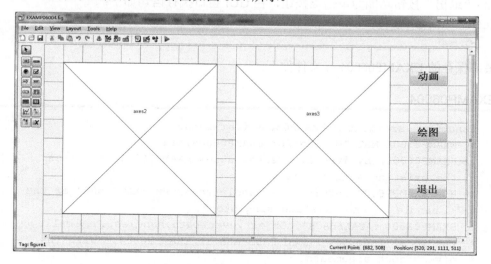

图6.17　3个按钮设置后的GUI界面

【第四步】　添加回调函数的执行语句。

(1)"动画"按钮功能为单击该按钮,在axes2窗口绘制动画,执行语句如下:

```
% (将axes2坐标系设置为当前坐标系)
axes(handles.axes2);
% (定义自变量x)
x=-700:0.1:700;
```

```matlab
% （显示外框）
box on
hold on
% （定义坐标系显示区间）
axis([-700 700 -0.24 1.1]);
% （绘制 sin(x)/x 函数的动画）
comet(x,sin(x)./x);
```

(2)"绘图"按钮功能为单击该按钮，在 axes3 窗口绘制图形，执行语句如下：

```matlab
% （将 axes3 坐标系设置为当前坐标系）
axes(handles.axes3);
% （定义自变量 x）
x=(-100:0.01:100)+eps;
% （定义因变量 y）
y=sin(x)./x;
% （绘制函数 y 的曲线）
plot(x,y);
% （定义 axes3 的坐标轴显示范围）
axis([-100 100 -0.24 1.1]);
% （在[30,0.5]位置显示文字'sin(x)/x'）
t1=text(30,0.5,'sin(x)/x');
% （设置文本 t1 的属性）
set(t1,'FontSize',20,'Color','b','FontWeight','bold')
```

(3)"退出"按钮功能为单击该按钮，在 GUI 窗口执行语句如下：

```matlab
% （关闭 GUI 窗口）
close
```

最终保存的 EXAMP06004.m 文件如下。

EXAMP06004

```matlab
function varargout = EXAMP06004(varargin)
% EXAMP06004 MATLAB code for EXAMP06004.fig
% EXAMP06004, by itself, creates a new EXAMP06004 or raises
%      the existing singleton*.
% H = EXAMP06004 returns the handle to a new EXAMP06004 or the
%      handle to the existing singleton*.
% EXAMP06004('CALLBACK',hObject,eventData,handles,...) calls
%      the local function named CALLBACK in EXAMP06004.M with the
%      given input arguments.
%
% EXAMP06004('Property','Value',...) creates a new EXAMP06004
%      or raises the existing singleton*. Starting from the left, property
%      value pairs are applied to the GUI before EXAMP06004_OpeningFcn
%      gets called. An unrecognized property name or invalid value makes
%      property application stop. All inputs are passed to
%      EXAMP06004_OpeningFcn via varargin.
%
% *See GUI Options on GUIDE's Tools menu. Choose "GUI allows
```

```matlab
%   only one instance to run (singleton)".
%
% See also: GUIDE, GUIDATA, GUIHANDLES
% Edit the above text to modify the response to help EXAMP06004
% Last Modified by GUIDE v2.5 15-Aug-2015 11:04:45
%
% Begin initialization code - DO NOT EDIT
gui_Singleton = 1;
gui_State = struct ('gui_Name', mfilename, ...
                    'gui_Singleton', gui_Singleton, ...
                    'gui_OpeningFcn', @EXAMP06004_OpeningFcn, ...
                    'gui_OutputFcn', @EXAMP06004_OutputFcn, ...
                    'gui_LayoutFcn', [] , ...
                    'gui_Callback', []);
if nargin && ischar (varargin{1})
    gui_State.gui_Callback = str2func (varargin{1});
end

if nargout
    [varargout{1:nargout}] = gui_mainfcn (gui_State, varargin{:});
else
    gui_mainfcn (gui_State, varargin{:});
end
% End initialization code - DO NOT EDIT
%
% --- Executes just before EXAMP06004 is made visible.
function EXAMP06004_OpeningFcn (hObject, eventdata, handles, varargin)
% This function has no output args, see OutputFcn.
% Choose default command line output for EXAMP06004
handles.output = hObject;
% Update handles structure
guidata (hObject, handles);
%
% --- Outputs from this function are returned to the command line.
function varargout=EXAMP06004_OutputFcn (hObject,eventdata,handles)
% Get default command line output from handles structure
varargout{1}=handles.output;
%
% --- Executes on button press in donghua.
function donghua_Callback (hObject,eventdata,handles)
% （将 axes2 坐标系设置为当前坐标系）
axes (handles.axes2);
% （定义自变量 x）
x=-700:0.1:700;
% （显示外框）
box on
hold on
% （定义坐标系显示区间）
axis ([-700 700 -0.24 1.1]);
% （绘制 sin (x)/x 函数的动画）
comet (x,sin (x) ./x);
%
```

```
% --- Executes on button press in huitu.
function huitu_Callback(hObject,eventdata,handles)
% （将 axes3 坐标系设置为当前坐标系）
axes(handles.axes3);
% （定义自变量 x）
x=(-100:0.01:100)+eps;
% （定义因变量 y）
y=sin(x)./x;
% （绘制函数 y 的曲线）
plot(x,y);
% （定义 axes3 的坐标轴显示范围）
axis([-100 100 -0.24 1.1]);
% 在[30,0.5]位置显示文字'sin(x)/x'.
t1=text(30,0.5,'sin(x)/x');
% （设置文本 t1 的属性）
set(t1,'FontSize',20,'Color','b','FontWeight','bold')
%
% --- Executes on button press in tuichu.
function tuichu_Callback(hObject,eventdata,handles)
close
```

在命令窗口运行 EXAMP06004 程序，然后先单击"绘图"按钮后再单击"动画"按钮，运行结果如图 6.18 所示。

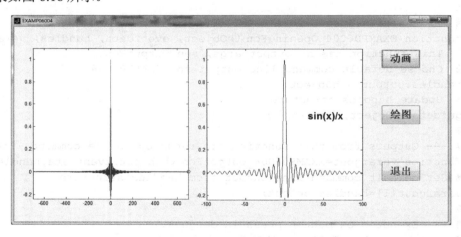

图 6.18　函数极限可视化最终 GUI 窗口

注意：
　　在定义 x 的值 "x=(-100:0.01:100)+eps" 中多了 "eps"，是为了避免 "x=0" 出现在分母上，使得 y 值趋向无穷大而难于绘制图形，并且 MATLAB 会给出警告提示。

注释：
　　函数 comet 彗星轨迹式绘图。
　　语法格式为：comet（y）　　显示彗星追踪式绘图。
　　　　　　　　comet（x,y）　　根据变量 x 和 y 以彗星追踪式绘图。
　　　　　　　　comet（x,y,p）　　根据变量 x 和 y 以 p*length（y）长度为彗星尾，绘制图形。

【例 6.5】 电路信号合成。

验证方波可用相应频率的基波及其奇次谐波来合成，设方波的宽度为 π，周期为 2π。

分析与求解：

一个以原点为奇对称中心的方波可以用奇次正弦波的叠加来逼近，即

$$y(t) = \sin t + \frac{1}{3}\sin 3t + \frac{1}{5}\sin 5t + \cdots + \frac{1}{2k-1}\sin(2k-1)t + \cdots$$

用 $y(t)$ 来描述波的函数，通过查看二维曲线和三维曲面来检验它的逼近程度和特征。

【第一步】 GUI 窗口包含组件有 2 个坐标系，一个坐标系绘制二维图像，另一个坐标系绘制三维图像；3 个按钮，分别为"二维图像"、"三维图像"和"退出"。分别选中按钮，单击菜单栏下面快捷工具栏中的"Property Inspector"按钮打开属性设置窗口，"二维图像"按钮属性设置如下："FontSize"设置为 20，"FontWeight"设置为"bold"，"ForegroundColor"设置为"蓝色"，"String"设置为"二维图像"，"Tag"设置为"two_pushbutton"；"三维图像"按钮属性设置如下："FontSize"设置为 20，"FontWeight"设置为"bold"，"ForegroundColor"设置为"蓝色"，"String"设置为"三维图像"，"Tag"设置为"three_pushbutton"；"退出"按钮属性设置如下："FontSize"设置为 20，"FontWeight"设置为"bold"，"ForegroundColor"设置为"蓝色"，"String"设置为"退出"，"Tag"设置为"quit_pushbutton"。GUI 初始化设计后窗口如图 6.19 所示。

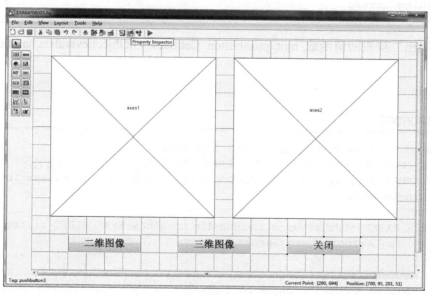

图 6.19 GUI 初始化设计后窗口示意图

【第二步】 编写回调函数。

（1）"二维图像"按钮的回调函数 two_pushbutton_Callback（hObject,evendata,handles）执行语句如下：

```
t=0:0.01:2*pi; y=sin(t);
axes(handles.axes1);
plot(t,y); pause(3);
y=sin(t)+sin(3*t)/3; plot(t,y); pause(3);
y=sin(t)+sin(3*t)/3+sin(5*t)/5+sin(7*t)/7+sin(9*t)/9;
```

```
plot(t,y); y=zeros(10,max(size(t))); x=zeros(size(t));
for k=1:2:19
      x=x+sin(k*t)/k;
      y((k+1)/2,:)=x;
end
pause(3); plot(t,y(1:9,:));
grid
line([0,pi+0.5],[pi/4,pi/4]);
text(pi+0.5,pi/4,'pi/4');
title('谐波合成的二维曲线')
```

（2）"三维图像"按钮回调函数 three_pushbutton_Callback（hObject,evendata,handles）执行语句如下：

```
t=0:0.01:2*pi; y=zeros(10,max(size(t))); x=zeros(size(t));
for k=1:2:19
      x=x+sin(k*t)/k;
      y((k+1)/2,:)=x;
end
halft=ceil(length(t)/2); pause(1);
axes(handles.axes2);
surf(t(1:halft),[1:10],y(:,1:halft));
shading interp
title('谐波合成的三维曲面')
```

（3）"退出"按钮回调函数 quit_pushbutton_Callback（hObject,evendata,handles）执行语句如下：

```
close
```

EXAMP06005

```
function varargout = EXAMP06005(varargin)
%  EXAMP06005 MATLAB code for EXAMP06005.fig
%  EXAMP06005, by itself, creates a new EXAMP06005 or raises the existing
%  singleton*.
%
%  H = EXAMP06005 returns the handle to a new EXAMP06005 or the handle to
%      the existing singleton*.
%
%   EXAMP06005('CALLBACK',hObject,eventData,handles,...) calls the local
%  function named CALLBACK in EXAMP06005.M with the given input
%  arguments. EXAMP06005('Property','Value',...) creates a new EXAMP06005 or
%  raises the existing singleton*. Starting from the left, property value pairs are
%  applied to the GUI before EXAMP06005_OpeningFcn gets called.  An
%  unrecognized property name or invalid value makes property application
%  stop. All inputs are passed to EXAMP06005_OpeningFcn via varargin.
%
%  *See GUI Options on GUIDE's Tools menu.  Choose "GUI allows only one
%  instance to run (singleton)".
%
```

```
%  See also: GUIDE, GUIDATA, GUIHANDLES
%  Edit the above text to modify the response to help EXAMP06005
%  Last Modified by GUIDE v2.5 15-Aug-2015 17:03:17

% -----Begin initialization code - DO NOT EDIT
gui_Singleton = 1;
gui_State = struct ('gui_Name', mfilename, ...
        'gui_Singleton', gui_Singleton, ...
        'gui_OpeningFcn', @EXAMP06005_OpeningFcn, ...
        'gui_OutputFcn',  @EXAMP06005_OutputFcn, ...
        'gui_LayoutFcn', [] , ...
        'gui_Callback', []);
if nargin && ischar (varargin{1})
gui_State.gui_Callback = str2func (varargin{1});
end

if nargout
[varargout{1:nargout}] = gui_mainfcn (gui_State, varargin{:});
else
gui_mainfcn (gui_State, varargin{:});
end
% -----End initialization code - DO NOT EDIT
%
% --- Executes just before EXAMP06005 is made visible.
function EXAMP06005_OpeningFcn (hObject, eventdata, handles, varargin)
%  This function has no output args, see OutputFcn.
%  Choose default command line output for EXAMP06005
handles.output = hObject;
%  Update handles structure
guidata (hObject, handles);
%
% --- Outputs from this function are returned to the command line.
function varargout = EXAMP06005_OutputFcn (hObject, eventdata, handles)
%  Get default command line output from handles structure
varargout{1} = handles.output;
%
% --- Executes on button press in two_pushbutton.
function two_pushbutton_Callback (hObject, eventdata, handles)
t=0:0.01:2*pi;
y=sin (t);
axes (handles.axes1);
plot (t,y);
pause (3);
y=sin (t) +sin (3*t) /3;
plot (t,y);
pause (3);
y=sin (t) +sin (3*t) /3+sin (5*t) /5+sin (7*t) /7+sin (9*t) /9;
plot (t,y);
y=zeros (10,max (size (t)));
x=zeros (size (t));
for k=1:2:19
    x=x+sin (k*t) /k;
```

```
        y((k+1)/2,:)=x;
end
pause(3);
plot(t,y(1:9,:));
grid
line([0,pi+0.5],[pi/4,pi/4]);
text(pi+0.5,pi/4,'pi/4');
title('谐波合成的二维曲线')
%
% --- Executes on button press in three_pushbutton.
function three_pushbutton_Callback(hObject, eventdata, handles)
t=0:0.01:2*pi;
y=zeros(10,max(size(t)));
x=zeros(size(t));
for k=1:2:19
    x=x+sin(k*t)/k;
    y((k+1)/2,:)=x;
end
halft=ceil(length(t)/2);
pause(1);
axes(handles.axes2);
surf(t(1:halft),[1:10],y(:,1:halft));
shading interp
title('谐波合成的三维曲面')
%
% --- Executes on button press in quit_pushbutton.
function quit_pushbutton_Callback(hObject, eventdata, handles)
close
```

在命令窗口运行 EXAMP06005 程序，然后分别单击"二维图像"按钮和"三维图像"按钮，结果如图 6.20 所示。

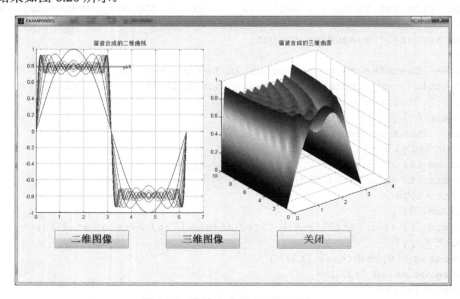

图 6.20 谐波合成后 GUI 运行界面

第 6 章 图形用户界面设计（GUI）

> 注释：
> （1）函数 line 绘制线。
> 语法格式：line（x,y） 以数组 x 和 y 在当前坐标系绘制二维线。
> 　　　　　line（x,y,z） 以数组 x、y 和 z 在当前坐标系绘制三维线。
> （2）函数 pause 等待用户反应。
> 语法格式：pause（n） 代表等待 n 秒后程序继续。
> 　　　　　pause 让程序停下来，等待用户的响应。

【例 6.6】 直接通过命令函数创建 GUI，它是关于圆形金属波导图形计算器测试软件运行界面。

分析与求解：

【第一步】 打开宽度为 1000 像素、高度为 750 像素的图形窗口，起始点坐标为（25,10），也就是离左边窗口距离为 25 像素，离底边距离为 10 像素，并将其句柄定义为 fig；然后借助句柄和函数 set 设置"defaultuicontrolfontsize"为 14，坐标系宽度占整个图形窗口的 0.7，高度占整个图形窗口的 0.7，起始点为（0.05，0.05），就是距离左边距离相当于图形窗口宽度的 0.05 倍，距离底边距离相当于图形高度的 0.05 倍。

【第二步】 文本框，图形句柄为 text1，类型属性设置为"toggle"，位置设置为"[100,600,850,70]"，也就是起始点为距离左边距离 100 像素，距离底边距离 600 像素，宽度为 850 像素，高度为 70 像素，显示的字符串为"圆形金属波导图形计算器测试软件封面"、字体设置为"仿宋体"，字号为 32，背景颜色设置为"[1,0.9,0.9]"，字体颜色设置为蓝色，加粗。

【第三步】 2 个按钮"圆形金属波导"和"退出程序"设置。"圆形金属波导"按钮属性设置如下：颜色设置为[1,0.3,0.9]，字号为 30，字体为"楷体_GB2312"、加粗；"退出程序"按钮属性设置如下：颜色设置为红色，字号为 28，字体为黑体、加粗。

【第四步】 文字显示控件，显示"临沂大学理学院软件开发中心"，字体为"楷体_GB2312"，字号为 26，背景颜色为[0.1,1,0.9]，字的颜色为红色、加粗。

EXAMP06006

```
function EXAMP06006
% Function EXAMP06006 MATLAB code for EXAMP06006 GUI.
% EXAMP06006 opens a figure to show the face page of a GUI.
% The GUI is a software to simulate the electromagnetic field
% in a circular metal waveguide.
%
% Records of revisions
% Date: 2015-08-15
% Programmer: Wang Yong-Long and Xia Chang-Long
% Description of changer: revision
%
% Clear variables in workspace.
% （清除工作空间中的变量）
clear
% Clear command window.
% （清除命令窗口）
```

```
clc
% Set a figure with position [25,10,1000,750].
% （设置图形窗口的位置为[25,10,1000,750]）
fig=figure('position',[25,10,1000,750]);
% Define 'defaultuicontrolfontsize' as '14' in fig.
% （定义图形窗口中控件的字号为14）
set(fig,'defaultuicontrolfontsize',14);
% Define an axes with position [0.05,0.05,0.7,0.7].
% （定义一个坐标系的位置为[0.05,0.05,0.7,0.7]）
axes('position',[0.05,0.05,0.7,0.7]);
% Define a string toggle with position [100 600 850 70], the string
% is "圆形金属波导图形计算器测试软件封面".
% （在位置[100 600 850 70]处定义一个字符串控件，字符为"圆形金属波导图形
%  计算器测试软件封面"）
htext1=uicontrol(fig,'style','toggle',...
     'position',[100 600 850 70],....
'string','圆形金属波导图形计算器测试软件封面');
% Set fontname as 仿宋体, fontsize as 32.
% （设置字体为仿宋体，字号为32）
set(htext1,'fontname','仿宋体','fontsize',32);
% Set background color as [1,0.9,0.9].
% （设置背景颜色为[1,0.9,0.9]）
set(htext1,'background',[1,0.9,0.9]);
% Set fontcolor as blue, fontweight as bold.
% （设置字的颜色为蓝色，字加粗）
set(htext1,'foreground','b','fontweight','bold');
%
% Define a text with "临沂大学理学院软件开发中心".
% （定义文本控件，文字为"临沂大学理学院软件开发中心"）
htext3=uicontrol(fig,'style','text',...
       'position',[730 100 170 160],....
       'string','临沂大学理学院软件开发中心');
% （设置字体为"楷体_GB2312"，字号为23，背景颜色为[0.1,1,0.9],
%  字的颜色为红色，字加粗）
set(htext3,'fontname','楷体_GB2312','fontsize',26);
set(htext3,'background',[0.1,1,0.9]);
set(htext3,'foreground',[1,0,0],'fontweight','bold');
% Create a button control with "圆形金属波导".
% （创建一个按钮标控件为"圆形金属波导"）
tb1=uicontrol(fig,'style','toggle',...
     'string','圆形金属波导','value',0,...
'position',[620 480 280 60],'callback','close,yxfzh');
% （设置字的颜色为[1,0.3,0.9]，字号为30，加粗，字体为楷体_GB2312）
set(tb1,'foreground',[1,0.3,0.9]);
set(tb1,'fontsize',30,'fontweight','bold');
set(tb1,'fontname','楷体_GB2312');
% Create a new pushbutton with "退出程序".
% （创建一个新按钮，标示为"退出程序"）
tb2=uicontrol(fig,'style','toggle',...
     'string','退出程序',...
'position',[660 360 240 60],'callback','close all');
% （设置字的颜色为红色，字号为28，加粗，字体为黑体
```

```
set(tb2,'foreground',[1,0,0]);
set(tb2,'fontsize',28,'fontweight','bold');
set(tb2,'fontname','黑体')

% Define a figure shown in the present axes.
%  (定义一个显示在当前坐标系的图形)
z=[eps:.01:10];phi=[0:.01:2*pi+0.1];
a=size(z);b=size(phi);
zz=ones(a(2),b(2));phii=zz;
for i=1:b(2)
    zz(:,i)=z;
end
for i=1:a(2)
    phii(i,:)=phi;
end
x=zz.*cos(phii);
y=zz.*sin(phii);
m=3;kc=5;
y1=0.5*(besselj(m-1,kc*z)-besselj(m+1,kc*z));
yy1=y1.^2;
y2=(cos(m*phi)).^2;
y3=besselj(m,z);
yy3=y3.^2./z.^2;
y4=(sin(m*phi)).^2;
y5=(yy1'*y2)+2*m^2*(yy3'*y4);
cla
surf(x,y,y5);
shading interp;
colormap jet;
axis off;
axis square;
```

运行结果如图 6.21 所示。

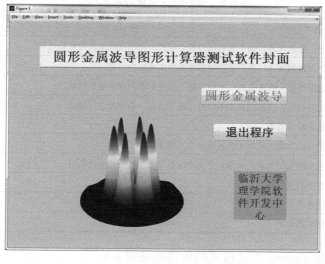

图 6.21 圆形金属波导图形计算器测试软件运行界面

> **注意：**
> 本例没有编写回调函数，单击"圆形金属波导"和"退出程序"2个按钮不会有相应语句执行，如果读者感兴趣可以自己编写。

6.4 常用 GUI 组件创建与设置

GUI 可以包括多个图形对象及多个接口容器对象，这些容器对象可成为图形用户对象或其他容器的接口容器。常见 GUI 图形对象的描述参见表 6.2。

表 6.2 常见 GUI 图形对象

用户交互对象	描述
uicontrol	创建用户交互控制对象
uimenu	创建用户交互菜单对象
uicontextmenu	创建用户交互上下文菜单对象
uitoolbar	在图形窗口创建工具条
uitoggletool	在图形窗口的工具条中创建双向切换按钮，为 uitoolbar 对象的子对象
uipushtool	在图形窗口的工具条中创建按钮，为 uitoolbar 对象的子对象
uitable	创建用户交互二维图形表格对象
uibuttongroup	创建用户交互容器对象
uicontainer	创建用户交互对象，包含其他一些用户交互对象并成为这些对象的父对象，在该容器对象中的子对象的位置和移动属性通常是相对于该容器对象而言的
uipanel	创建用户交互容器对象，与 uicontainer 比较相似，但 uipanel 对象可以设置边界和标题

uicontrol 对象基本构件的描述参见表 6.3。

表 6.3 uicontrol 对象的基本构件

基本构件	描述
checkbox	创建复选框对象，每个复选框对象由复选框和相应的标签对象构成
edit	创建可编辑文本，用户可以动态地修改文本框中内容
frame	创建组选框，组选框为一个透明的、带边框和阴影的矩形区域
listbox	创建列表框，产生的文本条目可以用于选择，但不能进行编辑
popupmenu	创建下拉式菜单
pushbutton	创建按钮对象，按钮上通常显示文本标签
radiobutton	创建单选按钮对象，通常一个文本标签和一个圆圈或菱形构件
slider	创建滑动框对象
text	创建静态文本
togglebutton	创建切换按钮

创建 uicontrol 和 uimenu 语法格式的描述参见表 6.4。

表 6.4 创建 uicontrol 和 uimenu 的语法格式

语 法 格 式	描 述
h=uicontrol（'Name', 'Value',…）	创建用户需要控件，并确定属性
h=uicontrol（parent, 'Name', 'Value',…）	在父对象中创建用户需要控件，并确定属性
h=uicontrol	在当前图形中创建按钮，并确定默认属性
uicontrol（uich）	将焦点转移到句柄对象 uich 上
h=uimenu（'Name', 'Value',…）	创建用户需要菜单，并赋予其属性，同时返回句柄
h=uimenu（Parent, 'Name', 'Value',…）	在父对象中创建用户需要菜单，并确定属性，同时返回句柄
uimenu（'Name', 'Value',…）	创建用户需要菜单，并赋予其属性
uimenu（parent, 'Name', 'Value',…）	在父对象中创建用户需要菜单，并确定属性

常见 uicontrol 对象属性的描述参见表 6.5。

表 6.5 常见 uicontrol 对象的属性

属 性 名	描 述
BackgroundColor	设置控件的背景颜色
BeingDeleted	设置控件能否被删除，默认为 off
BusyAction	设置回调程序终止
ButtonDownFcn	设置按下按键回调程序
CData	设置控件显示的图形颜色
Callback	设置控件动作
Children	设置控件有没有子对象
Clipping	该属性对控件没有影响，默认为 on
CreateFcn	在对象创建过程中执行回调程序
DeleteFcn	在对象删除过程中执行回调程序
Enable	设置允许或禁止控件
Extent	设置控件位置及大小，含有坐标（x, y）和宽度、高度
FontAngle	设置文字的倾斜角度
FontName	设置字体名称
FontSize	设置字体大小
FontUnits	设置字体大小的单位
FontWeight	设置字体
ForegroundColor	设置显示文本的颜色
HandleVisibility	处理句柄是否可以通过命令行或 GUI 图形界面对象接触
HitTest	设置通过鼠标单击是否可选
HorizontalAlignment	设置标签字符的排列
Interruptible	设置回调程序终止模式
KeyPressFcn	设置键盘按键按下回调程序
ListboxTop	设置列表框中最上面显示的序号
Max	设置最大值（取决于控件本身）
Min	设置最小值（取决于控件本身）
Parent	设置控件的父对象
Position	设置控件位置及大小，含有坐标（x, y）和宽度、高度

（续表）

属性名	描述
SelectionHighlight	设置选中控件时凸显或强调
SliderStep	设置滑动杆的步长
String	设置控件显示的文本
Style	设置控件的类型
Tag	设置控件的标签
TooltipString	设置控件的提示框内容
Type	设置控件的类
UIContextMenu	设置和 uicontrol 对象相关的上下文菜单对象
Units	设置位置向量的单位
UserData	设置用户指定的数据
Value	设置 uicontrol 对象的当前值
Visible	设置控件是否可见

控件数据处理函数的语法描述参见表 6.6。

表 6.6 控件数据处理函数的语法

语法	描述
value=getappdata（h, name）	获得应用程序中定义的数值
setappdata（h, name, value）	设置应用程序中的数值
rmappdata（h, name）	删除应用程序中的数值
isappdata（h, name）	判断是否为句柄函数值

【例 6.7】 函数回调编程。

EXAMP06007

```
function EXAMP06007
%-------主函数 EXAMP06007
% Function EXAMP06007 shows some examples to illustrate how to
% program callback function.
% EXAMP06007
%
% Records of revisions
% Date: 2015-08-15
% Description of changer: revision
%
% Generate two vectors.
% evalin: Evaluate expression in workspace.
%         （对工作空间的表达式估值）
evalin('base','data_x=0:pi/24:2*pi;')
evalin('base',...
       'data_y=exp(-(0:pi/24:2*pi)/3).*cos(2*(0:pi/24:2*pi));')
% Get the default background color of system.
% （获得系统背景颜色）
```

```matlab
panelColor=get(0,'DefaultUicontrolBackgroundColor');
%
% （打开图形窗口并设置其属性）
f=figure('Units','characters',...
         'Position',[30,30,120,35],...
         'Color',panelColor,...
         'HandleVisibility','callback',...
         'IntegerHandle','off',...
         'Renderer','painters',...
         'Toolbar','figure',...
         'NumberTitle','off',...
         'Name','WorkspacePlotter',...
         'ResizeFcn',@figResize);
% （创建底部面板图形对象）
botPanel=uipanel('BorderType','etchedin',...
         'BackgroundColor',panelColor,...
         'Units','characters',...
         'Position',[1/20,1/20,119.9,8],...
         'Parent',f,...
         'ResizeFcn',@botPanelResize);
% （创建右侧面板图形对象）
rightPanel=uipanel('BorderType','etchedin',...
         'BackgroundColor',panelColor,...
         'Units','characters',...
         'Position',[88,8,32,27],...
         'Parent',f,...
         'ResizeFcn',@rightPanelResize);
% （创建中间面板图形对象）
centerPanel=uipanel('BorderType','etchedin',...
         'Units','characters',...
         'Position',[1/20,8,88,27],...
         'Parent',f);
% （在中间面板添加坐标系）
a=axes('parent',centerPanel);
% （添加列表框和标签）
listBoxLabel=uicontrol(f,'Style','text',...
         'Units','characters',...
         'Position',[4,24,24,2],...
         'String','在workspace中选择两个变量',...
         'BackgroundColor',panelColor,...
         'Parent',rightPanel);
listBox=uicontrol(f,'Style','ListBox',...
         'Units','characters',...
         'Position',[4,2,24,20],...
         'BackgroundColor','white',...
         'Max',10,'Min',1,...
         'Parent',rightPanel,...
```

```matlab
            'Callback',@listBoxCallback);
%   (添加弹出菜单和标签)
popUpLabel=uicontrol (f,'Style','text',...
            'Units','characters',...
            'Position',[80,4,24,2],...
            'String','选择绘图模式',...
            'BackgroundColor',panelColor,...
            'Parent',botPanel);
popUp=uicontrol (f,'Style','popupmenu',...
            'Units','characters',...
            'Position',[80,2,24,2],...
            'BackgroundColor','white',...
            'String',{'Plot','Bar','Stem'},...
            'Parent',botPanel);
%   (添加hold和绘图对象)
holdToggle=uicontrol (f,'Style','toggle',...
            'Position',[45,2,24,2],...
            'Units','character',...
            'String','Hold State',...
            'Parent',botPanel,...
            'Callback',@holdToggleCallback);
plotButton=uicontrol (f,'Style','pushbutton',...
            'Units','character',...
            'Position',[10,2,24,2],...
            'String','绘图',...
            'Parent',botPanel,...
            'Callback',@plotButtonCallback);
%   (初始化列表框对象并保证hold按钮的属性值正确)
listBoxCallback (listBox,[])
holdToggleCallback (holdToggle,[])
% --------主函数EXAMP06007 (除了最后结束end)
%
  %  Programming of callback functions.
  %   (回调函数的编写)
  %----(1)调整图形大小的回调函数
  function figResize (src,evt)
      fpos=get (f,'Position');
      set (botPanel,'Position',...
         [1/20,1/20,fpos (3) -0.1,fpos (4) *8/35]);
      set (rightPanel,'Position',...
         [fpos (3) *85/120,fpos (4) *8/35,...
         fpos (3) *35/120,fpos (4) *27/35]);
      set (centerPanel,'Position',...
         [1/20,fpos (4) *8/35,fpos (3) *85/120,fpos (4) *27/35]);
  end
  %
  %----(2)调整底板大小的回调函数
```

```
function botPanelResize(src,evt)
    bpos=get(botPanel,'Position');
    set(plotButton,'Position',...
        [bpos(3)*10/120,bpos(4)*2/8,bpos(3)*24/120,2]);
    set(holdToggle,'Position',...
        [bpos(3)*45/120,bpos(4)*2/8,bpos(3)*24/120,2]);
    set(popUp,'Position',...
        [bpos(3)*80/120,bpos(4)*2/8,bpos(3)*24/120,2]);
    set(popUpLabel,'Position',...
        [bpos(3)*80/120,bpos(4)*4/8,bpos(3)*24/120,2]);
end
%
%----（3）调整右侧面板大小的回调函数
function rightPanelResize(src,evt)
    rpos=get(rightPanel,'Position');
    set(listBox,'Position',...
        [rpos(3)*4/32,rpos(4)*2/27,rpos(3)*24/32,rpos(4)*20/27]);
    set(listBoxLabel,'Position',...
        [rpos(3)*4/32,rpos(4)*24/27,rpos(3)*24/32,rpos(4)*2/27]);
end
%
%----（4）处理列表框的回调函数
function listBoxCallback(src,evt)
    %  evalin: Evaluate expression in workspace.
    %          （从工作空间获得向量）
    vars=evalin('base','who');
    set(src,'String',vars);
end
%
%----（5）绘图按钮的回调函数
function plotButtonCallback(src,evt)
    vars=get(listBox,'String');
    var_index=get(listBox,'Value');
    if length(var_index)~=2
        errordlg('必须选择两个变量','不正确的选择','modal')
        return
    end
    x=evalin('base',vars{var_index(1)});
    y=evalin('base',vars{var_index(2)});
    selected_cmd=get(popUp,'Value');
    axes(a);
    switch selected_cmd
    case 1
        plot(x,y)
    case 2
        bar(x,y)
    case 3
```

```
            stem(x,y)
        end
    end
%
%----(6)切换hold状态的回调函数
function holdToggleCallback(src,evt)
    button_state=get(src,'Value');
    if button_state==get(src,'Max')
        hold(a,'on')
        set(src,'String','Hold on');
    elseif button_state==get(src, 'Min')
        hold(a,'off')
        set(src,'String','Hold off');
    end
end
end % 主函数的结束标志
```

在命令窗口运行 EXAMP06007 程序，结果如图 6.22 所示。选中 Stem 绘图结果，如图 6.23 所示。

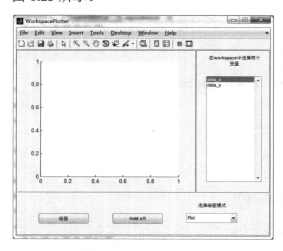

图 6.22　GUI 示意图

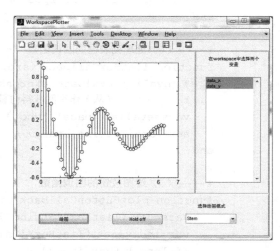

图 6.23　Stem 绘图结果

6.5　编译独立的应用程序

6.5.1　编译器的安装与配置

1. 编译器的安装

在安装 MATLAB R2013a 时，要选择安装组件，典型（T）使用默认设置，安装所有已许可的产品和自定义（U）指定所有安装选项（产品和快捷方式）；选择自定义模式，将提示用户选择安装组件，如图 6.24 所示。如果用户没有安装 MATLAB Compiler 4.18.1 组件，则可进行更新安装，在更新安装过程中选中该组件，完成安装。

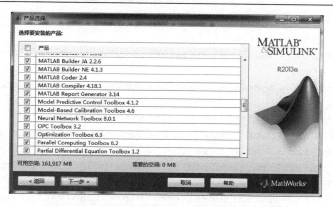

图 6.24　安装 MATLAB 时选择组件对话窗口

2. 配置编译器

借助例 6.8 介绍配置编译器过程。

【**例 6.8**】　配置 MATLAB 的编译器。

（1）通过 mex 函数对编译器进行配置。

```
>> mex -setup
Welcome to mex -setup. This utility will help you set up a default compiler.
For a list of supported compilers, see http://www.mathworks.com/ support/
compilers/R2013a/win64.html
Please choose your compiler for building MEX-files:
Would you like mex to locate installed compilers [y]/n?
```

输入 y 然后按回车键：

```
Select a compiler:
[1] Microsoft Visual C++ 2010 in D:\Program Files (x86)\Microsoft Visual
Studio 10.0
[0] None
Compiler:
```

以上程序显示表明：本机系统安装了 1 个编译器，只能选择这个编译器 Microsoft Visual C++2010。在 Compiler: 后面输入"1"然后按回车键：

```
Please verify your choices:
Compiler: Microsoft Visual C++ 2010
Location: D:\Program Files (x86)\Microsoft Visual Studio 10.0
Are these correct [y]/n?
```

需要用户进一步确认编译器的位置是否正确，如果正确，选择 y。

```
Are these correct [y]/n? y
***************************************************************
  Warning: MEX-files generated using Microsoft Visual C++ 2010 require
           that Microsoft Visual Studio 2010 run-time libraries be
           available on the computer they are run on.
           If you plan to redistribute your MEX-files to other MATLAB
           users, be sure that they have the run-time libraries.
```

```
************************************************************
Trying to update options file:
C:\Users\wyl\AppData\Roaming\MathWorks\MATLAB\R2013a\mexopts.bat
From template:
D:\PROGRA~2\MATLAB\R2013a\bin\win64\mexopts\msvc100opts.bat
Done...
************************************************************
   Warning: The MATLAB C and Fortran API has changed to support MATLAB
            variables with more than 2^32-1 elements. In the near future
            you will be required to update your code to utilize the new
            API. You can find more information about this at:
            http://www.mathworks.com/help/matlab/matlab_external/
            upgrading-mex-files-to-use-64-bit-api.html
            Building with the -largeArrayDims option enables the new API.
************************************************************
```

(2) 通过函数 mbuild 对编译器进行配置。

```
>> mbuild -setup
Welcome to mbuild -setup. This utility will help you set up a default compiler.
For a list of supported compilers, see
     http://www.mathworks.com/support/compilers/R2013a/win64.html
     Please choose your compiler for building shared libraries or COM components:
     Would you like mbuild to locate installed compilers [y]/n?y
```

输入 y 然后按回车键：

```
Would you like mbuild to locate installed compilers [y]/n? y
Select a compiler:
[1] Microsoft Visual C++ 2010 in D:\Program Files (x86)\Microsoft Visual Studio 10.0
[0] None
Compiler:
```

选择编译器"1"然后按回车键：

```
Compiler:1
Please verify your choices:
Compiler: Microsoft Visual C++ 2010
Location: D:\Program Files (x86)\Microsoft Visual Studio 10.0
Are these correct [y]/n?
```

进行确定：

```
Are these correct [y]/n? y
************************************************************
   Warning: Applications/components generated using Microsoft Visual C++
            2010 require that the Microsoft Visual Studio 2010 run-time
            libraries be available on the computer used for deployment.
            To redistribute your applications/components, be sure that the
            deployment machine has these run-time libraries.
************************************************************
Trying to update options file:
```

```
C:\Users\wyl\AppData\Roaming\MathWorks\MATLAB\R2013a\compopts.bat
From template:
D:\PROGRA~2\MATLAB\R2013a\bin\win64\mbuildopts\msvc100compp.bat
Done . . .
```

6.5.2 编译 exe 文件

mcc 的语法格式描述参见表 6.7。

表 6.7 mcc 的语法格式

语　法	描　述
mcc -m myfunction	为 myfunction.m 创建独立可执行文件 myfunction.exe
mcc -m -I /files/source -d /files/target myfunction	为文件夹/files/source 中的 myfunction.m 创建独立可执行文件 myfunction.exe,保存在文件夹/files/target 中
mcc -m myfunction1 myfunction2	将 myfunction1.m 和 myfunction2.m 创建成为可独立执行文件 myfunction1.exe
mcc -W lib:liba -T link:lib a0 a1	从 a0.m 和 a1.m 中生成共享链接库或动态链接库文件,文件名为 liba

完成上面配置,调用例 6.6 的程序 EXAMP06006,试运行,然后在 MATLAB 命令窗口输入如下命令语句:

```
>> mcc -m zhuye.m
```

就可得到 EXAMP06006.exe 文件。双击该文件打开窗口,如图 6.25 和图 6.26 所示。需要指出,编译文件不能是脚本文件,必须是函数 M 文件。

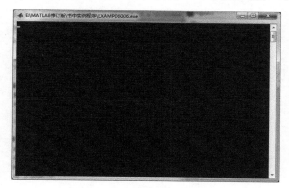

图 6.25 文件运行的 DOS 窗口

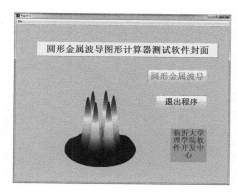

图 6.26 GUI 桌面

小　　结

本章首先介绍了如何启动 GUI,如何创建 GUI 以及构成 GUI 主要组元的属性与设置,并提供 3 个示例进行解释说明。随后提供了 3 个稍微复杂的实例,前 2 个实例借助 MATLAB 提供 GUI 创建模板完成,需要注意每个实例的各个组元属性设置及回调函数的编写,这为初学者提供了清晰的思路和可借鉴的经验;最后 1 个实例没有借助 MATLAB 提供的模板而是完全通过编程完成,这样更具灵活性,但是许多设置需要读者有清楚的认识。鉴于实例编程的需要,对常用 GUI 控件的创建与设置进行了总结,便于读者学习。最后对如何将 MATLAB 编写的 M 文件转化为 exe 可独立运行文件进行了简要介绍。

习 题

6.1 创建一个 GUI 窗口绘制方程 $y(x) = ax^2 + bx + c$ 图形,需要一个显示绘图结果的坐标系窗口,还能够输入 a、b、c 的值和 x 的取值范围(包含最小值和最大值),同时设置其他的所有组元属性。

6.2 创建一个 GUI 窗口,含有下拉菜单,下拉菜单中有背景颜色选项。

6.3 创建一个 GUI 窗口显示各类随机数的分布。程序需要创建一个含有 200 000 个随机数的样本,并能计算它们的分布情况,然后借助函数 hist 创建柱状图。注意设置和标注柱状图的属性。程序编写要支持标准分布、Gauss 分布、Rayleigh 分布,这些分布选择都显示在下拉菜单中。还有,该 GUI 窗口还为用户提供输入文本框,用户可以自己定义图形中显示柱的个数。

6.4 创建一个 GUI 窗口能够从外部文件读入数据,然后根据输入的 x 和 y 数据拟合出多项式,根据多项式绘出曲线。

6.5 随便修改前一个 GUI 窗口,让其具有编辑菜单功能。

第 7 章 Simulink 仿真系统

  Simulink 是 MATLAB 软件的一个软件包，是重要的组件之一，能够使用户和系统交互进行动力学系统建模、仿真和综合分析。Simulink 仿真系统支持完全的图形化界面，并且提供了标准的模型库，能够帮助用户在此基础上创建新的模型库，描述、模拟、评价和细化系统的行为，从而达到系统分析的目的。此外，还可以通过和其他工具箱产品来完成更多的分析任务。

  用户只需根据具体问题选择相应的组件，将相应组件从模型库复制到仿真系统的创建界面中，再将这些组件通过恰当的线合理地连接起来，完成模型创建，这就可以实现具体功能。用户可以借助模块分层设计流程，主要工作为模块结构的优化，不是算法设计。因此，Simulink 模型可以让用户明白具体环节的动态细节，也可以让用户清楚地理解系统各模块、各子系统、各分系统间的信息交换。

  Simulink 不仅仅可以解决线性模型，也可以解决非线性模型，如摩擦系数分析、空气阻力、齿轮滑动以及其他所有描述真实世界的现象。Simulink 还可将计算机变为模拟分析动力学系统的实验室，将许多的不可能转变为可能，比如自动离合系统的动作、飞机机翼的抖动、弱肉强食的过程、货币对经济的效用等的研究与分析模拟。

  Simulink 提供了图形用户界面，使得构建模型变得更直观、简单，只需鼠标的单击与拖放。Simulink 模块是分层次结构的，为用户寻找需要的模块提供了方便，提高了工作效率。所以 Simulink 是 MATLAB 软件一个非常重要的组成部分。关于 Simulink 相关产品可以通过网址 http://www.mathworks.com/products/simulink/related.html 进行了解。

  本章有 4 节内容。7.1 节介绍 Simulink 仿真系统的基础，7.2 节讲述 Simulink 模型操作和仿真系统的设置，7.3 节给出系统建模实例，7.4 节阐述如何创建仿真系统中的子系统。

7.1 Simulink 基 础

7.1.1 启动 Simulink

  启动 Simulink 有两种最常用的途径，一种是通过单击图 7.1 中被框中的工具栏 "Open Simulink block library" 图标实现的，另一种是在 MATLAB 命令窗口输入 simulink 然后按回车键。

  两种途径打开的 Simulink 模块库浏览器如图 7.2 所示。

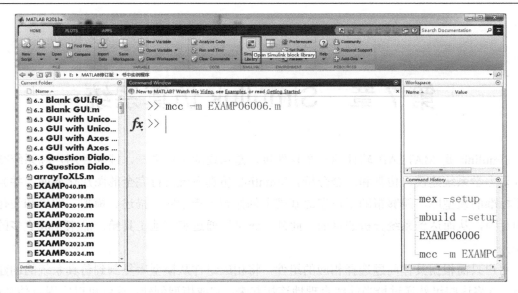

图 7.1　打开 Simulink Library Browser 的工具栏图标示意图

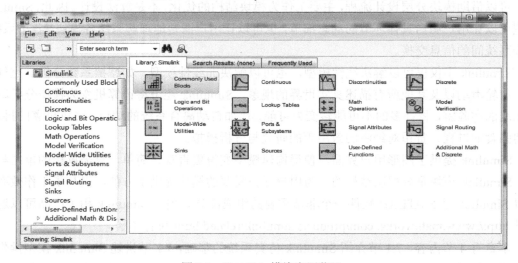

图 7.2　Simulink 模块库浏览器

在使用 Simulink 过程中，会经常用到 Simulink 模块库浏览器的菜单，参见表 7.1。

表 7.1　Simulink 模块库浏览器菜单

菜　单	子　菜　单	描　述
File	New	新建模型
	Open	打开已有模型
	Close	关闭打开模型
	Preferences…	属性设置
Edit	Find	寻找模块
	Add Selected Block to a New Model	向新模型中添加模块
	Select Results Block in Library View	选中库中模块

（续表）

菜　单	子　菜　单	描　述
View	Simulink Project	显示 Simulink 设计窗口
	Show Block Descriptions	显示模块描述
	Stay on Top	置顶
	Layout	图标显示格式，罗列格式（List）和平铺格式（Grid）
	Icon Size	图标大小设置含紧凑（Compact）、小图标（Small）、中等图标（Medium）和大图标（Large）4个选项
	Increase Font Size	加大字号
	Decrease Font Size	减小字号
	Show Parameters for Selected Block	显示选中模块的参数
	Refresh Tree View	更新窗口
Help	Help for the Selected Block	选中模块的帮助
	Library Browser Help	模块库浏览器帮助
	Simulink Help	Simulink 帮助

Simulink 模型的属性设置也会常常遇到，可以借助 File→Preferences 菜单进行设计，如图 7.3 所示。

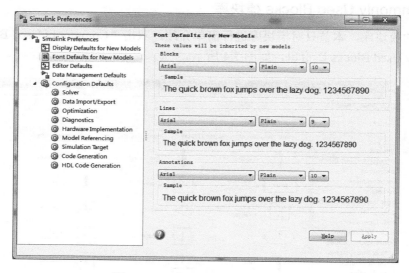

图 7.3　Simulink 界面属性设置窗口

7.1.2　Simulink 模块库浏览器

Simulink 模块库浏览器中有很多种模块库，参见表 7.2。

表 7.2　Simulink 模块库的分类

名　称	描　述
Commonly Used Blocks	常用模块库
Continuous	连续信号模块库

（续表）

名称	描述
Discontinuities	不连续信号模块库
Discrete	离散信号模块库
Logic and Bit Operations	逻辑和位操作模块库
Lookup Tables	查表模块库
Math Operations	数学运算模块库
Model Verification	模型确认模块库
Model-Wide Utilities	模型工具模块库
Ports & Subsystems	端口和子系统模块库
Signal Attributes	信号贡献模块库
Signal Routing	信号传输模块库
Sinks	信号接收器模块库
Sources	信号源模块库
User-Defined Functions	用户定义函数模块库
Additional Math & Discrete	附加数学和离散模块库

7.1.3 Commonly Used Blocks 模块库

如此多的模块库，本节只对模块库进行简单介绍，双击"Commonly Used Blocks"，打开 Commonly Used Blocks 库窗口，如图 7.4 所示。

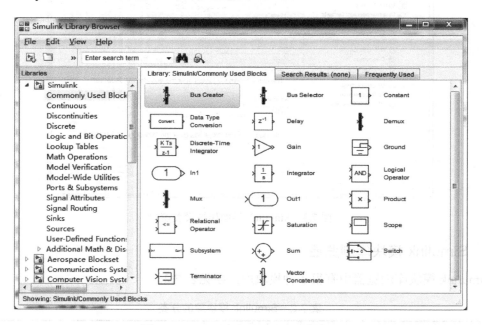

图 7.4　Commonly Used Blocks 库窗口

常用模块库中模块描述参见表 7.3。其他模块库格式类似，读者可以自己学习。

表 7.3　常用模块库中模块

模 块 名 称	描　　述	模 块 名 称	描　　述
Bus Creator	母线产生器	Mux	合并信号
Bus Selector	母线选择器	Out1	输出
Constant	常数	Product	相乘
Data Type Conversion	数据类型转换	Relational Operator	关系算符
Delay	延迟	Saturation	饱和度
Demux	拆分信号	Scope	示波器
Discrete-Time Integrator	离散时间积分器	Subsystem	子系统
Gain	增益	Sum	求和
Ground	接地	Switch	开关
In1	输入	Terminator	终端
Integrator	积分器	Vector Concatenate	矢量连接
Logical Operator	逻辑算符		

7.1.4　Simulink 模型窗口

当创建新的 Simulink 模型或打开一个已经存在的 Simulink 模型时，将会弹出 Simulink 模型窗口。此处，创建一个新 Simulink 模型，命名为 EXAMP07001，初始窗口如图 7.5 所示。

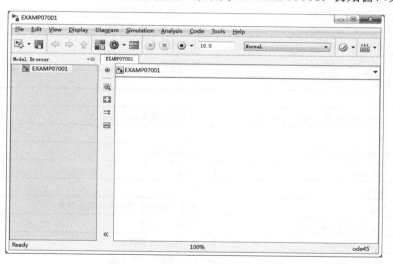

图 7.5　创建 Simulink 模型初始窗口

模型窗口同样含有菜单栏、工具栏、编辑框和状态栏等部分。其中菜单栏，主菜单栏有 10 个命令，分别为 File、Edit、View、Display、Diagram、Simulation、Analysis、Code、Tools 和 Help；工具栏有 ▦·（新建 Model、Chart、Library）、▤（保存）、▦（打开 Library Browser）、▦·（有 3 个选择，分别为 Model Configuration Parameters、Simulation Target For MATLAB & Stateflow、Model Propertes）、▦（打开 Model Explorer）、▦（打开 Stepping Options 设置）、▶（进行仿真）、▶（前进）、■（停止仿真）、▦·（有 5 个子菜单，分别为 Recond & Inspect Simulation Output、Configure Data Logging…、Log Chart Signals…、Help on Logging & Recording…、Simulation Data Inspectors…）、▦（仿真时间设置，仿真方式

选择有 Normal、Accelerator、Rapid Accelerator、Software-in-the-Loop（SIL）、Processor-in-the-Loop（PIL）、External）、 （可选项有 Model Advisor、Model Advisor Dashboard、Upgrade Advisor、Performance Advisor、Code Generation Advisor）、 ·（创建程序有 2 个选项：Build Model 和 Build Selected Subsystem）。状态栏的显示内容会显示模型窗口表示的状态信息：Ready 表示模型已经准备就绪；100%表示编辑框的显示比例；ode45 表示仿真选择的积分算法为 ode45。

Simulink 模型窗口主要菜单项及其子菜单描述参见表 7.4。

表 7.4 Simulink 模型窗口主要菜单项及其子菜单

菜 单	子 菜 单	描 述
File	New	新建 Simulink 模型
	Open	打开 Simulink 模型
	Close	关闭 Simulink 模型
	Save	保存
	Save as	另存为
	Source Control	设置 Simulink 和 SCS 的接口
	Export Model to Reports	导出 Simulink 模型报告
	Model Properties	设置 Simulink 属性
	Print	打印
	Simulink Preferences	设置 Simulink 界面属性
	Stateflow Preferences	设置 Stateflow 界面属性
	Exit MATLAB	退出 MATLAB
Edit	Undo Create Segment	撤销创建
	Cut	剪切
	Copy	复制
	Copy Current View to Clipboard	将当前视图复制到剪贴板
	Paste	粘贴
	Paste Duplicate Import	粘贴复制端口
	Select All	全选
	Delete	删除
	Find	搜索 Simulink 系统内的模块对象
	Find Referenced Variables…	搜索引用变量
View	Library Browser	打开模块库浏览器
	Model Explorer	打开模块资源管理器
	Simulink Project	打开 Simulink 工程
	Model Dependency Viewer	模块需要显示
	Model Browser	模型管理器
	Configure Toolbars	配置工具栏
	Toolbars	工具栏
	Status Bar	状态栏
	Explorer Bar	浏览器条
	Navigate	导航
	Zoom	工作区
	MATLAB Desktop	MATLAB 桌面

(续表)

菜　单	子　菜　单	描　述
Simulation	Update Diagram	更新图标
	Model Configuration Parameters	设置仿真心态的仿真参数和解法器等
	Data Display	数据显示
	Run	运行
	Step Forward	前进
	Stop	停止
	Output	输出
	Stepping Options	步长选择
	Debug	调试
Code		创建各种代码程序
Tools	Library Browser	打开 Simulink 模块库浏览器
	Model Explorer	打开 Simulink 模型管理器
	Report Generator	打开报告管理器
	MPlay Video Viewer	打开 MPlay 视窗
Help	Simulink	显示 Simulink 的帮助文件、示例、S 函数等
	Stateflow	显示 Stateflow 的帮助文件、示例、入门知识等
	Keyboard Shortcuts	显示 Simulink 的鼠标键盘快捷键
	Web Resources	显示相关网络资源
	Terms Of Use	显示使用条款

7.1.5　Simulink 建模仿真示例

【**例 7.1**】　用 Simulink 模拟正弦信号的产生与输出。

创建 Simulink 的步骤如下。

（1）创建新模型界面。

在 Simulink Library Browser 窗口，从菜单中选择 File→New→Model 或单击创建按钮，创建新模型界面。

（2）添加正弦信号产生模块和波形显示模块。

从 Simulink Library Browser 窗口找到 Sources 库，在 Sources 库中找到 Sine Wave，用鼠标拖入模型窗口释放；同样从 Simulink Library Browser 窗口找到 Sinks 库并找到 Scope，用鼠标拖入模型窗口释放，然后将两模块连接起来，其结果如图 7.6 所示。

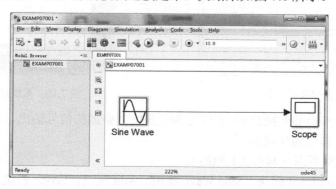

图 7.6　连接后的模块

(3) 模块属性设置。

在组件创建页面上双击正弦信号产生组件,打开属性设置窗口,如图 7.7 所示,可对组件属性进行设置;同样双击示波器模块,打开示波器初始窗口,如图 7.8 所示,可以借助上面的菜单对其属性进行设置。

图 7.7　正弦信号模块属性设置窗口

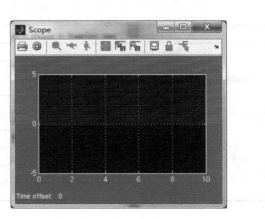

图 7.8　示波器初始窗口

(4) 运行仿真系统。

通过菜单选择 Simulation→Run 命令运行,或者单击 ▶ 按钮,模拟运行时间采用默认的 10.0s。对示波器属性设置窗口(如图 7.9 所示)进行设置,同时单击示波器窗口工具栏中的按钮 ,正弦波图形结果如图 7.10 所示。

图 7.9　示波器属性设置窗口

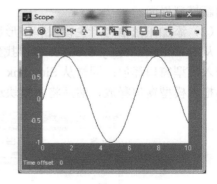

图 7.10　系统运行结果

【例 7.2】 用 Simulink 模拟 chirp 信号和正弦信号叠加的信号输出。

创建 Simulink 模型的步骤如下。

(1) 创建新模型界面(与例 7.1 相同)。

(2) 添加 Chirp Signal 模块、Sine Wave 模块、Add 模块、Scope 模块。

从 Sources 库找到 Chirp Signal 模块和 Sine Wave 模块,将它们拖入模型窗口;从 Sinks

库找到 Scope 模块,将其拖入模型窗口;从 Math Operations 库找到 Add 模块,将其拖入模型窗口,其结果如图 7.11 所示。

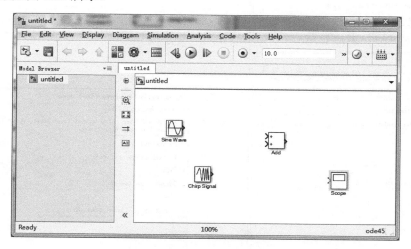

图 7.11　未连接模块的模型窗口

(3) 连接这些模块。

将 Sine Wave 和 Chirp Signal 两个模块作为信号源输入计算 Add 模块,用线连接;将计算后的信号作为信号源输入 Scope 模块,适当调整模块位置,如图 7.12 所示。

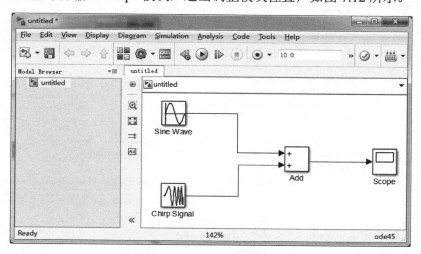

图 7.12　连接调整模块后的模型窗口

(4) 模块属性设置。

双击模块,根据需要设置其属性。Sine Wave 模块属性设置如图 7.13 所示,Chirp Signal 模块属性如图 7.14 所示,通过菜单 Simulink→Model Configuration Parameters 命令打开属性设置窗口,如图 7.15 所示。

(5) 运行系统。

通过菜单 File→Save as 命令保存为 EXAMP07002。通过对示波器属性设置窗口(如图 7.16 所示)进行设置,运行时间设置为 20.0s,通过菜单 Simulation→Run 命令运行该模型,示波器的窗口显示如图 7.17 所示。

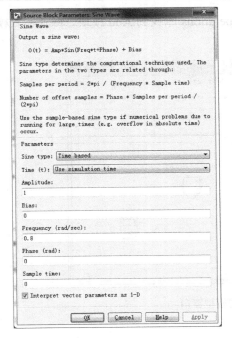
图 7.13 Sine Wave 模块属性设置

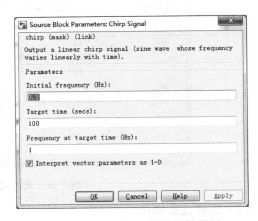
图 7.14 Chirp Signal 模块属性设置

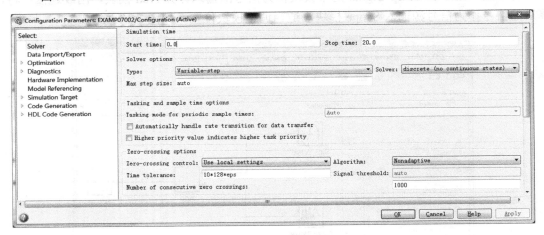
图 7.15 属性设置窗口

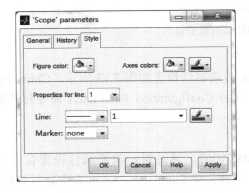
图 7.16 示波器属性设置

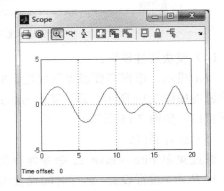
图 7.17 示波器结果图

7.2 Simulink 模型操作和仿真系统设置

7.2.1 Simulink 模型操作

Simulink 模型一般包括输入、输出和中间状态操作等。一般输入由信号源提供，中间状态操作借助各种模块共同完成（称为系统），输出由信号接收器完成。当然，根据任务需要也可以由两部分构成。

Simulink 对模型系统的仿真过程，可以看作是在用户给定的时间段内根据模型提供的信息计算系统的状态和输出。Simulink 仿真过程通常可分为模型编译阶段、连接阶段和仿真阶段。模型编译阶段完成系统评价模型参数的表达式、确定信号属性、对系统模型进行优化、确定模型运动的优先级及采样时间等任务；连接阶段主要按照执行次序安排方法的运行列表，同时将定位和初始化存储在每个模型中；仿真阶段主要包括仿真初始化和仿真迭代两个阶段。仿真初始主要初始化系统的状态和输出。在定义的时间间隔内，仿真迭代实现一个时间段重复运行一次，用于确定每个时间段计算模型的新输入、输出和状态。

1. 操作模块

在 Simulink 模块创建过程中，对模块的操作是必需的。首先，在模块库中选中模块并复制，然后，在模型创建窗口粘贴模块。这个操作也可以借助鼠标拖动完成。在 Simulink 模型创建窗口，可借助鼠标拖动调整模块图标的大小。如果有更多需要修改的模块属性、特点、外观等，可以先选中需要操作的模块图标，单击鼠标右键打开菜单栏，如图 7.18 所示。如果需要设置模块图标的文字属性，可以单击 Format 菜单打开次级，选中 Font Style 命令打开字体设置窗口，如图 7.19 所示。如果需要修改模块方向，选中 Rotate & Flip 命令，可实现对模块图标顺时针旋转、逆时针旋转、上下翻转和上下翻转图标与文字标识。

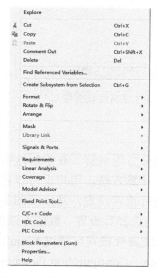

图 7.18　模块属性设置菜单栏

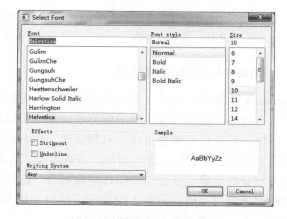

图 7.19　模块字体设置窗口

2. 操作连线

在 Simulink 中，信号由模块间连线传输，创建的多个模块最终需要通过连线将它们连接起来。在连线过程中,需要连接水平线、竖直线、斜线,有时还需要移动甚至删除连线。水平线和竖直线只需按住鼠标左键从起始点到终止点拖动光标,Simulink 会自动连线。斜线稍微复杂些,拖动光标时需要按下 Shift 键。移动首先要选中连线,然后借助鼠标拖动或者键盘上方向键进行移动,删除也要首先选中然后按 Delete 键即可。

当仿真系统比较复杂时,一个信号需要送到不同模块的输入端口,分支连线成为必然。将光标线移动到分叉点,然后按住鼠标右键拖动光标到分支线的终点释放鼠标,Simulink 自动完成分支线。

当所有模块间连线完成时,如果连线显得杂乱有失美观,可以通过选中需要移动位置的模块,然后借助键盘上的"上、下、左、右"键进行移动,直到移动到用户满意为止。

为了用户更容易理解和掌握所用信号线的信息,常常需要对信号线添加信号标识。为信号线添加标识时,只需双击需要添加标识的信号线,系统就会自动弹出标识的编辑框,在编辑框中输入需要的文字即为标识。标识编辑框在选中情况下,可以对其进行复制、修改和删除的基本操作。

信号线的属性也可以进行设置,在选中信号线的情况下,单击鼠标右键打开菜单,如图 7.20 所示。选中菜单中的 Properties 命令,打开信号线属性设置窗口,如图 7.21 所示。

图 7.20　信号线菜单

图 7.21　信号线属性设置窗口

7.2.2　Simulink 仿真系统设置

Simulink 仿真程序实质上是对描述系统的一组微分或差分方程求解。在求解这些微分方程或差分方程时,需要设置一种求解方法,系统提供默认的计算方法,用户也可以重新设置方法。当然,其他仿真参数系统也有默认值,用户也可以进行重新设置,常涉及的参数主要包括仿真系统的起始时间、终止时间、仿真步长、仿真容差的选择和设置、数值方法的选择、是否从外部获取数据、是否向外部传输数据等。这些仿真系统属性设置可通过 Simulink 仿真系统属性设置窗口完成,如图 7.22 所示。通过菜单 Simulation→Configuration Parameters…命令打开 Simulink 仿真系统属性设置窗口。

该窗口左边选项栏含有 Solver(求解器)、Data Import/Export(数据输入与输出)、Optimization(优化)、Diagnostics(仿真诊断)、Hardware Implementation(硬件需求)、Model Referencing(参考模型)、Simulation Target(仿真目标)、Code Generation(代码生成)和 HDL

Code Generation（HDL 代码生成）。其中，Optimization 含有 Signals and Parameters 和 Stateflow 共 2 个子选项；Diagnostics 含有 Sample Time、Data Validity、Type Conversion、Connectivity、Compatibility、Model Referencing、Saving 和 Stateflow 共 8 个子选项；Simulation Target 含有 Comments 和 Symbols 共 2 个子选项；Code Generation 含有 Report、Comments、Symbols、Custom Code 和 Debug 共 5 个子选项；HDL Code Generation 含有 Global Settings、Test Bench 和 EDA Tool Scripts 共 3 个子选项。

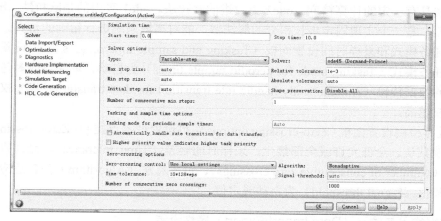

图 7.22 Simulink 仿真系统属性设置窗口

1. 求解器（Solver）属性设置

求解器属性设置窗口如图 7.22 所示，左侧为面板选择列表，右侧为可设置参数面板，含有仿真时间设置面板（Simulation time）、解法器设置面板（Solver options）、任务模式和采样时间设置面板（Tasking and sample time options）、过零检测设置面板（Zero-crossing options）。

（1）仿真时间设置面板（Simulation time）：设置仿真的起始时间和终止时间。

（2）解法器设置面板（Solver options）：可以设置步长类型、步长最大值、步长最小值、初始步长值、相对容错值、绝对容错值、模型保存设置、指定最小步数和求解方法。

Type 含有变步长（Variable-step）和固定步长（Fixed-step），默认选项为变步长。这两种计算器在当前仿真时间的基础上添加一个时间步长，进行求解计算。两者的主要区别在于，变步长的求解器的时间步长是根据求解仿真系统的动态特性来选择仿真步长的，而固定步长的求解器在求解时选择的步长是相等的时间步长。

在变步长情况下，仿真过程可根据误差限制自动改变步长，并提供误差控制和过零检测，Solver 含有选项 discrete、ode45、ode23、ode113、ode15s、ode23s、ode23t 和 ode23tb。默认解法器为 ode45。ode45 为 4/5 阶龙格库塔数值解法，适用于大多数连续或离散系统，但不适用于刚性（stiff）系统。它是单步解法器，计算 $y(t_n)$ 时仅需前一步 $y(t_{n-1})$ 的结果。离散求解器（discrete）能对离散系统进行精确求解。ode23 为 2/3 阶龙格库塔数值解法，在误差限要求不高和求解问题不太难的情况下会比 ode45 更有效，也是一个单步解法器。ode113 是一种阶数可变的多步解法器，也就是在计算当前时刻输出时，需要以前多个时刻的输出。在允许误差要求较严格的情况下，一般比 ode45 有效。ode15s 是一种基于数字微分公式的解法器，是一种多步解法器，适用于刚性系统。当要解决的问题较难，不能用 ode45 求解或者求解结果不理想时，可选择 ode115s。ode23s 是一种专用于刚性系统的单步解法器，在误差限制不严格情况下比 ode15s 效果好。ode23t 是一种实现梯形规则自由插值的解法器，适用于求解适度刚性问题。ode23tb 用于实现 TR-BDF2 数值解法，该解法具有两个阶段的隐式龙格库塔公式。在定步长情况下，仿真过程采用常数步长，无误差控制和过零检测，Solver 含有选项 discrete、ode8、ode5、ode4、ode3、ode2、ode1 和 ode14x。默认解法器为 ode3，使用的数值计算方法为固定步长的 2/3 阶龙格库塔数值解法。Ode8 采用 8 阶龙格库塔固定步长算法。ode5 是 ode45 的固定步长版，适用于大多数连续或离散系统，不适用于刚性系统。ode4 四阶龙格库塔法具有一定的计算精度。ode2 为改进的欧拉法。ode1 采用的是欧拉法。discrete 为离散求解器，是一个实现固定步长积分的解法器，能对离散系统进行精确的仿真。

（3）任务模式和采样时间设置面板（Tasking and sample time options）：可设置周期采样时间序列任务模式（Tasking mode for perodic sample times）为自动、单任务或多任务、自动处理速率的数据传输（Automatically handle rate transition for data transfer）、优先权越高指令任务优先权越高。

（4）过零检测设置面板（Zero-crossing options）：可设置过零点控制（Zero-crossing control）为可选系统默认、全部允许或全部禁止、采样时间容差（Time tolerance）、允许执行的仿真过零次数（Number of consecutive zero crossings）、过零算法（Algorithm）为自适应或非自适应、信号阈值（Signal threshold）为自动或定值。

2. 数据输入与输出（Data Import/Export）属性设置

选中"Data Import/Export"面板时，打开窗口如图 7.23 所示。数据输入与输出面板含有从工作空间加载数据面板（Load from workspace）、将数据保存到工作空间面板（Save to workspace）、输出数据设置面板（Save options）。

图 7.23　数据输入与输出属性设置窗口

（1）从工作空间加载数据面板（Load from workspace）：可设置加载的数据变量、加载变量的初始状态，同时可以通过"Edit Input"按钮打开加载设置窗口。

（2）将数据保存到工作空间面板（Save to workspace）：可设置时间变量名、状态变量名、输出变量名、终了状态变量名、信号日志变量名、信号日志格式、数据存储。

（3）输出数据设置面板（Save options）：可设置存储数据量为精细输出、额外输出或指定输出、精细因子（仅仅用于设置精细点输出的精细程度）、保存仿真结果为单独对象、记录和检查仿真输出。

3. 仿真诊断（Diagnostics）属性设置

选择 Diagnostics 选项，打开窗口如图 7.24 所示，可以设置仿真系统的仿真诊断选项。设置接发器诊断结果的报告方式，可选择沉默、示警、报错等 3 种方式。常规诊断设置项含有代数回路检测（Algebraic loop）、最小化代数回路（Minimize algebraic loop）、模块优先权干预（Block priority violation）、最小步长干预（Min step size violation）、采样点时间调节（Sample hit time adjusting）、接连不断过零的干预（Consecutive zero crossing violation）、未设置的采样时间继承性（Unspecified inheritability of sample time）、解法器数据矛盾（Solver data inconsistency）、自动的解法器参数选择（Automatic solver parameter selection）、无关的散射衍生信号（Extraneous discrete derivative signals）、状态变量冲突（State name clash）等。

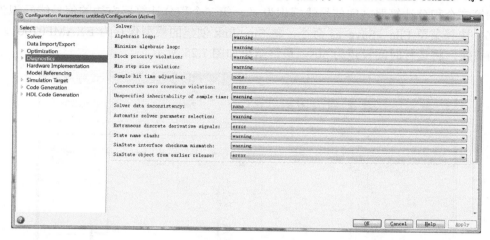

图 7.24 仿真诊断属性设置窗口

7.3 系统建模实例

【例 7.3】 借助积分器求微分方程 $x'' + 0.2x' + 0.4x = 0.2u(t)$，其中 $u(t)$ 为单位阶跃函数。

分析与求解：

将微分方程改写为

$$x'' = 0.2u(t) - 0.4x - 0.2x'$$

通过积分器求解微分方程的基本思想是通过积分器积分后得到低阶变量。如 x'' 通过一个积分器降阶为 x'，再通过一个积分器转变为 x，将其连到示波器上就能够显示出待求函数图像，连接到保存工作空间模块，可以在命令窗口输入相应变量进行显示。

设计思路如下，一般采取倒推法。

(1) 示波器模块与保存到工作空间模块的连接点为待求函数 x。

(2) 从 x 向后退一个积分计算器为 x'。

(3) 从 x' 向后退一个积分计算器为 x''。

(4) 然后根据 x'' 与 x'、x 和 $u(t)$ 的关系通过增益器"Grain"和加减组合器完成最终连接。

Simulink 模型创建步骤如下。

(1) 将 Sources 模块库中 Step 模块和 Clock 模块拖入模型窗口；将 Sinks 模块库中 Scope

模块和 To Workspace 模块拖入模型窗口；将 Commonly Used Block 模块库中 Integrator 模块、Gain 模块和 Bus Creator 模块拖入模型窗口；将 Math Operations 模块库中 Add 模块拖入模型窗口。

（2）复制 1 个 Integrator 模块和 2 个 Gain 模块，并且通过鼠标右键悬浮菜单 Rotate & Flip 命令翻转 2 个复制 Gain 模块。

（3）对三个 Gain 模块属性进行设置，双击打开设置窗口，分别将 Gain 属性设置为 0.2、0.4 和 0.2；模块 Add 的 List of signs 属性设置为+--。

（4）将所有模块用线连接起来，连线可以按住鼠标左键拖动完成，分叉线可以按住鼠标右键拖动完成。

（5）添加信号线标识，最上面一行，第一个积分计算器左边线标识为 x''，第一个积分计算器和第二个积分计算器之间信号线标识为 x'，第二个积分计算器右边的信号线标识为 x。

（6）选择菜单 File→Save as 命令，将 Simulink 创建的模型保存为 EXAMP07001，然后单击 按钮，将运行时间设定为 50s，最终结果如图 7.25 所示。

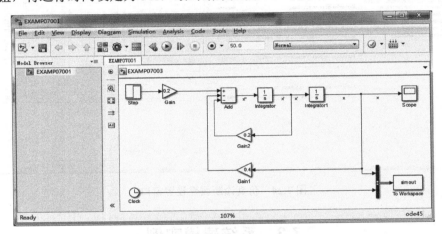

图 7.25　求解二次方程的 Simulink 模型

（7）双击 Scope 模块，对图像显示属性进行设置，如图 7.26 所示；Scope 图形窗口显示最终结果如图 7.27 所示。

如图 7.25 所示，显然仿真系统的运行结果以数组形式存储在工作空间，保存的变量名为 simout。

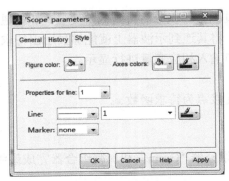

图 7.26　图形窗口属性设置

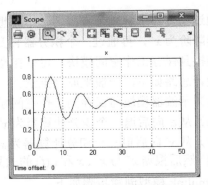

图 7.27　最终仿真结果

EXAMP07001

```
%  EXAMP07001 proves that the simulated data are saved in
%  workspace as simout.
%  (EXAMP07001 证明仿真数据保存在工作空间命名为 simout)
%  Load data from workspace.
%  (从工作空间加载数据)
t=simout.x.Time(:);
x=simout.x.Data(:);
%  Find the maximum in x.
%  (在变量 x 中寻找最大值)
[xm,km]=max(x);
%  Plot the figure of the final results.
%  (绘制最终结果图形)
plot(t,x)
hold on
%  Label the maximum of x.
%  (标注 x 的最大值位置)
plot(t(km),xm,'ro','markersize',15,'linewidth',2);
hold off
grid on
title('仿真结果')
```

在 MATLAB 命令窗口输入 EXAMP07001 执行命令得到如图 7.28 所示结果。

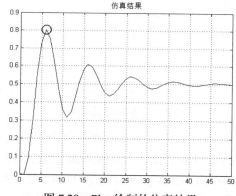

图 7.28 Plot 绘制的仿真结果

【例 7.4】 借助 Simulink 仿真求解微分方程：

$$x'' - k(1-x^2)x' + x = d(t)$$

其中 $k=10$，$d(t) = 10\sin(2t)$。初始条件为 $x_0 = \begin{bmatrix} x_{10} \\ x_{20} \end{bmatrix} = \begin{bmatrix} -2 \\ 2 \end{bmatrix}$。

分析与求解：

本例待求的二次方程可以转化为如下一次方程组：

$$\begin{cases} \dfrac{dx_1(t)}{dt} = x_2 & x_1(t_0) = x_{10} \\ \dfrac{dx_2(t)}{dt} = -x_1 + kx_2 - kx_1^2 x_2 + d(t) & x_2(t_0) = x_{20} \end{cases}$$

创建模型过程需要严格按照上面方程构建，先明确最终的输出为 x_1 和 x_2，然后基于各种

可能操作实现 x'_1 和 x'_2 的表示即可。

Simulink 模型创建步骤如下。

（1）将 Sources 模块库中 Signal Generator 模块和 Constant 模块拖入模型窗口；将 Sinks 模块库中 Scope 模块拖入模型窗口；将 Math Operations 模块库中 Add 模块、Product 模块、Gain 模块、Math Function 模块拖入模型窗口；将 Continuous 模块库或者 Commonly Used Block 模块库中 Integrator 模块拖入模型窗口；借助复制粘贴创建 2 个 Scope 模块、2 个 Gain 模块、2 个 Integrator 模块。

（2）单击鼠标右键选中菜单 Rotate & Flip 命令实现 2 个 Gain 模块翻转；双击打开 Add 属性设置窗口，将 List of signs 属性设置为-++-；双击 Product 模块打开属性设置窗口，将 Number of inputs 属性设置为 3；双击 Math Function 模块打开属性设置窗口，将 Function 属性设置为 square。

（3）通过拖动鼠标将所有模块连接在一起，根据需要调节模块位置，如图 7.29 所示。

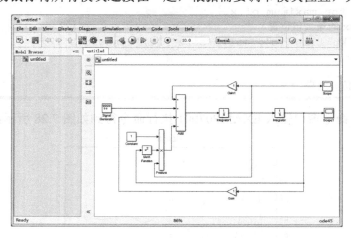

图 7.29 初步连线模型图

（4）模块的属性设置：双击 Signal Generator 图标打开属性设置对话框，Wave form 设置为 sine，Time（t）设置为 Use simulation time，Amplitude 设置为 10，"Frequency"设置为 1/pi，Units 设置为 Hertz，选中 Interpret vector parameter as 1-D；双击 Add 模块，List of signs 设置为++--，其他项为默认值；双击模块 Integrator 打开属性设置窗口，Initial condition 设置为-2，其他项为默认值；双击模块 Integrator 1 打开属性设置窗口，Initial condition 设置为 2，其他项为默认值；双击 Scope 模块打开窗口，采取默认值；双击 Scope1 模块，然后将鼠标移动到绘图区单击右键，从悬挂菜单中选择 Axes properties…命令，在打开窗口中将 y 的显示范围修改为"-10"和"10"；模块 Gain 属性采用默认值；双击模块 Gain1 打开属性设置窗口，将 Gain 设置为 10；双击模块 Math Function 打开属性设置窗口，将 Function 设置为 square，其他选项采用默认值；双击模块 Constant 打开属性设置窗口，将 Constant value 设置为 10；双击模块 Product 打开属性设置窗口，将 Number of inputs 设置为 3。通过菜单 Simulation→Configuration parameters 命令打开窗口，将 Stop time 设置为 40。通过菜单 File→Save as 命令将模型保存为 EXAMP07004，结果如图 7.30 所示。

（5）通过菜单 Simulation→Run 命令，或者工具栏 按钮运行仿真，Scope 给出结果如图 7.31 和图 7.32 所示。

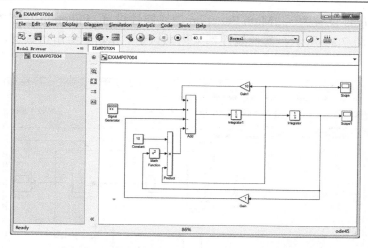

图 7.30　最终保存为 EXAMP07004 模型系统图

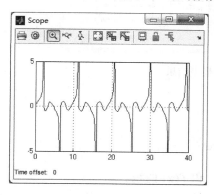

图 7.31　x2 仿真图像

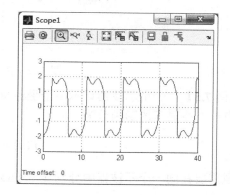

图 7.32　x1 仿真图像

提示：

　　有时需要对模块图标进行翻转和旋转操作，可以在选中该模块图标的情况下（选中时，模块图标 4 个角有小黑点），通过菜单 Rotate & Flip 完成。其他操作和 Windows 系统所带画图板中的操作类似。

　　通常连线，一般按住鼠标左键拖动到合适位置释放就可以；当需要画分支连线时，通常按住鼠标右键拖动到恰当位置即可。如果发现形状不满意，可以选中所画的线，然后选中某个标注点（一个小黑点），再按住鼠标左键拖动即可。

　　本例也可以采取例 7.3 中的思路进行创建模型，微分方程将改写为
$$x'' = k(1-x^2)x' - x + d(t)$$

设计模型如图 7.33 所示。

模块属性设置为：

Signal Generator 模块属性设置，Wave form 设置为 sine，Time（t）设置为 Use simulation time，Amplitude 设置为 10，"Frequency"设置为 1/pi，Units 选 Hertz，选中 Interpret vector parameter as 1-D；Add 模块属性设置，List of signs 设置为-++；Integrator 属性设置，Initial condition 设置为 2；Integrator1 属性设置，Initial condition 设置为-2；Product 模块属性设置，Number of inputs 设置为 3；Constant 模块属性设置，Constant value 设置为 10；Add1 模块属

性设置，List of signs 设置为-+；Math Function 模块属性设置，Function 设置为 square；Scope 模块属性设置如图 7.34 所示；其他属性都采用默认值。Scope 给出仿真解，结果如图 7.35 所示。

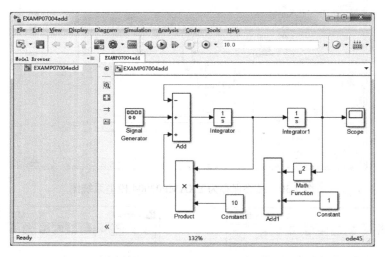

图 7.33　例 7.4 仿真模型结果

【例 7.5】　一个示波器显示多个坐标窗口。

有一个待显示图形函数为

$$f(x) = 2*\sin x e^{-\frac{1}{2}x^2}$$

现需要在同一个示波器中分别显示 $\sin x$、$e^{-\frac{1}{2}x^2}$ 和 $f(x)$ 曲线。

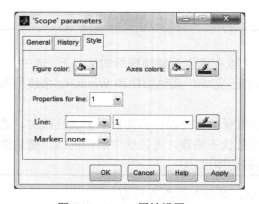

图 7.34　Scope 属性设置

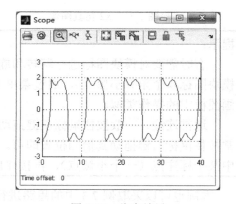

图 7.35　仿真结果

分析与求解：

x 数据可借助时钟产生，然后借助常数模块、乘法模块、指数函数模块、三角函数模块、示波器模块完成。

Simulink 模型创建步骤如下。

（1）将 Sources 模块库中的 3 个 Clock 模块、2 个 Constant 模块拖入模型窗口；将 Math Operations 模块库中 2 个 Product 模块、Sine Wave Function 模块、Math Function 模块拖入模型窗口；将 Sinks 模块库中 Scope 模块拖入模型窗口，如图 7.36 所示。

第 7 章　Simulink 仿真系统

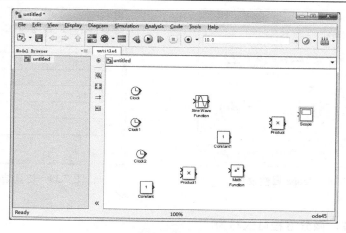

图 7.36　含有需要模块的模型窗口

（2）属性设置。Constant 模块，Constant Value 设置为-1/2；Constant1 模块，Constant Value 设置为 2；Product 模块，Number of inputs 设置为 3；Product1 模块，Number of inputs 设置为 3；Scope 模块，在 Scope 窗口单击"Parameters"按钮，在新打开的属性设置窗口将 Number of axes 设置为 3。

（3）连线，结果如图 7.37 所示。

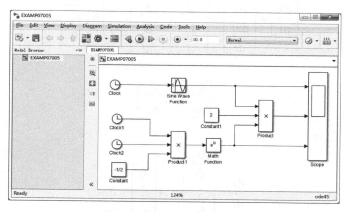

图 7.37　Simulink 模型结果

（4）将该仿真系统保存为 EXAMP07005，Stop time 设置为 30s，Scope 属性设置如图 7.38 所示，运行仿真结果如图 7.39 所示。

【例 7.6】　永久磁铁直流电动机。

永久磁铁直流电动机满足的微分线性方程组为

$$\begin{cases} \dfrac{\mathrm{d}i_a}{\mathrm{d}t} = -\dfrac{r_a}{L_a}i_a - \dfrac{k_a}{L_a}\omega_r + \dfrac{1}{L_a}u_a \\ \dfrac{\mathrm{d}\omega_r}{\mathrm{d}t} = \dfrac{k_a}{J}i_a - \dfrac{B_m}{J}\omega_r - \dfrac{1}{J}T_L \end{cases}$$

其中 $r_a = 1\Omega$，$L_a = 0.02\mathrm{H}$，$k_a = 0.3\mathrm{V}\times\mathrm{s/rad}$，$J = 0.0001\mathrm{kg}\times\mathrm{m}^2/\mathrm{rad}^2$，$B_m = 0.00005\mathrm{N}\times\mathrm{m}\times\mathrm{s/rad}^2$，$u_a = 25\mathrm{rect}(t)\mathrm{V}$。初始条件为 $\begin{bmatrix} i_{a_0} \\ \omega_{r_0} \end{bmatrix} = \begin{bmatrix} x_{10} \\ x_{20} \end{bmatrix} = \begin{bmatrix} 1 \\ 0 \end{bmatrix}$。

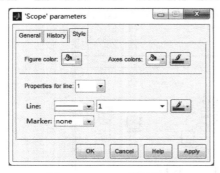
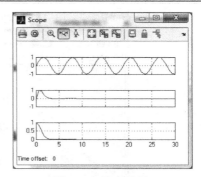

图 7.38　Scope 属性设置　　　　　图 7.39　仿真结果

分析与求解：

基于已知条件，微分方程可改写为

$$\begin{bmatrix} \dfrac{\mathrm{d}i_a}{\mathrm{d}t} \\ \dfrac{\mathrm{d}\omega_r}{\mathrm{d}t} \end{bmatrix} = \begin{bmatrix} -\dfrac{r_a}{L_a} & -\dfrac{k_a}{L_a} \\ \dfrac{k_a}{J} & -\dfrac{B_m}{J} \end{bmatrix} \begin{bmatrix} i_a \\ \omega_r \end{bmatrix} + \begin{bmatrix} \dfrac{1}{L_a} \\ 0 \end{bmatrix} u_a$$

假设 $x_1 = i_a$、$x_2 = \omega_r$、$u = u_a$，已知条件可转化为

$$\begin{bmatrix} \dfrac{\mathrm{d}x}{\mathrm{d}t} \end{bmatrix} = \begin{bmatrix} \dfrac{\mathrm{d}x_1}{\mathrm{d}t} \\ \dfrac{\mathrm{d}x_2}{\mathrm{d}t} \end{bmatrix} = \begin{bmatrix} -\dfrac{r_a}{L_a} & -\dfrac{k_a}{L_a} \\ \dfrac{k_a}{J} & -\dfrac{B_m}{J} \end{bmatrix} \begin{bmatrix} x_1 \\ x_2 \end{bmatrix} + \begin{bmatrix} \dfrac{1}{L_a} \\ 0 \end{bmatrix} u$$

初始条件为

$$\begin{bmatrix} i_{a_0} \\ \omega_{r_0} \end{bmatrix} = \begin{bmatrix} x_{10} \\ x_{20} \end{bmatrix}$$

通常方程组形式为

$$\dfrac{\mathrm{d}x}{\mathrm{d}t} = \begin{bmatrix} \dfrac{\mathrm{d}x_1}{\mathrm{d}t} \\ \vdots \\ \dfrac{\mathrm{d}x_n}{\mathrm{d}t} \end{bmatrix} = \boldsymbol{A} \begin{bmatrix} x_1 \\ \vdots \\ x_n \end{bmatrix} + \boldsymbol{B}u$$

本例有

$$\boldsymbol{x} = \begin{bmatrix} x_1 \\ x_2 \end{bmatrix},\ \boldsymbol{A} = \begin{bmatrix} -\dfrac{r_a}{L_a} & -\dfrac{k_a}{L_a} \\ \dfrac{k_a}{J} & -\dfrac{B_m}{J} \end{bmatrix},\ \boldsymbol{B} = \begin{bmatrix} \dfrac{1}{L_a} \\ 0 \end{bmatrix}$$

输出方程为 $y = \omega_r$

因此，永久磁体直流电动机的态空间（State-space）模型为

$$y = \omega_r = \boldsymbol{C}x + \boldsymbol{D} = \begin{bmatrix} 0 & 1 \end{bmatrix} \begin{bmatrix} i_a \\ \omega_r \end{bmatrix} + [0]u_a = \begin{bmatrix} 0 & 1 \end{bmatrix} \begin{bmatrix} x_1 \\ x_2 \end{bmatrix} + [0]u$$

创建 Simulink 仿真系统模型步骤如下。

（1）将 Sources 模块库中 Signal Generator 模块拖入模型窗口；将 Continuous 模块库中

State-Space 模块拖入模型窗口；将 Sinks 模块库中 Scope 模块拖入模型窗口。

（2）连线，设置模块属性。Signal Generator 模块，Wave form 设置为 square，Amplitude 设置为 25；State-Space 模块，Parameters A 设置为 [-1/0.02，-0.3/0.02; 0.3/0.0001，-0.000005/0.0001]，B 设置为[1/0.02; 0]，C 设置为[0，1]，D 设置为 0，Initial conditions 设置为[1，10]。仿真系统设置如图 7.40 所示。

（3）将以上仿真系统模型保存为 EXAMP07006，Stop time 设置为 2s，运行仿真结果如图 7.41 所示。

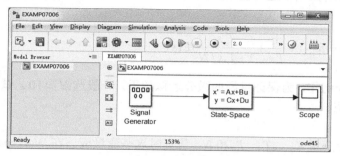

图 7.40　仿真系统模型

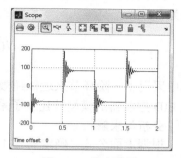

图 7.41　仿真结果

7.4　仿真系统中的子系统

在 Simulink 创建模型过程中，如果所研究的问题比较复杂，直接使用 Simulink 模块构成的模型较大，系统不容易辨认，可读性差。如果能够把整个模型按照实现功能或对应的物理器件划分成块进行研究，将有利于整个系统的概念研究，使整个模块更加简洁，可读性也更高。类似编程中所阐述的思想，将复杂系统分成多个较容易系统，然后分别对每个子系统进行创建模型，使其能够独立运行实现某些特定功能，如同用户自己编写的函数可以在 MATLAB 编程过程中调用一样，所创建的子系统也能够在创建模型过程中随便调用。使用子系统创建仿真系统具有以下优点：

（1）减少模型窗口中的模块个数。
（2）将一些功能相关的模块集成实现特定功能，可以重复使用。
（3）能够提高整个系统的运行效率和可靠性。
（4）便于构件分层的系统。

7.4.1　创建子系统

1. 组合相关模块创建子系统

当用户创建一些模块能够实现某些特定功能时，要把这些模块变成子系统，操作步骤如下所述。

（1）将需要放到子系统中的所有模块都选中。
（2）选择菜单 Edit→Create Subsystem 命令，或者单击鼠标右键，在弹出的快捷菜单中选择 Create Subsystem 命令，子系统建立。

2. 通过 Subsystem 模块创建子系统

通过 Subsystem 模块创建子系统，可以按照下面步骤进行操作。

(1) 打开 Simulink 模块库，将 Ports & Sybsystems 模块库中的 Subsystem 模块拖入模型窗口。

(2) 双击 Subsystem 模块，打开 Subsystem 窗口。

(3) 把要组合的模块拖入 Subsystem 窗口内，然后在窗口中加入 Import 模块表示从子系统外部到内部的输入，加入 Output 模块表示从子系统内部到外部的输出，把这些模块按顺序连接起来，子系统建立。

【例 7.7】 PID 控制器是在自动控制中经常使用的模块，在工程应用中其标准的数字模型为

$$U(s) = K_p(1 + \frac{1}{T_i s} + \frac{T_d s}{1 + T_d s/N})E(s)$$

其中采用了一阶环节来近似纯微分动作，为保证良好的微分近似效果，一般选 $N \geq 10$。试建立 PID 控制器的模型并建立子系统。

分析与求解：

根据上面公式，确定需要 4 个变量 K_p、T_i、T_d 和 N，仿真时这些变量应该在 Workspace 中赋值。

借助 Simulink 创建 PID 控制器模型步骤如下。

(1) 将 Sources 模块库中 In1 模块拖入模型窗口；将 Sinks 模块库中 Out1 模块拖入模型窗口；将 Math Operations 模块库中 Grain 模块和 Add 模块拖入模型窗口；将 Continuous 模块库中 2 个 Transfer Fcn 模块拖入模型窗口。

(2) 连线，设置各个模块属性。Gain 模块，Gain 属性设置为 Kp；Transfer Fcn 模块，Parameters 中 Denominator coefficients 设置为[Ti 0]；Transfer Fcn1 模块，Parameters 中 Numerator coefficients 设置为[Td 0]，Denominator coefficients 设置为[Td/N 1]；Add 模块，List of signs 设置为+++。模型保存为 EXAMP07007，结果如图 7.42 所示。

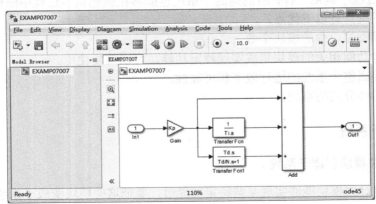

图 7.42 PID 模型窗口

创建子系统有两种方法，方法一步骤如下：

(1) 按住 shift 键，通过鼠标选中 Gain 模块、Transfer Fcn 模块、Transfer Fcn1 模块和 Add 模块。

(2) 单击鼠标右键打开悬挂菜单，选中 Create subsystem from selection 命令，子系统建立；也可以借助快捷键 Ctrl+G 创建子系统。

方法二步骤如下：

（1）将 Ports & Subsystems 模块库拖入建模窗口。

（2）双击 Subsystem 模块，打开子系统窗口。

（3）将 Math Operations 模块库中 Grain 模块和 Add 模块拖入模型窗口；将 Continuous 模块库中 2 个 Transfer Fcn 模块拖入模型窗口。

（4）连线，设置各个模块属性。Gain 模块，Gain 属性设置为 Kp；Transfer Fcn 模块，Parameters 中 Denominator coefficients 设置为[Ti 0]；Transfer Fcn1 模块，Parameters 中 Numerator coefficients 设置为[Td 0]，Denominator coefficients 设置为[Td/N 1]；Add 模块，List of signs 设置为+++。子系统建立，其模型窗口如图 7.43 所示。

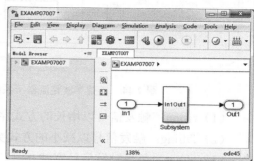

图 7.43 子系统创建的模型窗口

7.4.2 子系统的条件执行

子系统的执行可由输入信号来控制，用于控制子系统执行的信号称为控制信号，而由控制信号控制的子系统称为条件执行子系统。在一个复杂模型中，有的模块的执行依赖于其他模块，在此情况下，条件执行子系统十分有用。下面通过两个示例分别介绍使能子系统和触发子系统。

1. 使能子系统

使能子系统表示子系统由控制信号控制时，控制信号由负变为正时子系统开始执行，直到控制信号再次转变为负时结束。使能子系统控制信号可以是标量也可以是向量。如果控制信号是标量，当标量的值大于 0 时子系统执行；如果是向量，向量中的任意一个元素大于 0，子系统开始执行。使能子系统在外观上有一个"使能"控制信号输入口。"使能"指的是当且仅当"使能"输入口信号为正时，该模块开始接收 In 输入端的信号。

【例 7.8】 利用使能子系统构成一个正弦半波整流器。

创建子系统步骤如下：

（1）打开 Simulink Library Browser 窗口，创建一个仿真模型。

（2）将 Sources 模块库中的 Sine Wave 模块拖入模型窗口；将 Ports & Subsystems 模块拖入模型窗口；将 Sinks 模块库中的 Scope 模块拖入模型窗口。

（3）连线，设置模块属性。Scope 模块，Number of axes 设置为 2，结果保存为 EXAMP07008，如图 7.44 所示。

设置 Stop time 为 20s，运行该模型，仿真结果如图 7.45 所示。

2. 触发子系统

触发子系统是指当触发事件发生时开始执行的子系统。它在外观上有一个"触发"控件信号输入口，仅当触发输入信号所定义的某个事件恰巧发生时，该模块才开始接收 In 输入端的信号。子系统一旦被触发，其输出端口的值就保持不变，直到下次再触发才可能改变。触发信号也可以是向量，此时，只要向量中有一个分量信号发生"触发事件"，子系统就被触发。触发事件由系统内触发模块对话框定义，有 4 种触发事件形式可以选择。

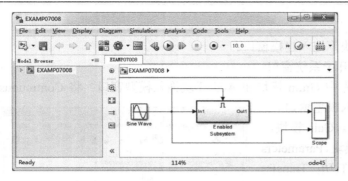 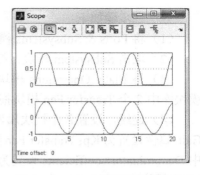

图 7.44　正弦半波整流器子系统　　　　图 7.45　正弦半波整流器仿真结果

（1）rising：触发信号以增长的方式穿越 0 时，子系统开始接收输入值。

（2）falling：触发信号以减小的方式穿越 0 时，子系统开始接收输入值。

（3）either：每当触发信号穿越 0 时，子系统开始接收输入值。

（4）function-call：这种触发方式必须和 S 函数配合使用。

在 Trigger 模块参数设置对话框中，有 Show output port 复选框，表示是否为 Trigger 模块添加一个输出端口，用来输出控制信号。选中后，下面参数 Output data type 被激活，在这里的输出控制信号的类型有 auto、int8 和 double 类型。

【例 7.9】　利用触发子系统获取零阶保持的采样信号实例。

创建模型步骤如下：

（1）将 Sources 模块库中的 Pulse Generator 模块和 Sine Wave 模块拖入建模窗口；将 Sinks 模块库中的 Scope 模块拖入建模窗口；将 Ports & Subsystems 模块库中的 Triggered Subsystem 模块拖入模型窗口。

（2）连线，模块属性设置。Pulse Generator 模块，Period 设置为 1，Pulse Width 设置为 50；Scope 模块，Number of axes 设置为 3，y 轴的显示范围定义为−1.5～1.5，背景设计颜色设置为白色，线条都设置为黑色；其他属性及其他模块输入/输出都采用默认值，如图 7.46 所示。

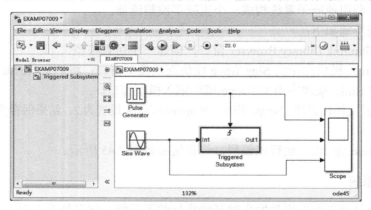

图 7.46　零阶保持的采样信号模型

（3）双击 Triggered Subsystem 模块，打开触发子系统窗口，在 In1 模块和 Out1 模块之间添加一个 Slider Gain 模块，所有属性都采用默认值，如图 7.47 所示。

（4）保存为 EXAMP07009，Stop time 设置为 20s，运行仿真结果如图 7.48 所示。

第 7 章 Simulink 仿真系统

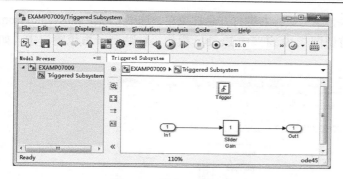
图 7.47 触发子系统窗口

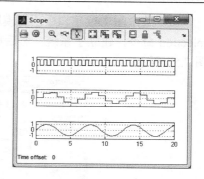
图 7.48 仿真结果

7.4.3 封装子系统

在使用子系统情况下，模型更加简洁，有利于提高问题研究的概念抽象能力以及面向对象的访问能力。但是，使用封装子系统创建时，子系统将直接从工作空间中获得变量的数值，容易发生变量冲突。同时当子系统比较多时，管理子系统将变得比较麻烦。当子系统需要修改时，也就是要修改子系统中模块属性时，需要对每个模块的属性都进行重新设置，这是一件相当烦琐的工作。Simulink 仿真系统为解决这个问题提供了办法，就是封装子系统。采用封装子系统方式创建的子系统，可以克服封装子系统的缺点，使得封装的子系统和 Simulink 模块库中模块具有相同的特点，存在自己的工作空间，也可以独立于基础模块的工作空间。

封装子系统的创建步骤如下：

（1）创建仿真系统模型；

（2）创建子系统；

（3）在选中子系统情况下，通过菜单 Diagram→Mask→Create Mask…命令或者快捷组合键 Ctrl+M，打开属性设置窗口，可以设置封装子系统的参数属性、模块描述、帮助说明等，完成子系统封装。

【例 7.10】 封装子系统创建。

创建封装子系统步骤如下：

（1）将 Sources 模块库中 2 个 In1 模块拖入子系统窗口；将 Sinks 模块库中 2 个 Out1 模块拖入子系统窗口；将 Continuous 模块库中 Derivative 模块和 Integrator 模块拖入子系统窗口；将 Ports & Subsystems 模块库中 Enable 模块拖入子系统窗口，连线结果如图 7.49 所示。

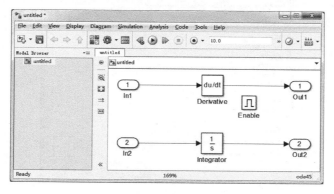
图 7.49 子系统创建窗口

（2）选中图 7.49 窗口中所有模块，单击鼠标右键打开悬挂菜单，选中 Create subsystem from selection 命令，建立子系统。然后剪切图 7.49 窗口中 Enable 模块，粘贴到子系统窗口中，如图 7.50 所示。

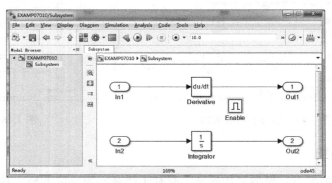

图 7.50　子系统窗口

将 untitled 模型保存为 EXAMP07010，仿真结果如图 7.51 所示。

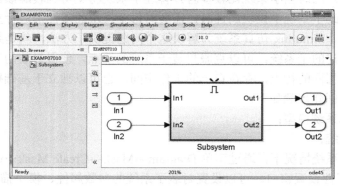

图 7.51　含有子系统的模型窗口

（3）在 EXAMP07010 模型窗口选中子系统 Subsystem，通过菜单 Diagram→Mask→Create Mask...命令打开封装子系统的属性设置窗口，如图 7.52 所示。属性设置有 4 个选项卡：图标和端口设置面板（Icon & Ports）、参数设置面板（Parameters）、初始化设置面板（Initialization）、封装模块描述和帮助文本设置页面（Documentation）。

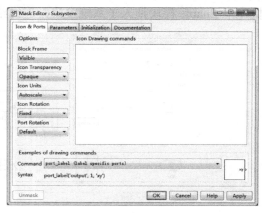

图 7.52　封装子系统属性设置对话窗口

小　　结

本章首先介绍了 Simulink 基础，包括 Simulink 启动，模块库浏览器的简单介绍，以 Commonly Used Blocks 模块库为例介绍了窗口结构，Simulink 模型窗口，通过两个简单示例阐述 Simulink 模型窗口的基本结构；然后简略介绍了 Simulink 模型操作和仿真系统的设置；随后给出 4 个建模示例，较详细地阐述了建模的基本过程和步骤；最后为了能够创建复杂模型的需要，对子系统进行了简单介绍，给出创建子系统的基本步骤、分析子系统的执行条件和封装的基本过程。

习　　题

7.1　借助 Simulink 工具完成下面动力学系统仿真：
$$x(t) = \frac{4}{\pi}[\cos(2\pi t) + \frac{1}{9}\cos(6\pi t) + \frac{1}{25}\cos(10\pi t)]$$

7.2　借助 Simulink 工具完成将摄氏度转换为华氏度，公式如下：
$$T_f = \frac{9}{5}T_c + 32$$

7.3　借助 Simulink 工具创建一个模型能够求解斜抛运动的水平运动距离与初始速度、仰角角度及时间的关系，重力加速度取 $9.8 m/s^2$。

7.4　借助 Simulink 工具创建一个模型，能够显示脉冲信号。

7.5　借助 Simulink 工具箱创建一个模型能够求解下面的微分方程组：
$$\begin{cases} x' + y - z = 1 \\ 2x - x' + y'' + z' = 0 \\ x + y' + z'' = 0 \end{cases}$$

7.6　借助 Simulink 工具创建一个模型，能够显示计算并显示如图 7.53 所示曲线，该曲线对应函数为 $f = x\sin x$。

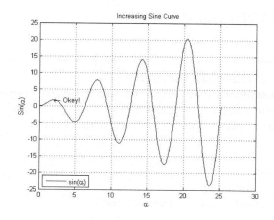

图 7.53　习题 7.6 图

第 8 章　MATLAB 在数字信号中的应用

数字信号处理是用数值计算的方法对信号进行处理的一门学科，MATLAB 语言非常适合数字信号处理。在通用计算机上编程实现各种复杂的算法，借助的主要工具就是 MATLAB。本章将讨论与数字信号处理相关的常用函数并自编一些函数，同时给出示例。本章示例简单，便于读者深入理解课程的基本理论和掌握具体的实现方法，借助 MATLAB 软件进行设计仿真。

本章有 4 节内容。8.1 节介绍时域离散信号和系统，8.2 节阐述离散时间傅里叶变换和 Z 变换函数，8.3 节讲述离散傅里叶变换和快速傅里叶变换，8.4 节对 IIR 滤波器的设计进行阐述。

8.1　时域离散信号和系统

本节介绍时域离散信号的表示方法和典型序列，以及线性常系数差分方程的各种 MATLAB 表示。时域表示方法很多，线性常系数差分方程（描述离散系统的输入/输出关系）和单位冲激响应（直接描述系统在时域的特性）都是系统对时域的描述。

8.1.1　信号、实现信号的基本运算及求解差分方程

基于 MATLAB 数值计算的特点，用它分析离散时间信号和系统较为方便。序列的表示里不包含位置信息，时域离散信号为有限长序列 $x(n)$，需要用两个向量参数 x 和 n 表示，x 是样值向量，n 是位置向量（作用相当于图形表示方法里的横坐标 n），n 和 x 的长度相等。下面列出常用的重要信号。

1. 单位采样序列

【例 8.1】　生成单位采样序列示例。

EXAMP08001

```
function [x,n]=EXAMP08001(n0,n1,n2)
% Function EXAMP08001 can generate unit sample sequence.
% （函数 EXAMP08001 能够产生单位采样序列）
% Call sequence: [x,n]=EXAMP08001(n0,n1,n2).
%
% Records of revisions:
% Date:2015-07-05
% Programmer: Yuan Hong
% Description of change: original code
%
% Generates x(n)=delta(n-n0);n1<=n<=n2
```

```
%  （产生单位采样序列）
n=[n1:n2];
x=[(n-n0)==0];
```

在命令窗口输入执行语句如下：

```
[x,n]=EXAMP08001(0,-3,3)
stem(x)
```

运行结果如下：

```
x =    0    0    0    1    0    0    0
n =   -3   -2   -1    0    1    2    3
```

单位脉冲序列示例如图 8.1 所示。

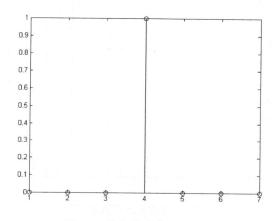

图 8.1　单位脉冲序列示例图

2. 单位阶跃序列

除了自编函数，还可以通过函数 ones 实现 ones（1,N），产生一个列向量。

【例 8.2】　用 MATLAB 生成长度为 21 的单位阶跃序列。

EXAMP08002

```
%  EXAMP08002.m
%  EXAMP08002 generates1-by-21 vector, all elements are 1.
%  （EXAMP08002 能够产生 1 乘 21 的矢量，所有元素都是 1）
%
%  Records of revisions:
%  Date:2015-07-05
%  Programmer: Yuan Hong
%  Description of change: original code
%
%  Clear all variables saved in workspace.
%  （清除所有保存在工作空间的变量）
clear
%  Clear command window.
%  （清除命令窗口）
clc
```

```
%   Define a 1-by-21 vector n.
%   (定义向量 n)
n=-10:10;
%   Define a simple vector xn is the same size as n.
%   (定义向量 xn)
xn=[zeros(1,10),ones(1,11)];
%   Plot the expectant figure by stem function.
%   (借助函数 stem 绘制预期图形)
stem(n,xn)
```

运行结果如图 8.2 所示。

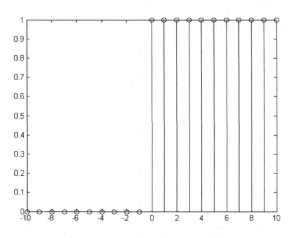

图 8.2　单位阶跃序列示例图

注意:
在 MATLAB 中，可以利用单位阶跃序列相减产生矩形序列 $R_N(n)=u(n)-u(n-N)$。

3. 复指数序列

$$x(n) = \begin{cases} e^{(\sigma+jw)n} & n \geqslant 0 \\ 0 & n < 0 \end{cases}$$

【例 8.3】 生成复指数序列示例。

EXAMP08003

```
function [n,x]=EXAMP08003(a,b,N)
%   Function EXAMP08003 generates a complex exponential sequence.
%   (函数 EXAMP08003 产生一个复数指数序列)
%   Call sequence: [n,x]=EXAMP08003(a,b,N)
%
%   Descriptions of variables:
%   n: a unit sequence
%   x: a complex exponential sequence
%   a: a constant parameter
%   b: a constant parameter
%   N: the number of elements of x
```

```
%
% Records of revisions:
% Date:2015-07-05
% Programmer: Yuan Hong
% Description of change: original code
%
% Define a unit sequence as n.
% （定义一个单位序列为 n）
n=[0:N];
% Define the expectant complex exponential sequence as x.
% （定义预期复数指数序列，记为 x）
x=exp((a+b*i)*n);
```

在命令窗口输入执行语句如下：

```
x=EXAMP08003(0.4,0.6,10)
stem(x)
```

运行结果如下：

```
x =
  Columns 1 through 4
   1.0000 + 0.0000i   1.2313 + 0.8423i   0.8064 + 2.0743i  -0.7543 + 3.2333i
  Columns 5 through 8
  -3.6523 + 3.3456i  -7.3151 + 1.0427i  -9.8851 - 4.8780i  -8.0622 -14.3328i
  Columns 9 through 11
   2.1466 -24.4384i  23.2286 -28.2818i  52.4235 -15.2556i
```

复指数序列示例图如图 8.3 所示。

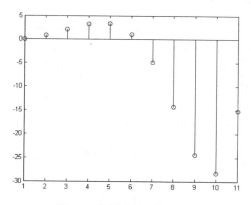

图 8.3 复指数序列示例图

4. 正弦序列

【例 8.4】 生成长度为 20 的正弦序列。

EXAMP08004

```
function EXAMP08004(m,n)
% EXAMP08004 generatesa set of sinusoidal sequence.
% （函数 EXAMP08004 产生一个正弦序列）
```

```
% Call sequence: EXAMP08004 (m,n)
%
% Descriptions of variables:
% m: the number of points in one period
% n: the number of the whole points
% N: a vectorfrom 0 to N
% xn: a sinusoidal sequence with respect to n
%
% Records of revisions:
% Date:2015-07-05
% Programmer: Yuan Hong and Wang Yong-Long
% Description of change: original code
%
% Define a vector N from 0 to n.
% （定义一个矢量 N 从 0 到 n）
N=0:n;
% Define a sinusoidal list xn with respect to N.
% （相应于定义一个正弦序列 xn）
xn=sin(2*pi/m*N);
% Plot the stems of xn.
% （绘制 xn 的 stem 图）
stem(N,xn)
```

在命令窗口输入执行语句如下：
EXAMP08004（20,30）
运行结果如图 8.4 所示。

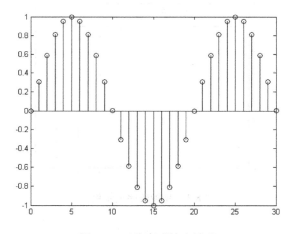

图 8.4 正弦序列输出波形

8.1.2 序列运算

1. 序列求和

对两个序列 f_1、f_2 相同位置的元素求和，需保证 f_1、f_2 的序列长度相同，否则用零补齐。

【例 8.5】 编写序列求和程序。

EXAMP08005

```
function [f,k]=EXAMP08005 (f1,f2,k1,k2)
% EXAMP08005 solves the sums of elements in f1 and f2 with same index.
% （EXAMP08005 函数对 f1 和 f2 相应元素求和）
% Call sequence: [f,k]=EXAMP08005 (f1,f2,k1,k2)
%
% Descriptions of variables:
% f1: a vector of data
% f2: a vector of data
% k1: a vector of index for f1
% k2: a vector of index for f2
% f: the final vector is the sum of f1 and f2
% k: a vector of index for f
%
% Records of revisions:
% Date:2015-08-05
% Programmer: Yuan Hong and Wang Yong-Long
% Description of change: original code
%
% Define a vector k ranging from the minimum in k1 or in k2 to
% the maximum.
% （定义一个矢量 k，从 k1 或 k2 中最小值到最大值）
k=min (min (k1),min (k2)) :max (max (k1),max (k2));
% Initialize two new sequences s1 and s2.
% （初始化两个新序列 s1 和 s2）
s1=zeros (1,length (k));
s2=s1;
% Transpose the corresponding elements in f1 to s1.
% （将 f1 中相应元素传递给 s1）
s1 (find ((k>=min (k1)) & (k<=max (k1)) ==1)) =f1;
% Transpose the corresponding elements in f2 to s2.
% （将 f2 中相应元素传递给 s2）
s2 (find ((k>=min (k2)) & (k<=max (k2)) ==1)) =f2 ;
% Solve the sum of s1 and s2.
% （计算 s1 和 s2 两个序列之和）
f=s1+s2;
% Plot the figure of f.
% （绘制 f 的图形）
stem (k,f)
```

在命令窗口输入执行语句如下：

```
>> k1=[-3,-2,-1,0,1,2,3,4];
>> k2=[1 2 3 4 5 6 7 8 9 10 11 12 13 14 15];
>> f1=[0.1 0.2 0.1 0.2 0.3 0.4 0.5 0.6];
```

```
>> f2=[0.1 0.2 0.3 0.4 0.5 0.6 0.7 0.8 0.9 1 1.1 1.2 1.3 1.4 1.5];
>> [f,k]=EXAMP08005(f1,f2,k1,k2)
```

运行结果如下：

```
f = Columns 1 through 5
    0.1000    0.2000    0.1000    0.2000    0.4000
  Columns 6 through 10
    0.6000    0.8000    1.0000    0.5000    0.6000
  Columns 11 through 15
    0.7000    0.8000    0.9000    1.0000    1.1000
  Columns 16 through 19
    1.2000    1.3000    1.4000    1.5000
k = Columns 1 through 8
    -3    -2    -1     0     1     2     3     4
  Columns 9 through 16
     5     6     7     8     9    10    11    12
  Columns 17 through 19
    13    14    15
```

两个序列求和示例图如图 8.5 所示。

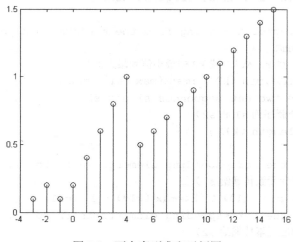

图 8.5 两个序列求和示例图

> **注意：**
> 用 find 函数寻找非零元素对应的下标区间示例，本章会经常用到。参加运算的两个序列必须有相同的维数。

2. 序列乘积

将 EXAMP08005 程序中 f=s1+s2 换成 f=s1.*s2，即可得到序列的乘积运算。得出的结论表明：序列的位置向量对序列的运算起到重要作用。

【例 8.6】 编写实现任意两个离散序列相乘的 MATLAB 子程序。

EXAMP08006

```
function [f,k]=EXAMP08006(f1,f2,k1,k2)
% EXAMP08006 solves the products of elements in f1 and f2 with same index.
% （EXAMP08006 函数对 f1 和 f2 相应元素求乘积）
% Call sequence: [f,k]=EXAMP08005(f1,f2,k1,k2)
%
% Descriptions of variables:
% f1: a vector of data
% f2: a vector of data
% k1: duration of f1
% k2: duration of f2
% f: the final vector is the product of f1 and f2
% k: duration of f
%
% Records of revisions:
% Date: 2015-08-05
% Programmer: Yuan Hong and Wang Yong-Long
% Description of change: original code
%
k=min(min(k1),min(k2)):max(max(k1),max(k2));
s1=zeros(1,length(k));s2=s1;
s1(find((k>=min(k1)) & (k<=max(k1))==1))=f1;
s2(find((k>=min(k2)) & (k<=max(k2))==1))=f2 ;
%sequence multiplication.
% （序列乘积）.
f=s1.*s2;
stem(k,f)
```

在命令窗口输入执行语句如下：

```
>> k1=[-3,-2,-1,0,1,2,3,4];
>> k2=[1 2 3 4 5 6 7 8 9 10 11 12 13 14 15];
>> f1=[0.1 0.2 0.1 0.2 0.3 0.4 0.5 0.6];
>> f2=[0.1 0.2 0.3 0.4 0.5 0.6 0.7 0.8 0.9 1 1.1 1.2 1.3 1.4 1.5];
>> [f,k]=EXAMP08006(f1,f2,k1,k2)
f =
  Columns 1 through 7
    0      0      0      0    0.0300  0.0800  0.1500
  Columns 8 through 14
  0.2400    0      0      0      0      0      0
  Columns 15 through 19
    0      0      0      0      0
k =
  Columns 1 through 13
   -3    -2    -1     0     1     2     3     4     5     6     7     8     9
  Columns 14 through 19
   10    11    12    13    14    15
```

两序列乘积示例图如图 8.6 所示。

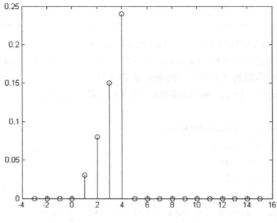

图 8.6 两序列乘积示例图

3. 能量和

【例 8.7】 用 MATLAB 语言编写求函数能量和 $E = \sum_{n=-\infty}^{\infty} |x(n)|^2 = \sum_{n=-\infty}^{\infty} x(n)x^*(n)$。

EXAMP08007

```
function Ex=EXAMP08007(x)
% EXAMP08007 solves the sum of elements square of x.
% （EXAMP08007 计算 x 的元素平方和）
% Calling sequence: Ex=EXAMP08007(x)
%
% Records of revisions:
% Date: 2015-08-05
% Programmer: Yuan Hong and Wang Yong-Long
% Description of change: original code
%
Ex=sum(x.*conj(x));
```

在命令窗口输入执行语句如下：

```
>> A=[1 2+3*i 4 2-i];
>> Energy=EXAMP08007(A)
Energy = 35
```

4. 序列的卷积

虽然 MATLAB 为计算两个有限长序列卷积提供了 conv 函数，但是卷积如果从 $n=0$ 开始，不需要输入序列的初始位置，调用方法 y=conv(x, h)，对于输出序列的位置信息无法获取。编写函数文件将 conv 加以扩展。

对于已知有限长序列 $\{x(n); n_{xb} \leqslant n \leqslant n_{xe}\}$ 和 $\{h(n); n_{hb} \leqslant n \leqslant n_{he}\}$，已经知道 $y(n)$ 的时间向量的起始点终点满足：$n_{yb} = n_{xb}+n_{hb}$，$n_{ye} = n_{xe}+n_{he}$。利用 conv 函数，编写有位置矢量输出为 y(n) 的 modconv.m 程序。

【例 8.8】 利用卷积函数 conv，扩展生成可获得有位置矢量（包含时间信息）输出的卷积函数。

EXAMP08008

```
function [y,ny]=modconv(x,nx,h,nh)
% MODCONVsolve convolution and polynomial multiplication with
% position vector.
% （MODCONV 计算卷积或多项式乘积，并且给出位置矢量）
% Calling sequence: [y,ny]=modconv(x,nx,h,nh)
%
% x: a signal
% h: a signal
% nx: the positions of elements in x
% nh: the positions of elements in h
% y: convolution multiplication of x and h
% ny: the positions of elements in y
%
% Records of revisions:
% Date: 2015-08-05
% Programmer: Yuan Hong
% Description of change: original code
%
nyb=nx(1)+nh(1);
nye=nx(length(x))+nh(length(h));
ny=[nyb:nye];
y=conv(x,h);
```

借助 conv 函数计算两个有限长序列的卷积，在命令窗口输入执行语句如下：

```
>> x=[3,11,7,0,-1,4,2];
>> h=[2,3,0,-5,2,1];
>> y=conv(x,h)
y = 6  31  47  6  -51  -5  41  18  -22  -3  8  2
```

借助作者编写函数 modconv 计算有限长序列卷积，在命令窗口输入执行语句如下：

```
>> x=[3,11,7,0,-1,4,2];nx=[-3:3];
>> h=[2,3,0,-5,2,1];ny=[-1:4];
>> [y,ny]=conv_m(x,nx,h,ny)
y = 6  31  47  6  -51  -5  41  18  -22  -3  8  2
ny =-4  -3  -2  -1  0  1  2  3  4  5  6  7
```

注意：
扩展后的卷积函数可以求出新序列在时间 n 的什么范围内有值。

5. 移位

$$x(n-m)=y(n)$$

右移 m 取正数，左移 m 取负数。设原序列用 x 和 m 表示，移位后的序列用 y 和 n 表示。命名移位函数为 sigshift。

【例 8.9】 借助 MATLAB 语言编写实现时域序列移位的函数。

EXAMP08009

```
function [y,n]=sigshift (x,m,n0)
%  SIGSHIFT implements y(n)=x(m+n0).
%  （SIGSHIFT 实现移位 y(n)=x(m+n0)）
%  Calling sequence: [y,n]=sigshift (x,nx,h,nh)
%
%  x: a signal
%  m: the positions of elements in x
%  y: a signal by shift
%  ny: the positions of elements in y
%
%  Records of revisions:
%  Date: 2015-08-05
%  Programmer: Yuan Hong
%  Description of change: original code
%
n=m+n0;
y=x;
```

将当前路径设置为含有 sigshift.m 文件的位置，在命令窗口输入执行语句如下：

```
>> x=[3,11,7,0,-1,4,2];m=[-3:3];
>> [y,n]= sigshift (x,m,2)
y =   3   11   7   0   -1   4   2
n =  -1    0   1   2    3   4   5
```

6. 系统的单位抽样响应和输出信号

【例 8.10】 设线性时不变系统的抽样响应 $h(n)=(0.9)^n u(n)$，输入 $x(n)=u(n)-u(n-10)$，求系统的输出 $y(n)$。

提示：系统的输出为输入和单位抽样响应的卷积，可利用子函数 modconv 求输出序列。

EXAMP08010

```
functionvarargout=EXAMP08010 (varargin)
%  EXAMP08010 is a program to solve example 10.
%  （EXAMP08010 为示例 8.10 的程序）
%  Calling sequence: EXAMP08010
%
%  Records of revisions:
%  Date: 2015-08-05
%  Programmer: Yuan Hong
%  Description of change: original code
%
n=-5:50;
x=stepseq(0,-5,50)-stepseq(10,-5,50);
h=((0.9).^n).*stepseq(0,-5,50);
%
subplot (311)
stem (n,x);
axis ([-5,50,0,2]);
```

第 8 章 MATLAB 在数字信号中的应用

```
  ylabel ('x (n) ');
%
subplot (312)
stem (n,h);
axis ([-5,50,0,2]);
ylabel ('h (n) ');
[y,ny]=modconv (x,n,h,n);
%
subplot (313)
stem (ny,y);
axis ([-5,50,0,8]);
ylabel ('y (n) ');
xlabel ('n');
end
%
%%---subfunction
function [x,n] = stepseq ( n0,ns,nf )
%   STEPSEQ implements   x=[ (n-n0) >=0].
%   （SIGSHIFT 实现阶跃 x=[ (n-n0) >=0])
%   Calling sequence: [x,n]=stepseq (n0,ns,nf)
%
%   n0: step point
%   ns: initial point
%   nf: final point
%   x: a signal
%   n: the positions of elements in x
%
n=[ns:nf];
x=[ (n-n0) >=0];
end
```

在命令窗口输入执行语句如下：

```
>> EXAMP08010
```

运行结果如图 8.7 所示。

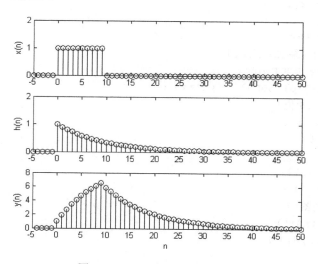

图 8.7 EXAMP08010 运行结果

> **注意:**
> 子函数的文件必须保存到主函数所在的文件夹,也可以子函数包含在主函数中。

8.2 离散时间傅里叶变换(DTFT)与 Z 变换函数

对信号和系统的分析可以在时域进行,也可以在频域展开。时间域分析比较直观,但有时会增加问题研究的难度。比如两个序列,一个波形变化快,一个波形变化慢,都混有噪声,用滤波器滤波噪声又不希望损伤信号。因为两种信号频谱结构不同,对滤波器的通带范围要求也不同。为了设计合适的滤波器,要求分析信号的频谱结构,需要把时间域转换到频率域。频域分析需要的数学工具傅里叶变换和 Z 变换在 MATLAB 语言中都有特定函数,使得问题大大简化。

8.2.1 函数 freqz

函数 freqz 是 MATLAB 软件自带函数,为数字滤波器的频率响应函数。

语法格式:[H, w]=freqz(b, a, N, 'whole') $[0, 2\pi]$ 内取 N 个频率点。

freqz(b, a) 不带输出向量的 freqz 函数,只自动画出幅频和相频曲线。

【例 8.11】 求 $x(n)$=[2, 3, 4, 3, 2]的离散时间傅里叶变换,并画出它的幅频特性及相频特性。

EXAMP08011

```
% (EXAMP08011 给出 x(n)=[2,3,4,3,2]的 DTFT,并画出它的幅频特性及相频特性)
% Calling sequence: EXAMP08011
%
clear all
clc
n=0:4;
x =[2,3,4,3,2];
k=0:1000;
w=(pi/500)*k;
% (用矩阵-向量乘法求 DTFT)
% Generates DTFT of x
X=x*(exp(-j*pi/500)).^(n'*k);
% (求幅度,求相位)
magX=abs(X);
angX=angle(X);
%
subplot(1,3,1)
stem(n,x,'.');
title('序列图')
ylabel('x(n)');
axis([0,5,0,6]);
%
```

```
subplot(1,3,2)
plot(w/pi,magX);
grid
xlabel('');
title('幅频特性');
ylabel('模值')
%
subplot(1,3,3)
plot(w/pi,angX);
grid
title('相频特性');
ylabel('弧度')
```

在命令窗口输入执行语句如下：

```
>> EXAMP08011
```

运行结果如图 8.8 所示。

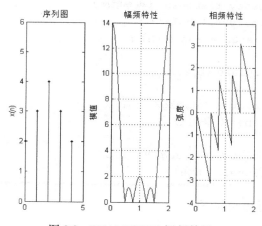

图 8.8　EXAMP08011 运行结果

8.2.2　函数 zplane

MATLAB 语言为绘制 $H(z)$ 的零极点图提供了 zplane 函数。$H(z)$ 的表达式为

$$H(z) = \frac{b(0) + b(1)z^{-1} + \cdots + b(M)z^{-M}}{a(0) + a(1)z^{-1} + \cdots + a(N)z^{-N}}$$

式中　　H ——频率响应；

$a(0)$ ——不能为零，且分子、分母按 z 的负幂次排列。

语法格式：zplane(b,a)　利用系统函数的 b,a 系数向量求零极点。该指令先利用 roots 函数找出由 b 及 a 构成的函数的零极点，然后再画出零极点，且自动设定坐标刻度，零点以 "○" 标记，极点以 "×" 标记。需要注意的是，a、b 作为系数时必须用行向量输入，如果是零极点向量，则必须取列向量形式输入，zplane 通过输入的是什么向量来判断用户输入的是系数或者是零极点。

【例 8.12】　已知：

$$H(z) = \frac{[1 - 1.8z^{-1} - 1.44z^{-2} + 0.64z^{-3}]}{[1 - 1.64853z^{-1} + 1.03882z^{-2} - 0.288z^{-3}]}$$

求 $H(z)$ 的零极点并画出零极点图。

EXAMP08012

```
% (EXAMP08012 绘画零极点图)
clc
clear all
% b: the coefficients of the numerator
b=[1,-1.8,-1.44,0.64];
% a: the coefficients of the denominator
a=[1,-1.64853,1.03882,-0.288];
% the system poles （系统极点）
rp = roots(a);
rz = roots(b);
% H: the frequency response （频率响应）
[H,w]=freqz(b,a,1024,'whole');
magX = abs(H);angX = angle(H);
%
subplot(311);zplane(b,a);
%
subplot(312);plot(w/pi,magX);grid
xlabel('');ylabel('幅值')
%
subplot(313);plot(w/pi,angX);grid
xlabel(' ');ylabel('相角')
```

在命令窗口输入执行语句如下：

>>EXAMP08012

运行结果如图 8.9 所示。

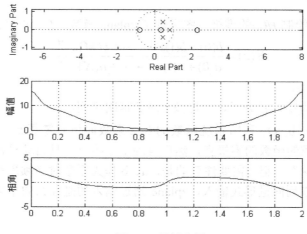

图 8.9 零极点图

【例 8.13】 三阶低通滤波器的差分方程表示如下：

$$y(n) - 1.76y(n-1) + 1.1829y(n-2) - 0.2781y(n-3)$$
$$= 0.0181x(n) + 0.0543x(n-1) + 0.0543x(n-2) + 0.0181x(n-3)$$

画出该滤波器的幅度和相位响应。

EXAMP08013

```
%   (EXAMP08013 画出系统函数的幅频和相频)
%
%   b: an array of filter coefficients
%   (滤波器系数数组)
b=[0.0181,0.0543,0.0543,0.0181];
%   a: an array of filter coefficients
%   (滤波器系数数组)
a=[1.000,-1.7600,1.1829,-0.2781];
m=0:length(b)-1;l=0:length(a)-1;
%   k: an index array for the frequencies
K=500;k=0:1:K;
w=pi*k/K;
%   num: numerator (分子)
%   den: denominator (分母)
num=b*exp(-j*m'*w);
den=a*exp(-j*l'*w);
%   H: Frequency response (频率响应)
H=num./den;
%mag and phase responses (幅频和相频响应)
magH=abs(H);angH=angle(H);
%
subplot(211);
plot(w/pi,magH);
grid;
axis([0,1,0,1]);
xlabel('');
ylabel('模值');
title('幅频响应');
%
subplot(212);
plot(w/pi,angH/pi);
grid;
xlabel('');
ylabel('弧度');
title('相频响应')
```

在命令窗口输入执行语句如下：

```
>> EXAMP08013
```

运行结果如图 8.10 所示。

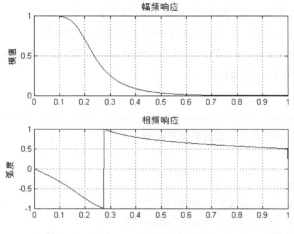

图 8.10 差分方程确定滤波器的幅频和相频

8.3 离散傅里叶变换及快速傅里叶变换

离散傅里叶变换（DFT）对于数字信号处理非常有用。快速傅里叶变换（FFT）和 DFT 是相同的一种变换，但比 DFT 计算次数少，更加快速有效。复习离散傅里叶变换定义：

$$X(k) = \text{DFT}(x(n)) = \sum_{n=0}^{N-1} x(n) e^{-j\frac{2\pi}{N}nk} = \sum_{n=0}^{N-1} x(n) W_N^{nk} \qquad k = 0, 1, \cdots, N-1$$

而 $X(k)$ 的 N 点离散傅里叶逆变换定义为

$$x(n) = \text{IDFT}[X(k)] = \frac{1}{N}\sum_{k=0}^{N-1} X(k) e^{j\frac{2\pi}{N}nk} = \frac{1}{N}\sum_{k=0}^{N-1} X(k) W_N^{-nk} \qquad n = 0, 1, \cdots, N-1$$

DFT 的定义还可用矩阵表示为

$$X = W_N x$$

式中　　X——N 点 DFT 频域的列向量，即

$$X = [X(0), X(1), \cdots X(N-2) X(N-1)]^T$$

x——时域序列的列向量，即

$$x = [x(0), x(1), \cdots, x(N-2), x(N-1)]^T$$

W_N——N 点 DFT 矩阵，定义为

$$W_N = \begin{bmatrix} 1 & 1 & 1 & \cdots & 1 \\ 1 & W_N^1 & W_N^2 & \cdots & W_N^{N-1} \\ 1 & W_N^2 & W_N^4 & \cdots & W_N^{2(N-1)} \\ \vdots & \vdots & \vdots & \ddots & \vdots \\ 1 & W_N^{N-1} & W_N^{2(N-1)} & \cdots & W_N^{(N-1)(N-1)} \end{bmatrix}$$

IDFT 的矩阵表示作为练习，读者可以自己写出。

8.3.1 几个扩展函数

1. 逆离散傅里叶变换

【例 8.14】 借助 MATLAB 语言编写逆离散傅里叶变换函数。

EXAMP08014

```
function xn=idft(Xk,N);
% FUNCTION IDFT is inverse discrete Fourier transformation.
% （IDFT 为逆离散傅里叶变换）
% Calling sequence: xn=idft(Xk,N)
%
%xn: N-point sequence over 0 <= n <= N-1
%Xk: DFT coefficient array over 0 <= k <= N-1
%N: the Length of DFT
%
% Records of revisions:
% Date: 2015-08-05
% Programmer: Yuan Hong
% Description of change: original code
%
n=[0:1:N-1];
k=[0:1:N-1];
WN=exp(-j*2*pi/N);
nk=n'*k;
WNnk=WN.^(-nk);
xn=(Xk * WNnk)/N;
```

2. 圆周移位

【例 8.15】 借助 MATLAB 语言编写圆周移位函数。

分析：编写函数是为了实现功能 $y(n)=x((n-m))_N R_N(n)$。

EXAMP08015

```
function y=cirshfit(x,m,N)
% CIRSHIFT calculates circular shift with y(n)=x((n-m))_N R_N(n).
% （CIRSHIFT 计算圆周移位，形式为 y(n)=x((n-m))_N R_N(n)）
% Calling sequence: y=cirshift(x,m,N)
%
%y: an output sequence containing the circular shift
%x: an input sequence
%m: sample shift
%N: the length of x
%
% Records of revisions:
% Date: 2015-08-15
% Programmer: Yuan Hong
```

```
%    Description of change: original code
%
if length(x)>N
    error('N must be greater or equal the length of x')
end
%    (将x补零到N长度)
x=[x zeros(1,N-length(x))];
n=[0:1:N-1];
n=mod(n-m,N);
y=x(n+1);
```

当 $x(n)=[1,2,3,4,5]$，借助 cirshift 函数求 $x((n-3))_5 R(n)$ 及 $x((n+3))_6 R_6(n)$。

将当前路径设置为含有 cirshift.m 文件的位置，在命令窗口输入执行语句如下：

```
>> clc
>> clear all
>> x=[1 2 3 4 5];
>> y1=cirshfit(x,3,5)
y1 =   3    4    5    1    2
>> y2=cirshfit(x,-3,6)
y2 =   4    5    0    1    2    3
```

3. 圆周卷积

两个长度分别为 N_1 和 N_2 的序列 $x_1(n)$ 和 $x_2(n)$ 的 L 点的圆周卷积和表示为

$$y(n) = x_1(n) \overset{L}{\otimes} x_2(n) \qquad L \geq \max[N_1, N_2]$$

圆周卷积和线性卷积之间的详细讨论关系可参考有关数字信号处理相关书籍。在时域和频域两个方面都可以用圆周卷积来计算线性卷积。

【例 8.16】 借助时域方法编写实现圆周卷积的函数。

EXAMP08016

```
function y=circonvt(x1,x2,N)
%   CIRCONVT calculates circular convolution between x1 and x2 in
%   time domain.
%   (CIRCONVT 在时域计算 x1 和 x2 的圆周卷积)
%   Calling sequence: y=circonvt(x1,x2,N)
%
%   y: an output sequence containing the circular convolution
%   x1: an input sequence
%   x2: another input sequence
%   N: the size of circular buffer
%   method:y(n)=sum(x1(m)*x2((n-m) mod N))
%
%   Records of revisions:
%   Date: 2015-08-15
%   Programmer: Yuan Hong
%   Description of change: original code
%
if length(x1)>N
```

```
        error ('N must be >= the length of x1');
    end
    if length (x2) >N
        error ('N must be >= the length of x2');
    end
    x1=[x1 zeros (1, N-length (x1))];
    x2=[x2 zeros (1, N-length (x2))];
    m=[0:1:N-1];
    x2=x2 (mod (-m,N) +1);
    H=zeros (N, N);
    for n=1:1:N
        H (n,:)=cirshfit (x2, n-1,N);
    end
    y=x1*H';
```

已知 $x_1(n) = [1,2,2]$、$x_2(n) = [1,2,3,4]$，借助 circonvt 函数求 4 点圆周卷积。

将当前路径设置为含有 circonvt.m 文件的位置，在命令窗口输入执行语句如下：

```
>> x1=[1,2,3]; x2=[1,2,3,4];
>> y=circonvt (x1,x2,4)
y = 18    16    10    16
```

当 $N \geq N_1 + N_2 - 1$ 时，圆周卷积 circonvt（x1,x2,N）的结果为线性卷积，可简化运算。线性卷积可以在频域上实现圆周卷积。最简单方法是采用以下语句即可以实现时域的圆周卷积，定理：时域序列作圆周卷积和，则离散频域中是相乘运算。

```
X1=fft (x1,N); % （对 x1 进行快速傅里叶变换）
X2=fft (x2,N); % （对 x2 进行快速傅里叶变换）
Y=X1*X2; % （求 X1 和 X2 乘积）
y=ifft (Y,N); % （求 X1 和 X2 乘积的逆快速傅里叶变换）
```

8.3.2 快速傅里叶变换

以上程序为了更好理解 DFT 的概念。实际应用中，便常用 FFT 提高运算效率。

fft 语法格式如下：

y=fft(x)　　向量 x 的离散傅里叶变换；如果 x 是矩阵，则计算该矩阵每一列的离散傅里叶变换；如果 x 是（n*d）维数组，则是对第一个非单元素的维进行离散傅里叶变换。

y=fft(x, n)　　n 点的离散傅里叶变换；如果 x 长度小于 n 将用零补齐，若 x 长度大于 n 截取。

y=fft(x, [], dim) 或 y=fft(x, n, dim)　　在参数 dim 指定的维上进行离散傅里叶变换；当 x 是矩阵时，dim 用来指定变换的实施方向，dim=1 表示变换按列进行，dim=2 表示变换按行进行。

函数 ifft 的参数应用与函数 fft 完全相同。

【例 8.17】 产生一个三角波序列 $x(n)$，长度为 $M=40$；计算 $N=64$ 时，$X(k) = DFT[x(n)]$，并图示 $x(n)$ 和 $X(k)$；对 $X(k)$ 在 $[0, 2\pi]$ 上进行 32 点抽样，得到 $X_1(k) = X(2k)$，k=0, 1,…, 31。求 $X_1(k)$ 的 32 点 IDFT，即 $x_1(n) = IDFT[X_1(k)]$；绘制 $x_1((n))_{32}$ 的波形图，观察 $x_1((n))_{32}$ 和 $x(n)$ 的关系，并加以说明。

EXAMP08017

```
% EXAMP08017 solves the questions in example 8.17.
%  (EXAMP08017 求示例 8.17 中的问题)
%
M=40;N=64;n=0:M;
xa=[0:floor(M/2)];
xb=ceil(M/2)-1:-1:0;
xn=[xa,xb];
% fft: fast discrete Fourier transform
Xk=fft(xn,64);
X1k=Xk(1:2:N);
% ifft: inverse fast discrete Fourier transform
x1n=ifft(X1k,32);
nc=0:4*N/2;
xc=x1n(mod(nc,N/2)+1);
%
subplot(321);
stem(n,xn,'x');
ylabel('x(n)');
title('40 点的三角波序列 x(n)');
%
subplot(322);
k1=0:N-1;stem(k1,abs(Xk),'.');
axis([0,65,0,401])
ylabel('|X(k)|');
title('64 点的 DFT[x(n)]')
%
subplot(323);
k2=0:N/2-1;
stem(k2,abs(X1k),'*');
axis([0,31,0,401])
ylabel('|x1(k)|');
title('隔点抽取 X(k) 得到的 X1(k)]');
%
subplot(324);
n1=0:N/2-1;
stem(n1,x1n,'.');
axis([0,31,0,21])
ylabel('x1(n)');
title('32 点的 IDFT[X2(k)=x1(n)]');
%
subplot(313);
stem(nc,xc,'p');
axis([0,131,0,21])
ylabel('x1((n)) 32');
title('x1(n) 的周期延拓序列');
```

运行结果如图 8.11 所示。

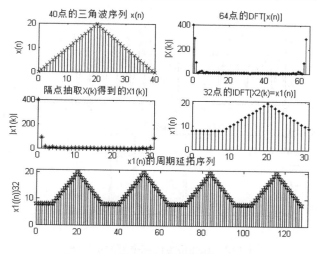

图 8.11 示例波形

【例 8.18】 设 $x(n)$ 是由任意两个正弦信号及白噪音的叠加,试用 FFT 对其作频谱分析。

EXAMP08018

```
%   EXAMP08018 first generates two sine signals and a white noise,
%   and then plays frequency analyzing.
%   (EXAMP08018 首先给出两个正弦信号及其白噪声,然后进行频谱分析)
%
N=256;
f1=0.1;f2=0.2;fs=1;
a1=5;a2=3;
w=2*pi/fs;
x=a1*sin(w*f1*(0:N-1))+a2*sin(w*f2*(0:N-1))+randn(1,N);
%
subplot(2,2,1);
plot(x(1:N/4));
title('原始信号');
f=-0.5:1/N:0.5-1/N;
% fft: fast discrete Fourier transform
X=fft(x);
% ifft: inverse fast discrete Fourier transform
y=ifft(X);
%
subplot(2,1,2);
plot(f,fftshift(abs(X)));
title('频谱响应');
%
subplot(2,2,2);
plot(real(x(1:N/4)));
title('时域信号')
```

运行结果如图 8.12 所示,该程序同时完成了傅里叶变换与傅里叶逆变换。

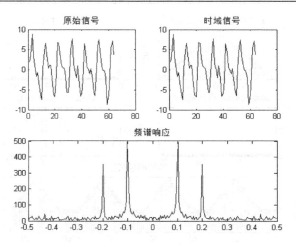

图 8.12　加噪信号的频率响应示例图

【例 8.19】　设 $x(n)$ 为长度 $N=5$ 的矩形序列，分析当 $N=8$、32、64 时 $x(n)$ 的频谱变化。

EXAMP08019

```
%  EXAMP08019analyses the frequency changing of a signal sequence with
%  N=8, 32, 64.
%  （EXAMP08019，在N=8、32、64时，分析信号序列的频谱变换）
%
x=[1,1,1,1,1];
%
N=8;
y1=fft(x,N);
n=0:N-1;
subplot(3,1,1);
stem(n,abs(y1),'xk');
axis([0,7,0,6]);
%
N=32;
y2=fft(x,N);
n=0:N-1;
subplot(3,1,2);
stem(n,abs(y2),'xk');
axis([0,32,0,6]);
%
N=64;
y3=fft(x,N);
n=0:N-1;
subplot(3,1,3);
stem(n,abs(y3),'xk');
axis([0,63,0,6]);
```

运行结果如图 8.13 所示

第8章 MATLAB 在数字信号中的应用

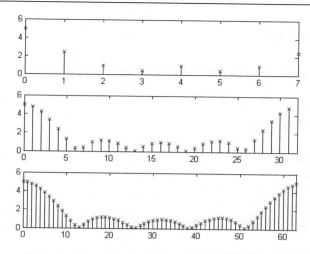

图 8.13 N=8、32、64 时矩形序列的频谱示例图

【例 8.20】 计算 $x(n) = [2,1,3,2,1,5,1]$ 与 $y(n) = [2,1,3,4]$ 的线性相关 $r_{xy}(m)$。

EXAMP08020

```
%  EXAMP08020solves the linear correlation sequence between two
%  signal sequences x=[2,1,3,2,1,5,1] and y=[2,1,3,4].
%  （EXAMP08020 计算两个信号序列 x=[2,1,3,2,1,5,1]和 y=[2,1,3,4]线性相关）
%
clc;clear all
x = [2,1,3,2,1,5,1];
y = [2,1,3,4];
N = length(x)+length(y)-1;
n = -length(y)+1:length(x)-1;
%xcorr: Cross-correlation function estimates
[rxy,mm] = xcorr(x, y);
X=fft(x,N);
Y=fft(y,N);
Rxy1 = X.*conj(Y);
rxy1 = ifft(Rxy1);
%
subplot(241)
stem(0:length(x)-1,x,'.');
grid;
title('x(n)');
xlabel('n');
%
subplot(242)
stem(0:length(y)-1,y,'.');
grid;
title('y(n)');
xlabel('n');
%
subplot(222)
stem(n,rxy(4:13),'.');
```

```
                grid;
                title('直接线性相关');
                xlabel('n');
                axis([-length(y)+1,length(x)-1,-inf,inf]);
                %
                subplot(223)
                stem(0:N-1,rxy1,'.');
                grid;
                title('圆周卷积');
                xlabel('n');
                %
                subplot(224)
                stem(n,[rxy1(length(x)+1:N),rxy1(1:length(x))],'.');
                grid;
                title('圆周相关转换为线性相关');axis([min(n),max(n),-inf,inf]);
                xlabel('n');
```

运行结果如图 8.14 所示。

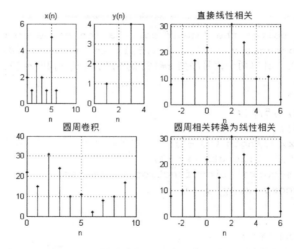

图 8.14 信号序列的线性相关示例图

【例 8.21】 利用 chirp 函数 Z 变换计算滤波器 h 在 100～200Hz 的频率特性，比较 CZT 和 FFT 函数。

EXAMP08021

```
                % EXAMP08021 supplies example to compare the usage between CZT
                % and FFT.
                %  (EXAMP08021 为比较 CZT 和 FFT 用法举例说明)
                %
                % fir1: FIR filter design using the window method
                %  (用窗函数法先确定一个 FIR 滤波器)
                h=fir1(30,125/500,boxcar(31));
                Fs=1000;
```

```
f1=100;
f2=200;
m=1024;
w=exp(-j*2*pi*(f2-1)/(m*Fs));
a=exp(j*2*pi*f1/Fs);
%
y=fft(h,m);
z=czt(h,m,w,a);
fy=(0:length(y)-1)'*Fs/length(y);
fz=(0:length(z)-1)'*(f2-f1)/length(z)+f1;
%
subplot(2,1,1)
plot(fy(1:500),abs(y(1:500)));
title('fft');
%
subplot(2,1,2)
plot(fz,abs(z));
title('czt');
```

运行结果如图 8.15 所示。

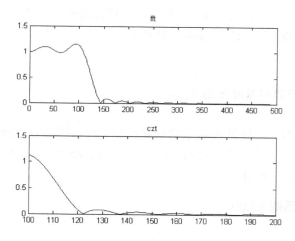

图 8.15 chirp 函数 Z 变换下的 CZT 和 FFT 比较示例图

8.4 IIR 滤波器的设计

IIR 数字滤波器的设计方法分为两类：间接法和直接法。间接法通过设计模拟滤波器的方法进行，用数字指标设计相应的过渡模拟滤波器，再用过渡模拟型转换成数字滤波器。直接法是直接在时域或频域设计数字滤波器。因为模拟滤波器的设计理论非常成熟，有很多性能好的典型滤波器可以使用（例如巴特沃斯、切比雪夫、椭圆等）。图表和公式现成可查，实际中也较多应用模拟滤波器的数字化仿真，故间接法应用很广。直接法需要求解联立方程，各种设计函数都有现成方法可供调用。只要掌握了基本设计原理，工程中应用直接法很容易。本节主要介绍模拟滤波器设计、数字滤波器的间接设计法和相应的 MATLAB

工具箱函数。

8.4.1 滤波器设计函数

本节以巴特沃斯滤波器设计函数为主要讨论对象，其他切比雪夫、椭圆、贝塞尔滤波器的指令函数使用方法基本一致。各种模拟滤波器的设计（包括低通、带通、高通、带阻）都归结为先设计一个归一化原型低通滤波器，然后通过模拟频带变换得到所需模拟滤波器。样本滤波器，即是所谓的归一化低通原型，就是将通带边缘频率归一化为 1。比如，对于巴特沃斯滤波器必定指的 3dB 衰减处的频率 Ω_c，归一化为 Ω_c =1。而对切比雪夫滤波器指的是通带边沿为某一衰减处的频率 Ω_p 归一化。两者的差别体现在具体设计里，需要注意。

1. 函数 buttord

巴特沃斯特滤波器函数，用来确定数字低通或模拟低通滤波器的阶次。

语法格式如下：

[N, Wn]=buttord（Wp, Ws, Rp, Rs） 对应数字滤波器，式中 Wp、Ws 分别是通带和阻带的截止频率，是归一化频率，其值在 0～1。对低通和高通滤波器，Wp、Ws 都是标量；对带通和带阻滤波器，Wp、Ws 是 1×2 的向量。Rp、Rs 是通带和阻带的衰减，单位为 dB。N 是阶次，Wn 是 3dB 对应的频率，它和 Wp 稍有不同。

[N, Wn]=buttord（Wp, Ws, Rp, Rs, 's'） 对应模拟滤波器，式中各个变量的含义与上面相同，但 Wp、Ws 及 Wn 的单位为 rad/s，是频率。

2. 函数 buttap

用来设计模拟低通原型滤波器 G（p）。

语法格式如下：

[z, p, k]=buttap（N） 计算 N 阶巴特沃斯特归一化模拟低通滤波器系统函数 G（p）的极点 z、零点 p 和增益 k。结构转换函数可以调用[B, A]=zp2tf（z, p, k）得到系统函数的分子和分母多项式系数向量 B 和 A。

3. 函数 lp2lp 和函数 lp2hp

语法格式如下：

[B, A]=lp2lp（b, a, Wo）或[B, A]=lp2hp（b, a, Wo） 实现模拟低通原型滤波器 G（p）分别转换为低通、高通滤波器。Wo 是低通或高通滤波器的截止频率；B、A 分别为待求滤波器系统函数的分子、分母系数向量。

4. 函数 lp2bp 和函数 lp2bs

语法格式如下：

[B, A]=lp2bp（b, a, Wo, Bw）或 [B, A]=lp2bs（b, a, Wo, Bw） 实现模拟低通原型滤波器 G（p）分别转换为带通及带阻滤波器。式中 b, a 分别是模拟低通原型滤波器 G（p）分子、分母多项式的系数向量；B、A 分别是转换后的 H（s）分子、分母多项式的系数向量；Wo 是带通或带阻滤波器的中心频率，Wo=sqrt（wp1*wp2）；Bw 是带宽，Bw=wp2-wp1。上下截止频率分别为 wp2、wp1。

5. 函数 bilinear

语法格式如下：

[Bz, Az]=bilinear（B, A, Fs）　　实现双线性变换，即由模拟滤波器 H（s）得到数字滤波器 H（z）。B、A 分别是 H（s）的分子、分母多项式的系数向量；Bz、Az 分别是 H（z）的分子、分母多项式系数向量；Fs 是抽样频率。

6. 函数 butter

巴特沃斯数字和模拟滤波器设计。实际上这个函数把 buttord、buttap、lp2lp 及 bilinear 等函数都包含进去，从而使设计过程更简单。

语法格式如下：

[B, A]=butter（N, Wn）　　设计数字带通滤波器。B、A 分别是系统函数 H（z）的分子、分母多项式的系数向量（系数函数按 z^{-1} 的升幂排列，首项为 z^0 项）；N 代表第 N 阶；Wn 是通带截止频率 0<Wn<1。

[B, A]=butter（N, Wn, 'high'）　　设计 3dB 截止频率为 Wn 数字高通滤波器。

[B, A]=butter（N, Wn, 'low'）　　设计低通滤波器。

[B, A]=butter（N, Wn, 'stop'）　　设计数字带阻滤波器。

[B, A]=butter（N, Wn, 's'）　　设计模拟滤波器。

7. 函数 cheb1ord

语法格式如下：

[N, wc] = cheb1ord（Wp, Ws, Rp, As）　　求切比雪夫 I 型滤波器的阶次。Wp、Ws 分别是通带和阻带的截止频率，其值为 $0 \leqslant wp(\text{或 ws}) \leqslant 1$，即它们是对 π 归一化的数字频率，当它们（Wp 或 Ws）的数值为 1 时，表示频率为抽样频率的 1/2。

与滤波器设计有关的函数还有 cheb1ap、cheby1、cheb2ord、cheb2ap、cheby2、ellipord、ellipap、ellip、besselap、besself 等分别为切比雪夫 I 型滤波器、切比雪夫 II 型滤波器、椭圆及贝塞尔滤波器设计函数。其格式与切比雪夫 I 型滤波器和巴特沃斯滤波器的参量调用方法类同。可借助 help 函数自己了解。

【例 8.22】　要求设计一个模拟低通滤波器，其通带截止频率 $f_p = 3000\text{Hz}$，通带最大衰减 $R_p = 2\text{dB}$，阻带截止频率 6000Hz，阻带最小衰减 $A_s = 30\text{dB}$，分别采用巴特沃斯、切比雪夫 I 型、切比雪夫 II 型、椭圆函数型 4 种类型来进行设计。

（1）巴特沃斯型。

EXAMP08022-Butterworth

```
%  EXAMP08022Butterworth designs Butterworth analog lowpass filter.
%  （EXAMP08022Butterworth 程序用来设计巴特沃斯低通模拟滤波器）
clc;
clear all
%  Define the given conditions in example 8.22.
OmegaP=2 *pi*3000;
OmegaS=2*pi*6000;
Rp=2;
```

```
As=30;
w0=[OmegaP,OmegaS];
% buttord: Butterworth filter Order selection.
[N,OmegaC]=buttord (OmegaP,OmegaS,Rp,As,'s')
% butter: Butterworth digital and analog filter design.
[b,a]=butter (N,OmegaC,'s')
[H,w]=freqs (b,a);
Hx=freqs (b,a,w0);
dbHx=-20*log10 (abs (Hx)/max (abs (H)))
plot (w/ (2 * pi) /1000,20 * log10 (abs (H)));
xlabel ('f (kHz)');
ylabel ('dB');
set (gca,'xtickmode','manual','xtick',[0,1,2,3,4,5,6,7]);
set (gca,'ytickmode','manual','ytick',[-40,-30,-20,-10,0]);
grid on;
```

运行结果如下:

```
N=6
OmegaC =2.1202e+04
b = 1.0e+25 *
    0    0    0    0    0    0    9.0825
a = 1.0e+25 *
    0.0000   0.0000   0.0000   0.0000   0.0000   0.0017   9.0825
dbHx =  0.9478   30.0000
```

巴特沃斯型模拟低通滤波器的频率响应示例图如图 8.16 所示。

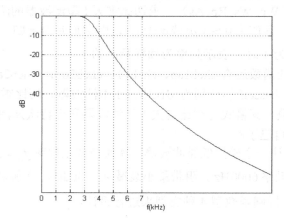

图 8.16　巴特沃斯型模拟低通滤波器的频率响应示例图

（2）切比雪夫 I 型。

EXAMP08022-ChebyshevI

```
% EXAMP08022ChebyshevI designs Chebyshev-I analog lowpass filter.
% （EXAMP08022ChebyshevI 程序用来设计切比雪夫 I 型低通模拟滤波器）
clc;
clear all
OmegaP=2*pi*3000;
```

```
OmegaS=2*pi*6000;
w0=[OmegaP,OmegaS];
Rp=2;
As=30;
[N,OmegaC]=cheb1ord(OmegaP,OmegaS,Rp,As,'s')
[b,a]=cheby1(N,Rp,OmegaC,'s')
[H,w]=freqs(b,a);
Hx=freqs(b,a,w0);
dbHx=-20*log10(abs(Hx)/max(abs(H)))
plot(w/(2*pi)/1000,20*log10(abs(H)))
xlabel('f(kHz)');
ylabel('dB');
set(gca,'xtickmode','manual','xtick',[0,1,2,3,4,5,6,7]);
set(gca,'ytickmode','manual','ytick',[-40,-30,-20,-10,-1]);
grid
```

运行结果显示如下：

```
N =4
OmegaC = 1.8850e+004
b =1.0e+016*
    0       0        0        0      2.0634
a = 1.0e+016 *  0.0000    0.0000    0.0000    0.0003    2.5976
dbHx =2.0000    37.4070
```

切比雪夫 I 型模拟低通滤波器的频率响应示例图如图 8.17 所示。

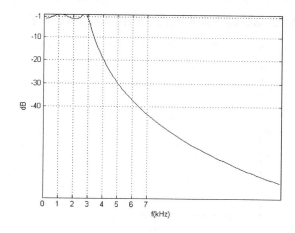

图 8.17　切比雪夫 I 型模拟低通滤波器的频率响应示例图

(3) 切比雪夫 II 型。

EXAMP08022-ChebyshevII

```
% EXAMP08022ChebyshevII designs Chebyshev-II analog lowpass filter.
% （EXAMP08022ChebyshevII 程序用来设计切比雪夫 II 型低通模拟滤波器）
clc;
clear all
OmegaP=2*pi*3000;
```

```
OmegaS=2*pi*6000;
w0=[OmegaP,OmegaS];
Rp=2;
As=30;
[N,OmegaC]=cheb2ord(OmegaP,OmegaS,Rp,As,'s')
[b,a]=cheby2(N,As,OmegaS,'s')
[H,w]=freqs(b,a);
Hx=freqs(b,a,w0);
dbHx=-20*log10(abs(Hx)/max(abs(H)))
plot(w/(2*pi)/1000,20*log10(abs(H)));
xlabel('f(kHz)');
ylabel('dB');
axis([0,15,-80,1])
set(gca,'xtickmode','manual','xtick',[0,2,3,6,8,10,15]);
set(gca,'ytickmode','manual','ytick',[-40,-30,-20,-10,-1]);
grid
```

运行结果如下:

```
N = 4
OmegaC = 3.1543e+04
b = 1.0e+17 *
    0.0000    0.0000    0.0000    0.0000    5.1099
a = 1.0e+17 *
    0.0000    0.0000    0.0000    0.0004    5.1099
dbHx = 0.4382   30.0000
```

切比雪夫 II 型模拟低通滤波器的频率响应示例图如图 8.18 所示。

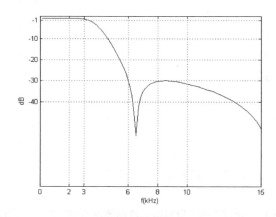

图 8.18 切比雪夫 II 型模拟低通滤波器的频率响应示例图

(4) 椭圆函数型。

EXAMP08022-Elliptic

```
% EXAMP08022Elliptic designs elliptic digital lowpass filter.
% （EXAMP08022Elliptic程序用来设计数字低通滤波器）
clc
clear all
```

```
OmegaP=2*pi*3000;
OmegaS=2*pi*6000;
w0=[OmegaP,OmegaS];
Rp=2;
As=30;
[N,OmegaC]=ellipord(OmegaP,OmegaS,Rp,As,'s')
[b,a]=ellip(N,Rp,As,OmegaC,'s')
[H,w]=freqs(b,a);
Hx=freqs(b,a,w0);
dbHx=-20*log10(abs(Hx)/max(abs(H)))
plot(w/(2*pi)/1000,20*log10(abs(H)));
xlabel('f(kHz)');
ylabel('dB');
set(gca,'xtickmode','manual','xtick',[0,2,3,4,5,6,10]);
set(gca,'ytickmode','manual','ytick',[-40,-30,-20,-10,-1]);grid;
```

运行结果如下:

```
N = 3
OmegaC = 1.8850e+04
b = 1.0e+12 *
        0    0.0000         0    2.6461
a = 1.0e+12 *
   0.0000    0.0000    0.0004    2.6461
dbHx = 2.0000   34.2134
```

椭圆函数模拟低通滤波器的幅频特性曲线示例图如图 8.19 所示。

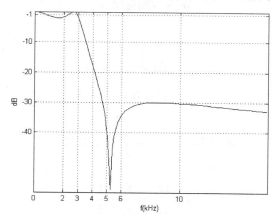

图 8.19 椭圆函数模拟低通滤波器的幅频特性曲线示例图

注意:
4 个不同滤波器各个参数指标不同,程序运行结果各不相同,可自行改变参数使得阻带衰减增大。

8.4.2 双线性变换法及冲激响应不变法设计 IIRDF

冲激响应不变法的基本思想:对 $h_a(t)$ 等间隔采样得到数字滤波器的单位脉冲响应 $h(n)$,

从而 $H_a(s)$ 转换成 $H(z)$。双线性变换法是从 s 域到 z 域的单值映射变换：$s=\dfrac{2}{T}\dfrac{1-z^{-1}}{1+z^{-1}}$，从原理上彻底消除频谱混叠。具体设计公式的推导可参考教材。下面举例说明两种变换方法，主要介绍直接调用 MATLAB 工具箱函数进行设计的方法。

1. 双线性不变换设计 IIR 数字滤波器

【例 8.23】 试用双线性变换法设计一个低通滤波器，技术指标分别为 $f_p=100\text{Hz}$、$f_{st}=300\text{Hz}$、$\alpha_p=3\text{dB}$、$\alpha_s=20\text{dB}$，抽样频率为 $F_s=100\text{Hz}$。

理论推导：先求数字频率 ω，由 2π 对应 F_s，有 $\omega_p=\dfrac{2\pi f_p}{F_s}=0.2\pi$，$\omega_{st}=\dfrac{2\pi f_{st}}{F_s}=0.6\pi$。

（1）用频率预畸公式 $\Omega=\tan\left(\dfrac{\omega}{2}\right)$ 将数字滤波器的技术要求转换为模拟滤波器的技术要求：

$$\Omega_p=\tan\left(\dfrac{\omega_p}{2}\right)=0.3249$$

$$\Omega_{st}=\tan\left(\dfrac{\omega_{st}}{2}\right)=1.37638$$

（2）设计低通滤波器 $G(s)$：

因为

$$\alpha_p=-20\lg(1-\delta_p)$$
$$\alpha_{st}=-20\lg(\delta_s)$$

所以

$$1-\delta_p=10^{-\frac{3}{20}}=0.70795$$

$$\delta_s=0.1$$

$$d=\left[\dfrac{(1-\delta_p)^{-2}-1}{\delta_s^{-2}-1}\right]^{1/2}=0.10027$$

$$k=\dfrac{\Omega_p}{\Omega_s}=\dfrac{f_p}{f_{st}}=\dfrac{1}{3}$$

由于

$$N\geqslant\dfrac{\log d}{\log k}=2.09$$

得滤波器阶数是 $N=2$，查归一化原型滤波器表得 $a_1=1.4142$、$a_2=1$，所以得归一化的二阶低通巴特沃斯滤波器 $G(p)$ 为

$$G(p)=\dfrac{1}{p^2+1.4142p+1}$$

去归一化，得到

$$G(s)=G(p)|_{p=\frac{s}{\Omega_p}}=\frac{0.3249^2}{s^2+0.4595s+0.3249^2}$$

（3）代入双线性变换法公式 $s=\frac{z-1}{z+1}$，有

$$H(z)=G(s)|_{s=\frac{z-1}{z+1}}=\frac{0.06745+0.1349z^{-1}+0.0645z^{-2}}{1-1.143z^{-1}+0.4208z^{-2}}$$

由本例可见，低阶滤波器的设计计算很麻烦，更高阶的滤波器的设计会更加复杂。而借助 MATLAB 函数，很容易实现双线性变换法设计的低通滤波器。

EXAMP08023

```
% EXAMP08023 designs IIR digital filter with bilinear invariance.
%  (EXAMP08023 借助双线性不变法设计 IIR 数字滤波器)
fp=100;
fst=300;
Fs=1000;
rp=3;
rs=20;
wp=2*pi*fp/Fs;
ws=2*pi*fst/Fs;
Fs=Fs/Fs;
wap=tan(wp/2);
was=tan(ws/2);
% buttord: Butterworth filter order selection.
[n,wn]=buttord(wap,was,rp,rs,'s');
% buttap: Butterworth analog lowpass filter prototype.
[z,p,k]=buttap(n);
% zp2tf: Zero-pole to transfer function conversion.
[bp,ap]=zp2tf(z,p,k)
% lp2lp: Lowpass to lowpass analog filter transformation.
[bs,as]=lp2lp(bp,ap,wap)
% bilinear: Bilinear transformation with optional
%           frequency prewarping
[bz,az]=bilinear(bs,as,Fs/2)
% freqz: Frequency response of digital filter.
[h,w]=freqz(bz,az,256,Fs*1000);
plot(w,abs(h));
grid on
```

运行结果如下：

```
bp = 0        0        1
ap = 1.0000   1.4142   1.0000
bs = 0.1056
as = 1.0000   0.4595   0.1056
bz = 0.0675   0.1349   0.0675
az = 1.0000  -1.1430   0.4128
```

双线性不变法设计数字滤波器的频响曲线示例图如图 8.20 所示。

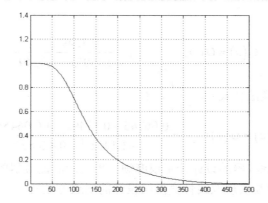

图 8.20 双线性不变法设计数字滤波器的频响曲线示例图

> **注意：**
> 由零极点得到系统函数分子、分母多项式系数向量，调用结构转换函数 zp2tf。

下面将 MATLAB 自带函数 freqz 扩展到 freqz_m，最大好处是可以得到滤波器的更多特性：幅度响应、以 dB 表示的幅度响应、相位响应、群延时响应。在以下例子里能更直观地看到差别。

【例 8.24】 编写 Modfreqz 函数实现 freqz 函数功能的扩展，使其可以得到滤波器具有更多响应，如幅度响应、以 dB 表示的幅度响应、相位响应、群延时响应。

EXAMP08024

```
function [db,mag,pha,grd,w]=Modfreqz(b,a)
% Function MODFREQZ computes z-domain frequency response.
% （函数 MODFREQZ 计算 z 域频率响应）
% Calling sequence: [db,mag,pha,grd,w]=Modfreqz(b,a)
%
% db: Relative magnitude in db over [0 to pi]
% mag: absolute magnitude over [0 to pi]
% pha: phase response in radians over [0 to pi]
% w: array of 500 frequency samples between [0 to pi]
% b: numerator polynomial coefficients of H(z)
% a: denominator polynomial coefficients of H(z)
%
% Records of revisions:
% Date: 2015-08-15
% Programmer: Yuan Hong
% Description of change: original code
%
[H,w]=freqz(b,a,1000,'whole');
H=(H(1:1:501))';
w=(w(1:1:501))';
mag=abs(H);
db=20*log10((mag+eps)/max(mag));
pha=angle(H);
% grpdelay: Group delay of digital filter.
grd=grpdelay(b,a,w);
```

【例 8.25】 用双线性不变法设计一个低通巴特沃斯数字滤波器,技术指标是 $w_p = 0.2\pi$、$w_s = 0.3\pi$、$R_p = 1\text{dB}$、$A_s = 15\text{dB}$。

EXAMP08025

```matlab
function varargout=EXAMP08025(varargin)
% Function EXAMP08025 designs Butterworth lowpass digital filter.
% (函数 EXAMP08025 设计低通巴特沃斯数字滤波器)
% Calling sequence: EXAMP08025
% Main Function.
%
% Records of revisions:
% Date: 2015-08-15
% Programmer: Yuan Hong
% Description of change: original code
%
% Express the known conditions in MATLAB language.
% (将已知条件表述为 MATLAB 语言)
wp=0.2*pi;
ws=0.3*pi;
Rp=1;
As=15;
T=1;
Fs=1/T;
OmegaP=(2/T)*tan(wp/2);
OmegaS=(2/T)*tan(ws/2);
% Analog Butterworth Prototype Filter Calculation:
% afd_butt: a subfunction given in this program.
%           (为一个子函数,本程序给出相应程序)
[cs,ds]=afd_butt(OmegaP,OmegaS,Rp,As);
%***Butterworth Filter Order = 6
%Bilinear transformation:
% bilinear: Bilinear transformation with optional frequency
%           prewarping.
[b,a]=bilinear(cs,ds,Fs);
% dir2cas: a subfunction given in this program.
%          (为一个子函数,本程序给出相应程序)
[C,B,A]=dir2cas(b,a)
[db,mag,pha,grd,w]=Modfreqz(b,a);
%
subplot(131)
plot(w/pi,mag)
title('Magnitude Response')
%
subplot(132)
plot(w/pi,db)
title('Magnitude in dB')
%
subplot(133)
plot(w/pi,pha/pi)
title('Phase Response ')
```

```matlab
end
%
%-----Subfunction afd_butt
function [b, a]=afd_butt(Wp, Ws, Rp, As);
  % AFD_BUTT designs Analog Lowpass Filter Design: Butterworth
  % Calling sequence: [b,a]=afd_butt(Wp, Ws, Rp, As);
  %
  % b: Numberator coefficients of Ha(s)
  % a: Denominator coefficients of Ha(s)
  % Wp: Passband edge frequeney in rad/sec; Wp>0
  % Ws: Stopband edge frequeney in tad/sec; Ws>Wp>0
  % Rp: Passband ripple in +dB; (Rp>0)
  % As: Stopband attenuation in +dB; (As> 0)
  %
  if Wp <= 0
      error(' Passband edge must be larger than 0');
  end
  if Ws <= Wp
      error('Stopband edge must be larger than Passband edge');
  end
  if (Rp <= 0)|(As<0)
      error('PB ripple and/or SB attenuation must be larger than 0');
  end
N=ceil((log10((10^(Rp/10)-1)/(10^(As/10)-1)))/(2*log10(Wp/Ws)));
  OmegaC=Wp/((10^(Rp/10)-1)^(1/(2*N)));
    % u_buttap: a subsubfunction designed in this program.
    %           （为一个子函数，本程序给出相应程序）
    [b, a]=u_buttap(N, OmegaC);
end
%
%===== subsubfunction u_buttap
function [b,a]=u_buttap(N,OmegaC);
    % Unnormalized Butterworth Analog Lowpass filter prototype
    % Calling sequence: [b,a]=u_buttap(N,OmegaC);
    %
    % b: numerator polynomial coefficients of Ha(s);
    % a: denominator polynomial coefficients of Ha(s)
    % N: order of the butterworth filter
    % OmegaC: cutoff freq in radians/sec
    [z,p,k]=buttap(N);
    p=p*OmegaC;
    k=k*OmegaC^N;
    B=real(poly(z));
    b0=k;
    b=k*B;
    a=real(poly(p));
end
%
%-----subfunction dir2cas
function [b0,B,A] = dir2cas(b,a);
  % DIRECT-form to CASCADE-form conversion (cplxpair version)
```

```
%   Calling sequence: [b0,B,A] = dir2cas (b,a)
%   b0: gain coefficient
%   B: K by 3 matrix of real coefficients containing bk's
%   A: K by 3 matrix of real coefficients containing ak's
%   b: numerator polynomial coefficients of DIRECT form
%   a: denominator polynomial coefficients of DIRECT form
b0=b (1) ;
b=b/b0;
a0=a (1) ;
a=a/a0;
b0=b0/a0;
M=length (b) ;
N=length (a) ;
if N>M
    b=[b zeros (1,N-M)];
elseif M>N
    a=[a zeros (1,M-N)];
    N=M;
else
    NM=0;
end
K=floor (N/2) ;
B=zeros (K,3) ;
A=zeros (K,3) ;
if K*2==N;
    b=[b 0];
    a=[a 0];
end
% cplxpair: Sort numbers into complex conjugate pairs.
broots=cplxpair (roots (b)) ;
aroots=cplxpair (roots (a)) ;
for i=1:2:2*K
    Brow=broots (i:1:i+1,:) ;
    Brow=real (poly (Brow)) ;
    B (fix ((i+1) /2) ,:) =Brow;
    Arow=aroots (i:1:i+1,:) ;
    Arow=real (poly (Arow)) ;
    A (fix ((i+1) /2) ,:) =Arow;
end
end
```

运行结果如下：

```
C = 5.7969e-04
B = 1.0000    2.0320    1.0323
    1.0000    1.9997    1.0000
    1.0000    1.9683    0.9687
A = 1.0000   -0.9459    0.2342
    1.0000   -1.0541    0.3753
    1.0000   -1.3143    0.7149
```

低通巴特沃斯数字滤波器的损耗函数曲线（dB）示例图如图 8.21 所示。

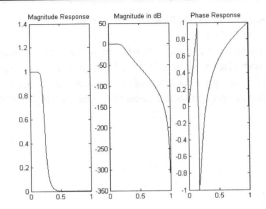

图 8.21 低通巴特沃斯数字滤波器的损耗函数曲线（dB）示例图

> **注意：**
> 调试过程中，要结合 IIR 滤波器的级联型结果计算二阶节的分子、分母。信号处理工具箱里自带函数为 sosfilt（sos, x）。用 G*sosfilt（sos, x）可求得级联结构输出。

2. 冲激响应不变法设计 IIR 数字滤波器

MATLAB 工具箱里函数 buttord 和 butter 设计巴特沃斯数字滤波器时用的就是双线性不变法。所以，上面例子的设计步骤可以借助 buttord、butter 函数完成，读者自己作为练习。

【例 8.26】 用例 8.25 的指标和冲激响应不变法设计数字滤波器。

EXAMP08026

```
function varargout=EXAMP08026(varargin)
% Function EXAMP08026 designs Butterworth lowpass digital filter.
% （函数 EXAMP08026 设计低通巴特沃斯数字滤波器）
% Calling sequence: EXAMP08026
% Main Function.
%
% Records of revisions:
% Date: 2015-08-15
% Programmer: Yuan Hong
% Description of change: original code
%
wp=0.2*pi;
ws=0.3*pi;
Rp=1;
As=15;
T=1;
Fs=1/T;
OmegaP=(2/T)*tan(wp/2);
OmegaS=(2/T)*tan(ws/2);
% afd_butt: a subfunction given in this program.
%            （为一个子函数，本程序给出相应程序）
[cs,ds]=afd_butt(OmegaP,OmegaS,Rp,As);
% imp_invr: a subfunction given in this program.
%            （为一个子函数，本程序给出相应程序）
```

```
    [b,a]=imp_invr(cs,ds,Fs);
% dir2par: a subfunction given in this program.
%          （为一个子函数，本程序给出相应程序）
[C,B,A]=dir2par(b,a)
[db,mag,pha,grd,w]=Modfreqz(b,a);
%
subplot(131)
plot(w/pi,mag)
title('Magnitude Response')
%
subplot(132)
plot(w/pi,db)
title('Magnitude in dB')
%
subplot(133)
plot(w/pi,pha/pi)
title('Phase Response ')
end
%
%-----Subfunction afd_butt
function [b, a]=afd_butt(Wp, Ws, Rp, As)
  % AFD_BUTT designs Analog Lowpass Filter Design: Butterworth
  % Calling sequence: [b,a]=afd_butt(Wp, Ws, Rp, As);
  %
  % b: Numberator coefficients of Ha(s)
  % a: Denominator coefficients of Ha(s)
  % Wp: Passband edge frequeney in rad/sec; Wp>0
  % Ws: Stopband edge frequeney in tad/sec; Ws>Wp>0
  % Rp: Passband ripple in +dB; (Rp>0)
  % As: Stopband attenuation in +dB; (As> 0)
  %
  if Wp <= 0
      error(' Passband edge must be larger than 0');
  end
  if Ws <= Wp
      error('Stopband edge must be larger than Passband edge');
  end
  if (Rp <= 0)|(As<0)
      error('PB ripple and/or SB attenuation must be larger than 0');
  end
  N=ceil((log10((10^(Rp/10)-1)/(10^(As/10)-1)))/(2*log10(Wp/Ws)));
  OmegaC=Wp/((10^(Rp/10)-1)^(1/(2*N)));
  % u_buttap: a subsubfunction designed in this program.
  %           （为一个子函数，本程序给出相应程序）
  [b, a]=u_buttap(N, OmegaC);
end
%
%===== subsubfunction u_buttap
function [b,a]=u_buttap(N,OmegaC)
    % Unnormalized Butterworth Analog Lowpass filter prototype
    % Calling sequence: [b,a]=u_buttap(N,OmegaC)
```

```matlab
    %
    % b: numerator polynomial coefficients of Ha(s);
    % a: denominator polynomial coefficients of Ha(s)
    % N: order of the butterworth filter
    % OmegaC: cutoff freq in radians/sec
    [z,p,k]=buttap(N);
    p=p*OmegaC;
    k=k*OmegaC^N;
    B=real(poly(z));
    b0=k;
    b=k*B;
    a=real(poly(p));
end
%
%-----subfunction imp_invr
function [b,a]=imp_invr(c,d,T)
    % Impulse Invariance TransFormation from Analog to Digital Filter
    % Calling sequence: [b,a]=imp_invr(c,d,T)
    %
    % b: Numerator polynomial in z^(-1) of the digital filter
    % a: Denominator polynomial in z^(-1) of the digital filter
    % c: Numerator polynomial in s of the analog filter
    % d: Denominator polynomial in s of the anglog filter
    % T: Sampling (transformation) parameter
    %
    % residue: Partial-fraction expansion (residues)
    [R,p,k]=residue(c,d);
    p=exp(p*T);
    [b,a]=residuez(R,p,k);
    b=real(b');
    a=real(a');
end
%
%-----subfunction dir2par
function [C,B,A]=dir2par(b,a)
    % DIRECT-form to PARALLEL-form conversion.
    % Calling sequence: [C,B,A]=dir2par(b,a)
    % C: Polynomial part when length(b) <= length(a)
    % B: k by 2 matrix of real coeffcients containing bk's
    % A: k by 3 matrix of real coeffcients containing ak's
    % a: numerator polynomial coefficients of DIRECT form
    % b: denomininator polynomial coefficients of DIRECT form
    M=length(b);
    N=length(a);
    % residuez: Z-transform partial-fraction expansion.
    [r1,p1,C]=residuez(b,a);
    % cplxpair: Sort numbers into complex conjugate pairs.
    p=cplxpair(p1,10000000*eps);
    % cplxcomp: a subsubfunction given in this program.
    I=cplxcomp(p1,p);
    r=r1(I);
```

```
            K=floor(N/2);
            B=zeros(K,2);
            A=zeros(K,3);
            if K*2==N
                for i=1:2:N-2
                    Brow=r(i:1:i+1,:);
                    Arow=p(i:1:i+1,:);
                    [Brow,Arow]=residuez(Brow,Arow,[]);
                    B(fix((i+1)/2),:)=real(Brow);
                    A(fix((i+1)/2),:)=real(Arow);
                end
                [Brow,Arow]=residuez(r(N-1),p(N-1),[]);
                B(K,:)=[real(Brow) 0];
                A(K,:)=[real(Arow) 0];
            else
                for i=1:2:N-1
                    Brow=r(i:1:i+1,:);
                    Arow=p(i:1:i+1,:);
                    [Brow,Arow]=residuez(Brow,Arow,[]);
                    B(fix((i+1)/2),:)=real(Brow);
                    A(fix((i+1)/2),:)=real(Arow);
                end
            end
        end
%
%===subsubfunction
function I=cplxcomp(p1,p2)
% （向量重新排序后的序号计算）
% Calling sequence: I = cplxcomp(p1,p2)
%
% （计算复数极点p1变为p2后留数的新序号）
% （本程序必须用在CPLXPAIR程序之后以便重新频率极点向量及其相应的留数向量）
%
% p2 = cplxpair(p1)
%
% （设一个空的矩阵）
I=[];
% （逐项检查改变排序后的向量p2）
for j=1:1:length(p2)
    % （把该项与p1中各项比较）
    for i=1:1:length(p1)
        % （看与哪一项相等）
        if (abs(p1(i)-p2(j))< 0.0001)
            % （把此项在p1中的序号放入I）
            I=[I,i];
        end
    end
end
% （最后的I表示了p2中各个元素在p1中的位置）
I=I';
end
```

运行结果如下：

```
C = []
B = 1.9193    -0.6258
   -2.2162     1.1537
    0.2969    -0.4630
A = 1.0000   -0.9732    0.2454
    1.0000   -1.0412    0.3575
    1.0000   -1.2645    0.6863
```

冲激响应不变法设计 IIR 数字滤波器的损耗函数曲线示例图如图 8.22 所示。

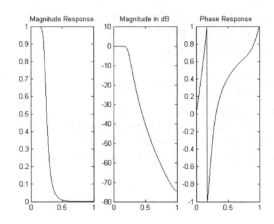

图 8.22　冲激响应不变法设计 IIR 数字滤波器的损耗函数曲线示例图

> **注意：**
> 　　冲激响应不变法和双线性不变法设计数字滤波器时，实现语句不同，二阶节系数 C、A、B 不同，幅度响应（dB）图更加直观明显。

8.4.3　MATLAB 自带函数设计各类数字滤波器

借助于第 1 节的部分 MATLAB 函数和自编函数，一般按照阶次、零极点和增益、系统函数分子、分母多项式的系数来设计。系统函数按 z^{-1} 的升幂排列，首项是 z^0。

【例 8.27】　直接设计法设计一个巴特沃斯型数字低通滤波器。技术指标为 $f_p=1\text{kHz}$、$f_{st}=1.5\text{kHz}$、$R_p=1\text{dB}$、$A_s=15\text{dB}$，采样频率 $F_s=10\text{kHz}$。

EXAMP08027

```
% （EXAMP08027 设计一个巴特沃斯型数字低通滤波器）
%
clc
clear all
Fs=10000;
wp=2*1000/Fs;
ws=2*1500/Fs;
Rp=1;
As=15;
```

```
[N,wc]=buttord(wp,ws,Rp,As);
[b,a]=butter(N,wc);
% dir2par: a function given by authors.
[C,B,A]=dir2par(b,a);
w0=[wp*pi,ws*pi];
Hx=freqz(b,a,w0);
[H,w]=freqz(b,a);
dbHx=-20*log10(abs(Hx)/max(abs(H)));
% Modfreqz: a function written by authors.
[db,mag,pha,grd,w]=Modfreqz(b,a);
plot(w/pi,db);title('Magnitude (dB)');
xlabel('\omega/\pi');ylabel('dB');
axis([0,0.4,-20,2]);
set(gca,'xtickmode','manual','xtick',[0,0.1,0.2,0.3,0.4]);
set(gca,'ytickmode','manual','ytick',[-20,-As,-Rp,0]);grid;
```

运行结果如图 8.23 所示。

【例 8.28】 用直接法设计一个切比雪夫Ⅰ型数字带通滤波器，指标为 $f_{p1}=20\text{kHz}$、$f_{p2}=30\text{kHz}$、$R_p=2\text{dB}$、$f_{st1}=10\text{kHz}$、$f_{st2}=45\text{kHz}$、$A_s=20\text{dB}$ 抽样频率 $F_s=100\text{kHz}$。

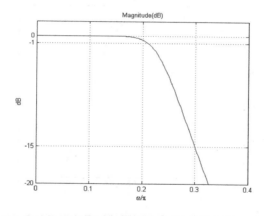

图 8.23 巴特沃斯型数字低通滤波器的幅度响应示例图

EXAMP08028

```
% （EXAMP08028 设计一个切比雪夫Ⅰ型数字带通滤波器）
%
clc
clear all
Fs=10^5;
wp1=2*20000/Fs;
wp2=2*30000/Fs;
ws1=2*10000/Fs;
ws2=2*45000/Fs;
wp=[wp1,wp2];
ws=[ws1,ws2];
Rp=2;
As=20;
```

```
[N,wc]=cheb1ord(wp,ws,Rp,As);
[b,a]=cheby1(N,Rp,wc);
[C,B,A]=dir2par(b,a);
w0=[ws1*pi,wp1*pi,wp2*pi,ws2*pi];
Hx=freqz(b,a,w0);
[H,w]=freqz(b,a);
dbHx=-20*log10(abs(Hx)/max(abs(H)));
% (Modfreqz 函数在前面已经由作者给出)
[db,mag,pha,grd,w]=Modfreqz(b,a);
plot(w/pi,db)
title('magnitude in dB')
xlabel('\omega/\pi')
ylabel('dB')
axis([0,1,-30,3])
set(gca,'xtickmode','manual','xtick',[0.1,0.2,0.5,0.6,0.8,1.0]);
set(gca,'ytickmode','manual','ytick',[-30,-As,-10,-2,0]);
grid
```

运行结果如图 8.24 所示。

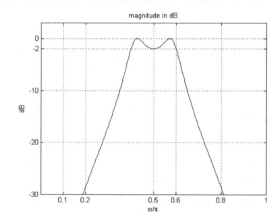

图 8.24 切比雪夫 I 型数字带通滤波器的幅度响应示例图

注意:
运行结果里 $N=2$，对于数字带通滤波器，得到的系统函数是 $2N$ 阶的。切比雪夫 I 型在通带内等波纹形。

【例 8.29】 用直接法设计数字椭圆带阻滤波器，技术指标为 $w_{p1}=0.3\pi$、$w_{p2}=0.7\pi$、$A_s=30\text{dB}$、$R_p=2\text{dB}$、$w_{st1}=0.4\pi$、$w_{st2}=0.6\pi$。

EXAMP08029

```
% （EXAMP08029 设计数字椭圆带阻滤波器）
%
clc
clear all
wp1=0.3;
```

```
wp2=0.7;
ws1=0.4;
ws2=0.6;
wp=[wp1,wp2];
ws=[ws1,ws2];
Rp=2;
As=30;
[N,wc]=ellipord(wp,ws,Rp,As);
[b,a]=ellip(N,Rp,As,wc,'stop')
[C,B,A]=dir2par(b,a);
w0=[wp1*pi,ws1*pi,ws2*pi,wp2*pi];
Hx=freqz(b,a,w0);
[H,w]=freqz(b,a);
dbHx=-20*log10(abs(Hx)/max(abs(H)));
% （Modfreqz函数在前面已经由作者给出）
[db,mag,pha,grd,w]=Modfreqz(b,a);
plot(w/pi,db)
title('magnitude in dB')
xlabel('\omega/\pi')
ylabel('dB')
axis([0,1,-60,4])
grid
```

运行结果如下：

```
b =0.2409   -0.0000    0.5800   -0.0000    0.5800    0.0000    0.2409
a =1.0000   -0.0000    0.2161    0.0000    0.6177   -0.0000   -0.1921
```

椭圆型数字带阻滤波器的幅度响应示例图如图 8.25 所示。

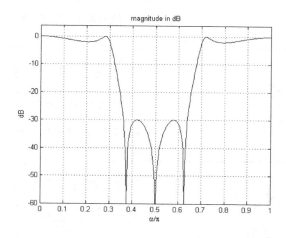

图 8.25　椭圆型数字带阻滤波器的幅度响应示例图

注意：
　　上例椭圆滤波器的阶次是 $N=3$，对比巴特沃斯和切比雪夫型，在相同的幅度指标下，椭圆阶次最小。

8.4.4 基于数字频带变换法设计数字滤波器

低通数字滤波器转到数字各型（低通、高通、带通、带阻）滤波器的变换公式，要求实现有理代换。利用相应的数字频带变换关系，求出各个参量，步骤基本分为先数字化，再做数字频带变换的方案。表 8.1 列出了 6 种变换公式和参数关系式。

表 8.1 通带截止频率 θ_p 低通原型数字滤波器的（Z）转为各型数字滤波器（z）

变换类型	变 换 公 式	变换参数的公式
低通|低通	$Z^{-1} = \dfrac{z^{-1} - \alpha}{1 - \alpha z^{-1}}$	$\alpha = \sin\left(\dfrac{\theta_p - \omega_p}{2}\right) / \sin\left(\dfrac{\theta_p + \omega_p}{2}\right)$ ω_p 为待求的低通滤波器的通带截止频率
低通|高通	$Z^{-1} = -\left(\dfrac{z^{-1} + \alpha}{1 + \alpha z^{-1}}\right)$	$\alpha = -\cos\left(\dfrac{\theta_p + \omega_p}{2}\right) / \cos\left(\dfrac{\theta_p - \omega_p}{2}\right)$ ω_p 为待求的高通滤波器的通带截止频率
低通|带通	$Z^{-1} = -\dfrac{z^{-2} - \dfrac{2\alpha}{k+1} z^{-1} + \dfrac{k-1}{k+1}}{\dfrac{k-1}{k+1} z^{-2} - \dfrac{2\alpha k}{k+1} z^{-1} + 1}$	$\alpha = \cos\left(\dfrac{\omega_{p_2} + \omega_{p_1}}{2}\right) / \cos\left(\dfrac{\omega_{p_2} - \omega_{p_1}}{2}\right) = \cos\omega_0$ $k = \cot\left(\dfrac{\omega_{p_2} - \omega_{p_1}}{2}\right) \tan\dfrac{\theta_p}{2}$ ω_{p_2}、ω_{p_1} 分别为带通滤波器通带截止频率的上、下截止频率，ω_0 为通带中心频率
低通|带阻	$Z^{-1} = \dfrac{z^{-2} - \dfrac{2\alpha}{1+k} z^{-1} + \dfrac{1-k}{1+k}}{\dfrac{1-k}{1+k} z^{-2} - \dfrac{2\alpha}{1+k} z^{-1} + 1}$	$\alpha = \dfrac{\cos\left(\dfrac{\omega_{p_2} + \omega_{p_1}}{2}\right)}{\cos\left(\dfrac{\omega_{p_2} - \omega_{p_1}}{2}\right)} = \cos\omega_0$ $k = \tan\left(\dfrac{\omega_{p_2} - \omega_{p_1}}{2}\right) \tan\dfrac{\theta_p}{2}$ ω_{p_2}、ω_{p_1} 分别为待求的带阻滤波器的上、下截止频率，ω_0 为通带中心频率
低通|多带通	$Z^{-1} = -\dfrac{z^{-N} + \sum\limits_{k=1}^{N} d_k z^{-(N-k)}}{1 + \sum\limits_{k=1}^{N} d_k z^{-k}}$	$\cos\left(\dfrac{\theta_p}{2} + \dfrac{N}{2}\omega_{i1}\right) + \sum\limits_{i=1}^{N} d_k \cos\left(\dfrac{\theta_p}{2} + \dfrac{N}{2}\omega_{i1} - k\omega_{i1}\right) = 0$ $\cos\left(\dfrac{\theta_p}{2} - \dfrac{N}{2}\omega_{i2}\right) + \sum\limits_{i=1}^{N} d_k \cos\left(\dfrac{\theta_p}{2} - \dfrac{N}{2}\omega_{i2} - k\omega_{i2}\right) = 0$ ω_{i2}、ω_{i1} 分别为第 i 个通带的上、下截止频率，$1 \leqslant i \leqslant N/2$，$N/2$ 为 $(0\sim\pi)$ 的通带数
低通|多带阻	$Z^{-1} = \dfrac{z^{-N} + \sum\limits_{k=1}^{N} d_k z^{-(N-k)}}{1 + \sum\limits_{k=1}^{N} d_k z^{-k}}$	$\sin\left(\dfrac{\theta_p}{2} + \dfrac{N}{2}\omega_{i1}\right) + \sum\limits_{i=1}^{N} d_k \sin\left(\dfrac{\theta_p}{2} + \dfrac{N}{2}\omega_{i1} - k\omega_{i1}\right) = 0$ $\sin\left(\dfrac{\theta_p}{2} - \dfrac{N}{2}\omega_{i2}\right) + \sum\limits_{i=1}^{N} d_k \sin\left(\dfrac{\theta_p}{2} - \dfrac{N}{2}\omega_{i2} + k\omega_{i2}\right) = 0$ ω_{i2}、ω_{i1} 分别为第 i 个阻带的上、下通带的截止频率，$1 \leqslant i \leqslant N/2$，$N/2$ 为 $(0\sim\pi)$ 的阻带数

【例 8.30】 编写实现 Z 域到 z 域的频带变换的函数。

EXAMP08030

```
function [bz,az]=zmapping(bZ,aZ,Nz,Dz)
% Function ZMAPPING is designed for frequency band transformation
% from Z-domain to z-domain
% （函数 ZMAPPING 设计为实现 Z 域到 z 域的频带变换）
% 语法格式为 [bz,az] = zmapping(bZ,aZ,Nz,Dz)
%
bzord=(length(bZ)-1)*(length(Nz)-1);
azord=(length(aZ)-1)*(length(Dz)-1);
bz=zeros(1,bzord+1);
for k=0:bzord
    pln=[1];
    for i=0:k-1
        pln=conv(pln,Nz);
    end
    pld=[1];
    for i=0:bzord-k-1
        pld=conv(pld,Dz);
    end
    bz=bz+bZ(k+1)*conv(pln,pld);
end
az=zeros(1,azord+1);
for k=0:azord
    pln=[1];
    for i=0:k-1
        pln=conv(pln,Nz);
    end
    pld=[1];
    for i=0:azord-k-1
        pld=conv(pld,Dz);
    end
    naz=az+aZ(k+1)*conv(pln,pld);
end
az1=az(1);
az=az/az1;
bz=bz/bz1;
```

【例 8.31】 首先设计一个归一化的数字巴特沃斯低通滤波器。技术指标为 $\Omega_p = 2\pi \times 3000\text{rad/s}$、$\Omega_{st} = 2\pi \times 4500\text{rad/s}$、$A_s = 15\text{dB}$、$R_p = 2\text{dB}$、$F_s = 2 \times 10^4 \text{s}$。再用 zmapping 函数实现数字低通到高通的（截止频率为 $\Omega_p = 2\pi \times 4000\text{rad/s}$）转换。

EXAMP08031

```
% （EXAMP08031 设计一个归一化的数字巴特沃斯低通滤波器，借助 zmapping 函数实现
% 低通到高通转换）
%
clc
clear all
Fs=2*10^4;
Rp=2;
As=15;
```

```
wp=2*pi*3000/Fs;
ws=2*pi*4500/Fs;
OmegaP=2*Fs*tan(wp/2);
OmegaS=2*Fs*tan(ws/2);
[N,OmegaC]=buttord(OmegaP,OmegaS,Rp,As,'s');
[b,a]=butter(N,OmegaC,'s');
[bz,az]=bilinear(b,a,Fs);
wc=0.4*pi;
wclp=2*atan(OmegaC/(2*Fs));
alpha=-(cos((wc+wclp)/2))/(cos((wc-wclp)/2))
Nz=-[alpha,1];
Dz=[1,alpha];
[bhp,ahp]=zmapping(bz,az,Nz,Dz)
[Hlp,wlp]=freqz(bz,az);
%
subplot(221);
plot(wlp/pi,20*log10(abs(Hlp)));
xlabel('\omega/\pi')
ylabel('dB')
axis([0,1,-150,20])
grid
[Hhp,whp]=freqz(bhp,ahp);
%
subplot(222)
plot(whp/pi,20*log10(abs(Hhp)))
xlabel('\omega/\pi')
ylabel('dB')
axis([0,1,-150,20])
grid
```

运行结果如下:

```
N = 4;
OmegaC = 2.2274e+04;
bz = 0.0237  0.0947  0.1421  0.0947  0.0237;
az =  1    -1.3846  1.0981  -0.3958  0.0612;
bhp = 0.1672 -0.6687  1.0031  -0.6687  0.1672;
ahp =   1   -0.7821  0.6800  -0.1827  0.0301;
alpha = -0.423
```

数字低通和数字高通滤波器示例图如图 8.26 所示。

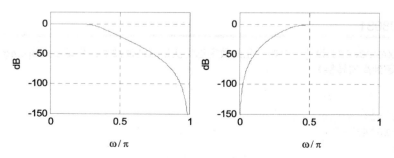

图 8.26 数字低通和数字高通滤波器示例图

第8章　MATLAB在数字信号中的应用

【例 8.32】 用模拟频带变换法设计一个模拟切比雪夫 I 型带通滤波器。

技术指标为 $f_{p1}=200\text{Hz}$、$f_{p2}=300\text{Hz}$、$R_p=2\text{dB}$、$f_{st1}=100\text{Hz}$、$f_{st2}=400\text{Hz}$、$A_s=20\text{dB}$。

EXAMP08032

```
% （EXAMP08031 设计一个归一化的数字巴特沃斯低通滤波器，借助 zmapping 函数实现
%  低通到高通转换）
%
clc
clear all
OmegaP1=2*pi*200;
OmegaP2=2*pi*300;
OmegaS1=2*pi*100;
OmegaS2=2*pi*400;
Rp=2;
As=20;
OmegaZ=sqrt(OmegaP1*OmegaP2);
B=OmegaP2-OmegaP1;
ws1=(OmegaS1^2-OmegaZ^2)/(OmegaS1*B);
ws2=(OmegaS2^2-OmegaZ^2)/(OmegaS2*B);
OmegaS=min(abs(ws1),abs(ws2));
OmegaP=1;
g=sqrt((10^(As/10)-1)/(10^(Rp/10)-1));
OmegaR=OmegaS/OmegaP;
N=ceil(log10(g+sqrt(g*g-1))/log10(OmegaR+sqrt(OmegaR*OmegaR-1)))
[z0,p0,k0]=cheb1ap(N,Rp);
a0=real(poly(p0));
b0=k0*real(poly(z0));
[b,a]=lp2bp(b0,a0,OmegaZ,B)
[H,w]=freqs(b,a,2000);
w0=[OmegaS1,OmegaP1,OmegaP2,OmegaS2];
Hx=freqs(b,a,w0);
dbHx=-20*log10(abs(Hx)/max(abs(H)))
plot(w/(2*pi)/1000,20*log10(abs(H)));
xlabel('f(KHz)');
ylabel('dB');
axis([0,0.6,-45,1])
set(gca,'xtickmode','manual','xtick',[0,0.1,0.2,0.3,0.4,0.5,0.6,0.7])
set(gca,'ytickmode','manual','ytick',[-40,-30,-20,-10,-2,0])
grid
```

运行结果如下：

```
N = 3
b =1.0e+007 *
    8.1085  -0.0000  0.0000  -0.0000
a =1.0e+019 *
0.0000  0.0000  0.0000  0.0000  0.0000  0.0003  1.3290
dbHx = 51.3855  1.9998  1.9998  32.4803
```

切比雪夫Ⅰ型带通滤波器示例图如图 8.27 所示。

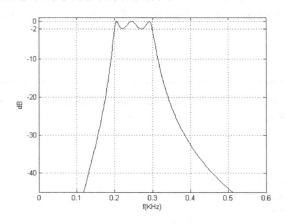

图 8.27　切比雪夫Ⅰ型带通滤波器示例图

用数字频带变换法设计数字滤波器的具体数学变换公式复杂，特别当阶数高时计算繁复，利用 MATLAB 可以方便实现数字频域变换。工程实际中可直接调用工具箱里现有函数用来设计数字滤波器。

【例 8.33】　按照下列基本步骤设计合适滤波器滤除含噪语音信号。

（1）语音信号的采集。

利用 Windows 下的录音机录制一段音频信号，时间在 1s 内。然后在 MATLAB 软件平台下，利用函数 wavread 对语音信号进行采样，得到采样序列 $x(n)$。

（2）语音信号的频谱分析。

画出语音信号 $x(n)$ 的时域波形，对语音信号进行傅里叶变换得到 $X(k)$，画出相应的幅频图。

（3）语音信号加噪后的频谱分析。

利用函数 awgn （或者采用其他白噪声）对语音信号 $x(n)$ 加噪得到 $y(n)$，对信号 $y(n)$ 进行快速傅里叶变换后，绘制含噪信号相应的幅频曲线图。

（4）设计滤波器。

根据 $X(k)$ 和 $Y(k)$ 的幅度曲线给出各滤波器的性能指标，设计 IIR 模拟和数字滤波器等，求出滤波器传递函数分子和分母多项式系数向量 **B** 和 **A**。设计 IIR 滤波器，利用函数 freqz 或者 Modfreqz 绘制各滤波器的频率响应。

（5）用滤波器对信号进行滤波。

各滤波器分别对加噪信号 $y(n)$ 进行滤波，得到滤波后信号 $z(n)$，分析 $z(n)$ 的幅频特性 $Z(k)$，绘图。FIR 滤波器利用函数 fftfilt 对信号进行滤波，IIR 滤波器利用函数 filter 对信号进行滤波。

（6）滤波效果分析。

选用低通、带通、带阻、高通等滤波器对同一个语音信号进行滤波，对比分析滤波前后时域、频域波形和滤波以后声音效果。

选用不同的模拟滤波器转数字滤波器的变换方法对同一个语音信号进行滤波，对比分析前后差异。

选用同一个滤波器，浮动调节技术指标对同一个语音信号进行滤波，比较滤波以后语音信号的频谱和声音效果,看哪个滤波器滤除噪声更好。

选用不同的滤波器对同一个语音信号进行滤波,对比分析滤波前后频域和声音效果。

注意:若采用巴特沃斯带通滤波器,放大后的频谱分析可以看出滤波效果较好。限于篇幅,只列出一种滤波器。读者可采用之前例子里的各种数字滤波器,自己录制语音信号加噪以后滤波,这是本科生数字信号处理课程设计中的一个部分实验,最直接简单地用软件MATLAB 设计的方法作最基本的实际工程应用。

滤波前后部分频带内的幅度曲线对比示例图如图 8.28 所示。

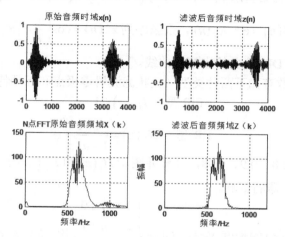

图 8.28　滤波前后部分频带内的幅度曲线对比示例图

小　　结

本章是 MATLAB 语言和数字信号处理紧密结合的实验仿真和应用问题。依次在时域离散信号和系统的基本理论、时域离散信号和系统的频域分析、离散傅里叶变换和快速算法、模拟和数字滤波器的设计和分析 4 个部分进行展开。每部分小节都结合基本内容,介绍了相应的 MATLAB 信号处理工具箱函数和涉及的自编函数,给出部分理论仿真和例题求解程序及运行结果。最后一个例子是对经典数字信号处理中设计的滤波器的一个最基本的应用。

习　　题

8.1　给定两个离散序列:$x1(n)=\{-2,-1,0,1,2\}$,$x2(n)=\{2,1,1\}$,试用 MATLAB 绘出它们的波形及 $x1(n)+x2(n)$ 的波形。

8.2　给定因果稳定线性时不变系统的差分方程描述:

$$\sum_{k=0}^{N} a_k y(n-k) = \sum_{k=0}^{M} b_k x(n-k)$$

对于输入序列 $x(n)$,求输出序列 $y(n)$。

（1）$x(n)$

（2）$x(n)=u(n)$

(3) $x(n)=R_{30}(n)$

(4) $x(n)=e^{j\frac{\pi}{4}n}u(n)$

8.3 求一个因果线性移不变系统 $y(n)=0.81y(n-2)+x(n)-x(n-2)$ 的单位抽样响应 $h(n)$，单位阶跃响应 $g(n)$，然后画出 $H(e^{j\omega})$ 的幅频和相频特性。

8.4 求解差分方程 $y(n)-0.4y(n-1)-0.45y(n-2)=0.45x(n)+0.4x(n-1)-x(n-2)$，其中，$x(n)=0.8^n\varepsilon(n)$，初始状态 $y(-1)=0$、$y(-2)=1$、$x(-1)=1$、$x(-2)=2$。

8.5 已知复正弦序列 $x1(n)=e^{j\frac{\pi}{8}n}R_N(n)$，余弦序列 $x2(n)=\cos\left(\frac{\pi}{8}n\right)R_N(n)$，对序列分别求当 $N=16$ 和 $N=8$ 时的 DFT，并绘出幅频曲线。

8.6 梳状滤波器系统函数有如下两种类型（梳状滤波器零极点和幅频特性）：

$$\text{FIR 型} \quad H_1(z)=1-z^{-N}$$

$$\text{IIR 型} \quad H_2(z)=\frac{1-z^{-N}}{1-a^N z^{-N}}$$

分别令 $N=8$，$a=0.8$、0.9、0.98，计算并图示 $H_1(z)$ 和 $H_2(z)$ 的零、极点图及幅频，说明极点位置对滤波器的影响。

8.7 已知有限长序列 $x(n)$ 的长度 $N=4$，$x(n)=\begin{cases}1 & n=0\\2 & n=1\\-1 & n=2\\3 & n=3\end{cases}$，用 FFT 求 $X(k)$，用 IFFT 求 $x(n)$。

8.8 一个数字系统的抽样频率 $F_s=2000\text{Hz}$，试设计一个为此系统使用的带通数字滤波器 $H(z)$，采用巴特沃思滤波器。要求：

（1）通带范围为 300～400Hz，在带边频率处的衰减不大于 3dB。

（2）在 200Hz 以下和 500Hz 以上衰减不小于 18dB。

8.9 设计低通数字滤波器，要求在通带内频率低于 $0.2\pi\text{rad}$ 时，允许幅度误差在 1dB 以内，在频率 $0.3\sim1\pi\text{rad}$ 的阻带衰减大于 15dB。用脉冲响应不变法设计数字滤波器（$T=1$），模拟滤波器采用切比雪夫 I 型原型滤波器。

8.10 设计巴特沃斯高通滤波器，要求通带截止频率为 0.8π，通带内衰减不大于 $1dB$，阻带起始频率为 0.6π，阻带内衰减 $15dB$，设 $T=1$。

8.11 设计巴特沃思带通滤波器，指标：通带上下截止频率分别为 0.4π、0.3π，通带最大衰减 3dB，阻带上下初始频率为 0.5π、0.2π，阻带内最小衰减为 18dB。

8.12 用窗函数法设计一个线性相位 FIR 带通滤波器，技术指标：下阻带截止频率 $f_{st_1}=2\text{kHz}$，上阻带截止频率 $f_{st_2}=6\text{kHz}$，通带下截止频率为 $f_{p_1}=3\text{kHz}$，通带上截止频率 $f_{p_2}=5\text{kHz}$，通带最大衰减为 0.2dB，阻带最小衰减为 55dB，抽样频率为 $f_s=20\text{kHz}$。

参 考 文 献

[1] 王永龙. Introduction and Applications in MATLAB（手稿）[M]. 2007.
[2] Rudra Pratap. Getting Started with MATLAB 5: A Quick Introduction for Scientists and Engineers [M]. London: Oxford University，1999.
[3] 程卫国，冯峰，王雪梅，等．MATLAB 5.3 精要、编程及高级应用[M]. 北京：机械工业出版社，2000.
[4] Adrian Biran，Moshe Breiner．MATLAB 6 for Engineers[M]．New York: Prentice Hall，2002.
[5] Edward B. Magrab. MATLAB 原理与工程应用[M]. 高会生，李新叶，胡智奇等，译. 北京：电子工业出版社，2002.
[6] Delores M. Etter，David C. Kuncicky，Dug Hull．Introduction to MATLAB 6（第 2 版）[M]．New Jersey: Pearson Education Inc，2004.
[7] 飞思科技产品研发中心．MATLAB7 基础与提高[M]. 北京：电子工业出版社，2005.
[8] John H. Mathews, Kurtis D. Fink. Numerical Methods Using MATLAB（第 4 版）[M]. 北京：电子工业出版社，2005.
[9] Stephen J. Chapman．MATLAB Programming for Engineers（第 2 版）[M]．北京：科学出版社，2005.
[10] 彭芳麟. 数学物理方程的 MATLAB 解法与可视化[M]. 北京：清华大学出版社，2004.
[11] 陈晓平，李长杰．MATLAB 及其在电路与控制理论中的应用[M]．合肥：中国科学技术大学出版社，2004.
[12] 钟季康，鲍鸿吉. 大学物理习题计算机解法 MATLAB 编程应用[M]. 北京：机械工业出版社，2007.
[13] 王永龙，夏昌龙，刘朋. 基于 MATLAB 编程给出圆形波导中能流密度分布图仿真[J]. 临沂师范学院学报，2008，30（3）：46-50.
[14] 王永龙，夏昌龙，樊三强，等. 圆波导管壁电流分布图仿真[J]. 临沂师范学院学报，2009，31（3）：46-50.
[15] 张性辉. 在大学物理教学中使用 MATLAB 制作图像和动画的几个实例[J]. 大学物理，2004，23（9）：59-62.
[16] 刘耀康. 用计算机绘制点电荷对的电场线[J]．大学物理，2005，24（8）：59-61.
[17] 周莉英，董慎行. 空间静电场的计算机模拟——轴测投影法[J]. 大学物理，2005，24（8）：56-58.
[18] 邵斌，贺黎明. 基于 MATLAB 平台的量子力学三维图形设计及动画生成工具[J]. 大学物理，2005，24（11）：55-58.
[19] 陈垚光，毛涛涛，王正林，等. 精通 MATLAB GUI 设计[M]. 北京：电子工业出版社，2008.
[20] 周建兴，岂兴明，矫津毅，等. MATLAB 从入门到精通[M]. 北京：人民邮电出版社，2008.
[21] 李国朝. MATLAB 基础及应用[M]. 北京：北京大学出版社，2011.
[22] 陆君安，尚涛，谢进. 偏微分方程的 MATLAB 解法[M]. 武汉：武汉大学出版社，2001.
[23] Sergey E. Lyshevski. Engineering and Scientific Computations Using MATLAB [M]．New Jersey: John Wiley & Sons, Inc., Hoboken, 2003.
[24] 李庆扬，王能超，易大义. 数值分析[M]. 长沙：华中科技大学出版社，2006.
[25] 王沫然．MATLAB 与科学计算[M]. 北京：电子工业出版社，2003.
[26] Richard L. Burden，J. Douglas Faires．Numerical Analysis（第 7 版）[M]. 北京：高等教育出版社，2009.
[27] 原思聪．MATLAB 语言与应用技术[M]. 北京：国防工业出版社，2011.
[28] Cleve B．Moler．MATLAB 数值计算[M]．俞文健，译. 北京：机械工业出版社，2006.

[29] 刘师少. 计算方法[M]. 北京：科学出版社，2005.
[30] Gerald Recktenwald. 数值方法和 MATLAB 实现与应用[M]. 吴卫国，译. 北京：机械工业出版社，2004.
[31] John H. Mathews, Kurtis D. Fink. 数值方法（MATLAB 版）（第3版）[M]. 陈渝，译. 北京：电子工业出版社，2002.
[32] Jeffery J. Leader. 数值分析与科学计算（影印版）[M]. 北京：清华大学出版社，2008.
[33] 何汉林，梅家斌. 数值分析[M]. 北京：科学出版社，2007.
[34] 刘卫国. MATLAB 程序设计教程[M]. 北京：中国水利水电出版社，2008.
[35] 王月明. MATLAB 基础与应用教程[M]. 北京：北京大学出版社，2012.
[36] 丁玉美，高西全. 数字信号处理[M]. 西安：西安电子科技大学出版社，2005.
[37] Vinay K. Ingle, John G. Proakis. Digital Signal Processiong Using MATLAB[M]. 北京：科学出版社，2006.
[38] 程佩青. 数字信号处理教程（第 3 版）[M]. 北京：清华大学出版社，2013.
[39] A．V．奥本海姆. 信号与系统[M]. 刘树棠，译，西安：西安交通大学出版社，1985.
[40] 胡广书. 数字信号处理——理论，算法与实现（第 2 版）[M]. 北京：清华大学出版社，2003.